Ultrasonic Methods in Evaluation of Inhomogeneous Materials

NATO ASI Series

Advanced Science Institutes Series

A Series presenting the results of activities sponsored by the NATO Science Committee, which aims at the dissemination of advanced scientific and technological knowledge, with a view to strengthening links between scientific communities.

The Series is published by an international board of publishers in conjunction with the NATO Scientific Affairs Division

A	**Life Sciences**	Plenum Publishing Corporation
B	**Physics**	London and New York
C	**Mathematical and Physical Sciences**	D. Reidel Publishing Company Dordrecht, Boston, Lancaster and Tokyo
D	**Behavioural and Social Sciences**	Martinus Nijhoff Publishers Boston, Dordrecht and Lancaster
E	**Applied Sciences**	
F	**Computer and Systems Sciences**	Springer-Verlag Berlin, Heidelberg, New York
G	**Ecological Sciences**	London, Paris, Tokyo
H	**Cell Biology**	

Series E: Applied Sciences – No. 126

Ultrasonic Methods in Evaluation of Inhomogeneous Materials

edited by

Adriano Alippi

University of Rome
La Sapienza
Italy

Walter G. Mayer

Georgetown University
Washington DC
USA

1987 **Martinus Nijhoff Publishers**
Dordrecht / Boston / Lancaster
Published in cooperation with NATO Scientific Affairs Division

Proceedings of the NATO Advanced Study Institute on "Ultrasonic Methods in Evaluation of Inhomogeneous Materials", Erice, Italy, October 15-25, 1985

Library of Congress Cataloging in Publication Data

NATO Advanced Study Institute on "Ultrasonic Methods in Evaluation of Inhomogeneous Materials" (1985 : Erice, Sicily)
Ultrasonic methods in evaluation of inhomogeneous materials.

(NATO ASI series. Series E, Applied sciences ; no. 126)
"Proceedings of the NATO Advanced Study Institute on "Ultrasonic Methods in Evaluation of Inhomogeneous Materials", Erice, Italy, October 15-25, 1985"--T.p. verso.
"Published in cooperation with NATO Scientific Affairs Division."
Includes index.
1. Ultrasonic testing--Congresses. 2. Inhomogeneous materials--Testing--Congresses. I. Alippi, Adriano. II. Mayer, Walter G. III. Title. IV. Series.
TA417.4.N38 1985 620.1'1274 87-1711

ISBN-13: 978-94-010-8099-6 e-ISBN-13: 978-94-009-3575-4
DOI: 10.1007/978-94-009-3575-4

Distributors for the United States and Canada: Kluwer Academic Publishers, P.O. Box 358, Accord-Station, Hingham, MA 02018-0358, USA

Distributors for the UK and Ireland: Kluwer Academic Publishers, MTP Press Ltd, Falcon House, Queen Square, Lancaster LA1 1RN, UK

Distributors for all other countries: Kluwer Academic Publishers Group, Distribution Center, P.O. Box 322, 3300 AH Dordrecht, The Netherlands

Softcover reprint of the hardcover 1st edition 1987

NATO Advanced Study Institute
on
ULTRASONIC METHODS IN EVALUATION OF INHOMOGENEOUS MATERIALS
Erice, 15 - 25 October, 1985

SCIENTIFIC COMMITTEE

Prof. A. Alippi
University of Rome, ITALY

Prof. W.G. Mayer
Georgetown University,
Washington, USA

Prof. L. Bjørnø
Technical University,
Lyngby, DENMARK

Prof. R.W.B. Stephens
Chelsea College, London, UK

LECTURERS

Prof. L. Adler
Ohio State University
Columbus, Ohio, USA

Prof. B. A. Auld
Stanford University
CA, USA.

Dr. G. Busse
University of Stuttgart
FRG

Prof. J.D.N. Cheeke
Sherbrooke University
Quebec, CANADA

Dr. J.J. Gagnepain
C.N.R.S.
Besançon, FRANCE

Dr. J.A. Gallego Juarez
Centro de Fisica Aplicada
Madrid, SPAIN

Prof. W.G. Mayer
Georgetown University
Washington, DC, USA

Dr. R. Schmitt
Fraunhofer Institut
Saarbrücken, FRG

Dr. J.D. Skinner
General Electric Company
Wembley, Middlesex, UK

Prof. G. Socino
University of Perugia
ITALY

Prof. B.R. Tittmann
Rockwell International
CA, USA

Prof. A. Zarembowitch
University Pierre et Marie Curie
Paris, FRANCE

TABLE OF CONTENTS

III. MATERIALS

PREFACE

The purpose of the School, the content of which is reflected in this book, is to bring together experiences and knowledge of those acousticians who are particularly sensible to materials and their properties, specifically to those materials that may be called inhomogeneous.

The two things together: acoustics and inhomogeneity, define factually a dimensionless parameter, λ/a, which is the ratio between the sound wavelength and the spatial length of the material where its physical characteristics notably change. An implicit definition is, therefore, at hand for an inhomogeneous medium, which has the characteristic of a conditioned definition and sets a looser constraint to the otherwise strict statement of invariance under translations.

Composite, biological, porous, stratified materials are in the list of inhomogeneous materials, whose technological or structural interest has grown greatly in recent times. Ultrasonic waves offer a means for their investigation, which is valuable for it can be nondestructive, continuous in time, spatially localized, dependent on the size of inhomogeneities.

Which ultrasonic method for which material characteristics or properties is a problem to be optimized. The overview articles together with the contributed papers offer a wide range of possibilities where sound waves act as a powerful means of investigation and it could be left to the reader's or scientist's own experience to foresee additional ways of how applying what can be inferred as a suggestion from these topics. The invited lectures and the advanced research papers from participants were classified into three main groups; the first one devoted to general classes or properties of materials that can be investigated with ultrasounds, the second one dealing with ultrasonic methods aimed at the investigation of inhomogeneous materials, and a third one where materials are treated and investigated with ultrasonic methods.

The actual presentation of topics followed a temporal sequence of invited papers in the morning sessions and of advanced research papers in the afternoon ones, while the book order follows the subdivision in three sections, in each one of which the advanced research papers follow the presentation of the invited lectures.

I wish now to express my sincere gratitude to all the contributors for writing and presenting valuable papers, and to all the participants to the School for making the Institute a scientific success.

Adriano Alippi

INAUGURAL ADDRESS

Ladies and Gentlemen:

or dear colleagues and friends, if you permit me to address you more familiarly, I would like to take few minutes of your time at the beginning of this course, just to give it an official opening.

Welcome, then, to Erice; welcome to this second course of the School on Physical Acoustics, on Ultrasonic Methods in the Evaluation of Inhomogeneous Materials. This course is following the first one, held here in 1982, that dealt with the Fundamentals of Acoustic Wave Propagation, and it therefore gives the concrete sense of a series to the School. Those of you - and there are few among the lecturers and few, with my deepest satisfaction, among the participants - those of you who had the adventure of attending that December "Summer school" will probably remember that at that time we were, in a certain sense, experimenting our course: problems such as the opportunity for presentation of advanced research papers, their relative room with respect to the main lectures, and even type and time of lectures to match the attendants' requests, were all to be tested and solved duly. We did that on a subject that could be considered as introductory to the courses to come: as I said, on the fundamentals of acoustical wave propagation. At that time, we set the bases for the present course, that should have been dealing with a more specific, though still broadband, subject. The merging of personal experiences and active discussions focused the attention on ultrasounds and materials, not however to be connected in the field of non-destructive testing evaluation, rather in that of defining physical characteristics. And the interest resulted into leaning, on one side, towards the ultrasonic methods used to get physical information on the materials, and, on the other, towards the classes of materials that could be investigated through ultrasounds.

The participation to the School with advanced research papers was then highly recommended and, finally, an Advanced Study Institute was organized sponsored by the NATO organization. We are deeply grateful for such sponsorship, which permitted to extend the participation, implement the organization and publish the Proceedings.

I should also acknowledge the European Physical Society and the National Science Foundation for granting some additional scholarships, that permitted us to gain an even ampler international participation, and the Ettore Majorana Centre for Scientific Culture, which contributed with a rich financial support.

The Institute of Acoustics of the Italian National Council of Researches has to be thanked for allowing the partipation of Mrs. L. Covi, whose qualities for acting as secretary of the School I greatly appreciated.

I do also acknowledge the Institutions and Organizations that sponsor the Ettore Majorana Centre: the Italian Ministry of Education, the Italian Ministry of Scientific and Technological Research and the Regione Siciliana. A warm thank has to be tributed to the local Organization, whose efficiency and promptness allowed old problems to be solved.

A grateful thought to the friends of the Scientific Committee, specifically to Walter Mayer, the colleague Director, whose great collaboration has been always thoughtful and patient!

It is on theirs and his behalf, my friends, that I want to mention finally your contribution to the success of this course. It is your presence and courtesy: those of the lecturers and those of the participants, the effort of those who communicate and of those who receive knowledge, that make the success of an academe. And, being on a historically Greek territory, let the spirit of the Academe fill our days here in Erice, and let science blend here with humanity.

Prof. A. Alippi

Erice, October 16th, 1985

I. GENERAL

ANISOTROPIC MATERIAL ORIENTATION WITH ULTRASONICS

Walter G. Mayer

ABSTRACT

Anisotropic solids, like single crystal structures, are considered and a short theoretical review is given to show how one can calculate the velocities of the sets of longitudinal and shear waves in the various directions of such an anisotropic structure.

The use of acousto-optic experimental techniques is discussed and examples are given of applications of these methods to determine ultrasonic velocities in anisotropic solids, and thus the elastic constants of the material.

The use of ultrasonic surface waves for the determination of single crystal surface orientation is discussed.

Experimental extensions from surface orientation of anisotropic solids to localized flaws on or near the surface of otherwise homogeneous solids are considered. Some theoretical background information and experimental evidence are presented.

1. INTRODUCTION

Anisotropic substances, like single crystals, may be thought of as being quite homogeneous inasmuch as, in principle, they do not have any irregularities in the commonly accepted sense. Their structure is quite regular, consisting of an array of unit cells; their symmetries and their alignments give a single crystal a measure of "homogeneity" so that a single crystal sample has identical properties anywhere in its volume, provided one defines homogeneity as absence of structural differences from one location to another.

But, of course, single crystals fall into the category of inhomogeneous substances because the mechanical properties do vary with orientation. Lattice spacings change as a function of direction as do elastic properties and moduli. For this reason, ultrasonic velocities are not the same in all directions, although they are the same in all locations of the sample as long as directions are not changed. One might use true invariance of velocities (both with respect to location and direction) as a working definition of homogeneity.

To be sure, the usual concept of an imhomogeneity would include a change in velocity in a localized volume within the sample, whereas in the case of an anisotropic substance the change in ultrasonic velocities would not be caused by the presence of a volume element with different properties, but by a change in propagation direction.

As will become apparent later, the observation of changes in the velocity of ultrasonic waves can be used to determine both kinds of "inhomogeneities", that is, the overall crystal orientation and local changes in the mechanical structure or properties of otherwise truly homogeneous substances. This will be illustrated in greater detail for surfaces rather than the bulk of solids, both single crystal and isotropic.

Changes in velocity, either as a function of crystal orientation in anisotropic solids, or caused by local irregularities in otherwise isotropic materials can, in principle, always be used as an indicator of the existence of inhomogeneities. There are various experimental methods which one can use to determine such velocity changes. The present paper considers only one of these: the interaction of ultrasonic waves and light, usually referred to as acousto-optic interactions. This techniques allows one to determine all the appropriate sound velocities in a given direction of a transparent single crystal through an interpretation of the light diffraction pattern created by the ultrasonic waves and to use the light in the orders of this diffraction pattern to create a visual image of the ultrasonic beam, its direction and its structure.

Moreover, the acousto-optic interaction makes it possible to image ultrasonic beam reflections from surfaces of samples immersed in a transparent liquid like water, and since reflection phenomena often carry information about the mechanical properties of the surface, they can be used to obtain information about surface inhomogeneities. In order to illustrate some uses of this method, a short review of the relations between anisotropic structure and sound velocities and their determination via acousto-optics is in order.

2. WAVE VELOCITIES AND ANISOTROPY

Mechanical waves propagating in an isotropic solid may either be longitudinal or transverse, and their respective velocities are given by

$$V = (M/\rho)^{1/2}, \tag{1}$$

where M is the appropriate modulus, compressional or shear, and ρ is the density of the substance. If, however, the substance is a single crystal, the modulus is replaced by the appropriate combination of elastic constants and the resulting velocities of the possible waves depend on the crystallographic direction of wave propagation, often expressed in terms of direction cosines, l, m, and n, with respect to an x, y, z coordinate system. Since one longitudinal and two distinct shear waves may exist in any given crystallographic direction the relations between wave velocities, density of the crystal, wave propagation direction cosines, and elastic constants will result in a determinant as shown in Fig. 1. The values of the λ entered in the determinant are defined, in general terms, in Fig. 1. Many of the elastic constants, C, will be zero for a relatively simple crystal structure like the cubic and the determinant will have zero off-diagonal terms when the wave vectors are along a crystal axis. In this simplest of all cases the determinant reduces to a cubic equation whose three

$$\begin{vmatrix} \lambda_{11}-\rho V^2 & \lambda_{12} & \lambda_{13} \\ \lambda_{12} & \lambda_{22}-\rho V^2 & \lambda_{23} \\ \lambda_{13} & \lambda_{23} & \lambda_{33}-\rho V^2 \end{vmatrix} = 0$$

$$\boxed{\begin{aligned}\lambda_{ab} = {} & l^2 C_{1a1b} + m^2 C_{2a2b} + n^2 C_{3a3b} + lm(C_{1a2b} + C_{2a1b}) \\ & + ln(C_{1a3b} + C_{3a1b}) + mn(C_{2a3b} + C_{3a2b})\end{aligned}}$$

$$\lambda_{11} = l^2 C_{11} + m^2 C_{66} + n^2 C_{55} + 2lm C_{16} + 2ln C_{15} + 2mn C_{56}$$

$$\vdots$$

$$\lambda_{23} = l^2 C_{56} + m^2 C_{24} + n^2 C_{34} + lm(C_{25} + C_{46}) + ln(C_{36} + C_{45}) + mn(C_{23} + C_{44})$$

FIGURE 1. Determinant relating wave velocities to direction cosines and elastic constants.

roots are the wave velocities. When the waves propagate in directions which do not coincide with any of the three principal axes, i.e., l, m, and n are not either 0 or 1, more terms appear in the velocity determinant and the roots of that cubic equations are different from the previous simple result. This implies that all velocities are functions of crystal orientation and a plot of the three velocities as a function of crystal orientation will clearly show the same rotational symmetry as is present in the crystal.

Suppose velocities are measured for the longitudinal wave whose propagation vector is in the x-y plane, the values one obtains will repeat in every quadrant, that is, there will be four-fold symmetry; if the velocity profile is measured in the (111) plane, the symmetry will be three-fold. Obviously, if a crystal with more complex symmetry relations is used than is exhibited by the cubic system, the velocity determinant will be more complicated.

Conversely, if one has measured the longitudinal and the two transverse wave velocities in a given direction of an anisotropic sample one can determine at least some of the elastic constants. Similar measurements in other directions will yield different velocity values and thus it will be possible to determine the values of the remaining constants.

This relationship between symmetries of sample and velocity profiles enables one to use the orientation-dependent velocities to determine the orientation of the sample. In order to do this one would need to find the longitudinal and transverse wave velocities in a plane which, e.g., is perpendicular to a face of the crystal sample and relate the resulting velocity profile and its symmetry to the values predicted from the equations in Fig. 1.

This appears to be an almost impossible task, considering that many velocity measurements of both longitudinal and shear waves would have to be taken before orientation-dependent velocity values and profiles can be established. Fortunately, acousto-optic methods can make this task possible.

3. ACOUSTO-OPTIC INTERACTION

An ultrasonic wave is a pressure wave which, when propagating in a transparent medium, changes the refractive index for light, where the relative change is proportional to the amplitude of the ultrasonic wave. This change has the same periodic structure as the ultrasonic wavelength. If an initially plane light wave travels through this pressure wave, the originally uniform phase of the light will be changed periodically and the light will exit the ultrasonic field with a corrugated wave front. This process is shown schematically in Fig. 2a. This corrugation of the wavefronts will give rise to constructive and destructive interferences, resulting in the formation of a light diffraction pattern. This pattern can be focused behind the volume of interaction as indicated in Fig. 2b.

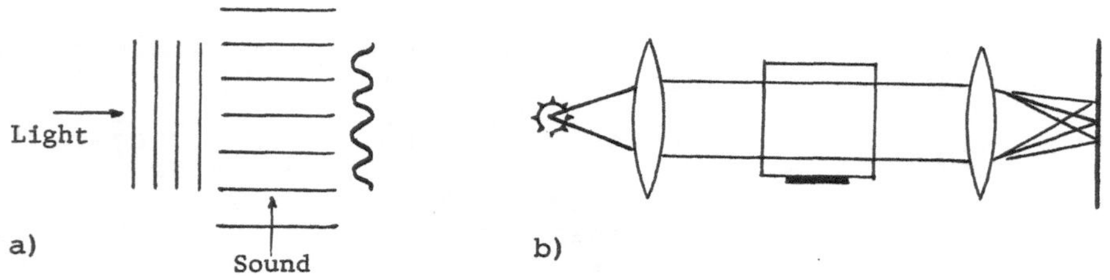

FIGURE 2. a) Change from plane to corrugated light wavefront caused by ultrasonic wave. b) Experimental setup to focus resulting diffraction pattern in plane behind interaction region.

A theoretical analysis of the process was first published by Raman and Nath [1] for progressive ultrasonic waves of relatively low MHz frequencies and a more general theory appeared more recently [2]. Both theories establish relations of importance to the present consideration which is that the angular spacing, θ, of the diffraction orders is given by

$$\sin\theta = \lambda/\lambda^* , \tag{2}$$

where λ and λ^* are the wavelengths of light and sound, respectively. A strong diffraction pattern will be formed if the light and sound propagation directions are at right angles to each other. The pattern will disappear rapidly when this condition is violated by a fraction of a degree.

To make use of this, consider a transparent single crystal mounted in the arrangement shown in Fig 2b in such a manner that the transducer generates a strong ultrasonic field in the crystal, with all generated waves multiply reflected in many directions in the sample. But the incident light will only be diffracted by those waves whose propagation direction is in the plane normal to the light direction. Since all these waves have velocities given by the equations in Fig. 1, their respective wavelength will vary correspondingly. Each partial wave will obey Eq. (2) and a multiplicity of light diffraction orders will be formed, the direction of each individual set corresponding to the direction of travel of the partial wave.

The whole set visible on a suitably placed screen then shows the same symmetry as that of the crystallographic plane in which the longitudinal and mode-converted shear waves traveled. A representative diffraction pattern produced in this fashion is shown in Fig. 3. From the multiple pattern one can not only determine the value of some of the elastic

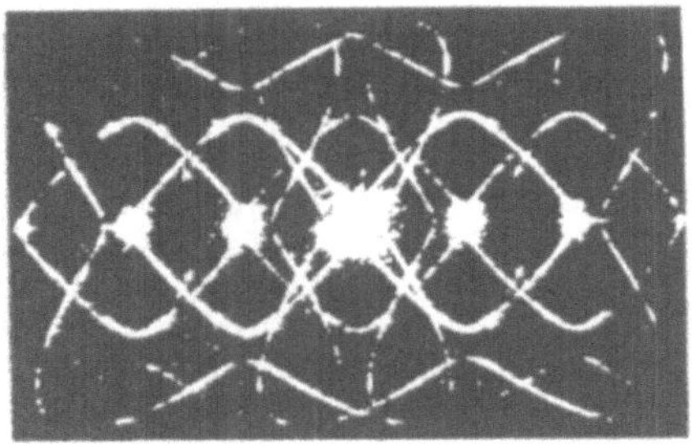

FIGURE 3. Multiple diffraction pattern produced by longitudinal and transverse ultrasonic waves in a transparent crystal. (Photograph supplied by E. Haussuehl, private communication)

constants but clearly one sees immediately what symmetry plane is pictured, and therefore, one has determined a direction of a normal to the plane - one has a measure of the crystallographic orientation.

4. SURFACE ORIENTATION

The method described above was useful to determine the orientation of a crystal sample by means of the direction and the spacing of multiple diffraction patterns. A somewhat simpler technique is available to determine the crystallographic orientation of a crystal surface. It is based on the evaluation of ultrasonic beam reflections at a crystal surface and water interface.

Just as bulk wave velocities in anisotropic substances are orientation dependent, so are wave velocities on surfaces of single crystals. Farnell [3] has calculated the orientation-dependent velocities of such surface waves for a number of cubic materials. The velocity profile, again, reflects the symmetry of the planes in question. A surface wave velocity in a given direction on the crystal plane depends on the magnitudes of the elastic constants of the substance, thus on the velocities of the various bulk waves in the appropriate directions.

The velocity of a surface wave on a sufficiently thick homogeneous solid (essentially the velocity of the Rayleigh wave) depends on the values of the longitudinal and the shear wave velocities. If the substance is immersed in water and an ultrasonic beam is incident onto the surface at an angle A, some of the incident energy will be refracted into the solid at an angle B for the resulting bulk longitudinal wave and some energy will be refracted into a bulk shear wave at an angle C. The process is governed by Snells' law. But if the refraction angle is such that the generated wave travels along the boundary between liquid and solid, the refraction angle is 90 degrees, the angle at which the Rayleigh wave travels. In this case, Snells' law can be written as

$$(\sin A)/VL = 1/VR , \qquad (3)$$

where VL is the sound velocity in water and VR is the Rayleigh wave velocity. The particular angle of incidence where this surface wave is excited is referred to as the Rayleigh angle.

The same consideration holds when the substance is anisotropic except here VR is not a constant but varies with direction on the crystal surface.

The angle of reflection of the incident beam is, of course, always the negative of the angle of incidence for all real angles of incidence, including the Rayleigh angle. But is was first shown by Schoch [4] that the reflected beam is "displaced" along the surface for ultrasonic beams incident at the Rayleigh angle. This and other nonspecular reflection phenomena were recently explained on a theoretical basis by Ngoc and Mayer [5].

This then implies that one should be able to determine the Rayleigh wave velocity if one could readily observe at what incidence angle nonspecular reflection occurs. Clearly, in the case of anisotropic substances, the Rayleigh angle is a function of crystal orientation, and finding it would determine a crystallographic direction along the surface. Figure 4 is an example of the critical angle (for Rayleigh waves and for pseudo-surface waves) distribution around the [110] direction on a single crystal copper surface.

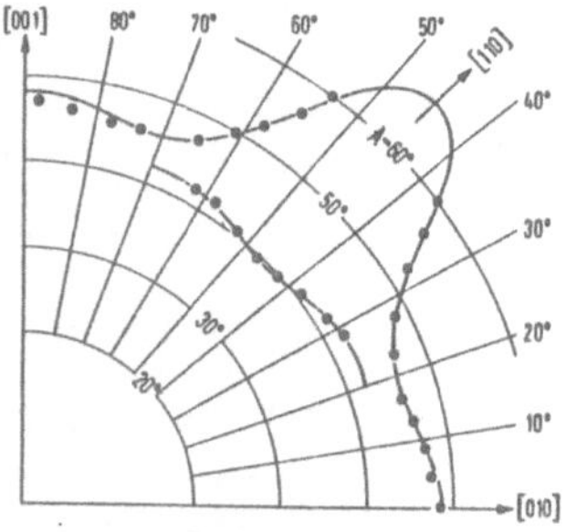

FIGURE 4. Rayleigh angle (outer curve) and critical angle for pseudo-surface waves (inner curve) on single crystal copper surface in water.

The observation of nonspecular reflections can again be done through the use of acousto-optic techniques. In this case one uses the expanded light beam to travel through the volume of water located above the crystal surface where the ultrasonic beam impinges. Thus the incident and the reflected beam create a diffraction pattern. But instead of interpreting the diffraction pattern one blocks its central order and allows the light of the higher orders to form, by means of a simple optical lens arrangement, to form a dark-field illumination image behind the plane of the diffraction pattern.

This image, frequently called a Schlieren image, then shows the location of the incident and the reflected beam. If the beam is reflected specularly, as is depicted in Fig. 5a, the incidence was not at the Rayleigh angle. But if nonspecular

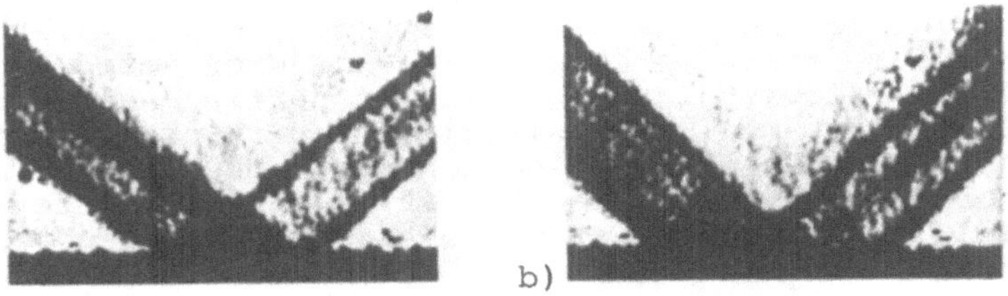

a) b)

FIGURE 5. Schlieren image of a) specular reflection and b) nonspecular reflection at critical angle for surface waves on solid.

reflection phenomena are visible, one knows that the incidence angle was at the critical angle for the excitation of a surface wave. The nonspecular reflection shown in Fig. 5b is of the simple "displacement" type, others are also possible, like dark strips in the reflected beam, a long trailing field in the reflection, etc.

Now, if one knows the value of the critical angle for a given anisotropic surface, as for instance for the [110] direction on copper shown in Fig.4, one simply sets the incidence angle of the ultrasonic beam at that angle and slowly rotates the copper sample under the beam without tilting the sample with respect to the plane of the incident and reflected ultrasonic beam. The image of the ultrasonics incident and reflected beam will show pure specular reflection for all rotational positions of the sample surface - until the [110] direction is directly lined up with the plane of the incident and reflected ultrasonic beams. At this precise setting the reflection will be nonspecular - thus the orientation of the sample surface is accomplished.

5. ACOUSTO-OPTIC DETECTION OF SURFACE INHOMOGENEITIES

This method, as outlined in Section 4, is based on the fact that nonspecular reflection phenomena can be observed when the angle of incidence corresponds to the Rayleigh angle. This angle of incidence depends on the surface wave velocity which in turn is influenced by the longitudinal and bulk wave velocities. This implies that the Rayleigh angle will change if any of the bulk wave velocities change, that is, if the elastic properties in the sample change. Any local change of the elastic properties of a sample quite naturally constitutes an inhomogeneity. It thus follows, that if nonspecular Rayleigh angle reflection is observed for all portions of a solid surface, the surface will be homogeneous as far as the values of the bulk wave velocities are concerned. The Rayleigh angle would change if there were a change in the shear modulus or any other modulus, perhaps somewhere on the surface.

This clearly suggests that one should be able to find local inhomogeneities on the surface of a solid flat surface by scanning an ultrasonic beam across the area, with the angle of incidence set at the Rayleigh angle which results in observable nonspecular reflection. If the scanning beam impinges on a section of the surface which is different (in its elastic properties) from the rest of the sample, nonspecular reflection will disappear and the beam will be reflected specularly, provided the velocity inhomogeneity is sufficiently pronounced.

Small changes in the elastic properties may not cause a total shift to specular reflection but may only change the profile of the reflected beam so it is merely "different" from the profile one observes at true Rayleigh angle nonspecular reflection.

A Schlieren image observation of the reflected beam structure may reveal even small changes in the beam profile if the surface of the solid sample is scanned and if the area under the beam is elastically different from the rest of the sample. Figure 6 is an example of the images obtained when

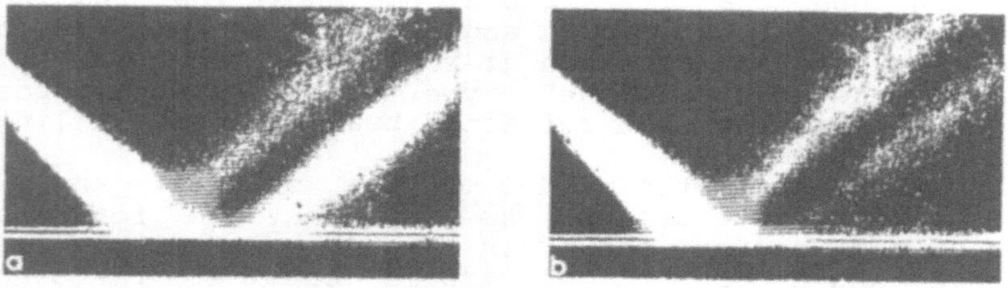

FIGURE 6. a) 2 MHz beam reflected at the Rayleigh angle from a brass/water interface. b) Changed reflected beam profile when irradiated area has slightly different elastic properties. Incidence in both cases from the left at identical angles.

the incidence is at the Rayleigh angle for brass in water, a) in the figure, which does not change as the sample is scanned until an area is reached where an inhomogeneity exists that gives rise to the image shown in b).

It is possible to calculate [6] how much deviation from the shear velocity must exist (as compared to the shear velocity in the homogeneous part of the sample) so that an adjustment by a given fraction of a degree in the incidence angle will restore the profile shown in Fig. 6b to the initial profile depicted in Fig. 6a.

Based on the theory of reflections of bounded beams near Rayleigh angle [7] one can incorporate in the calculations [8] some possible shapes of the inhomogeneities from which one can predict how the nonspecularly reflected beam changes its profile when certain dimensions and elastic properties of the inhomogeneity change.

This acousto-optic technique coupled with theoretical calculations can be used to find local inhomogeneities and to make some predictions about their structure and extent.

6. CONCLUSION

The use of acousto-optic techniques was described to show how a light diffraction pattern produced by ultrasonic waves in a transparent crystal can be interpreted to reveal the crystallographic orientation of an anisotropic solid, and how elastic constants can be determined. It was shown that the reflection of ultrasonic beams, imaged via a dark-field illumination technique (or Schlieren technique), can be useful in the determination of crystallographic directions on surfaces of anisotropic solids.

An application of this technique to isotropic solids makes it possible to not only measure the Rayleigh wave velocity of the solid but allows one to find localized inhomogeneities in the surface of the solid and to make some predictions about their structure and size.

Acknowledgment: Portions of the work described in this paper were supported by the Office of Naval Research, U.S. Navy.

REFERENCES

[1] C.V. Raman and N.S. Nath, Proc. Indian Acad. Sci., A3, 75 (1936)
[2] W.R. Klein, B.D. Cook, and W.G. Mayer, Acustica 15, 67 (1965)
[3] G.W. Farnell, in Physical Acoustics, W.P. Mason, ed., Vol. 6, Ch. 3 (Academic Press, New York, 1970)
[4] A. Schoch, Ergebn. Exakt. Naturw. 23, 160 (1950)
[5] T.D.K. Ngoc and W.G. Mayer, IEEE Trans. SU-27, 229 (1980)
[6] W.G. Mayer, Ultrasonics 19, 109 (1981)
[7] T.D.K. Ngoc and W.G. Mayer, J. Acoust. Soc. Am. 67, 1149 (1980)
[8] T.D.K. Ngoc and W.G. Mayer, J. Nondestr. Eval. 3, 93 (1982)

THIN FILM AND LAYER CHARACTERIZATION

LASZLO ADLER

1. INTRODUCTION

The problems of elastic wave propagation in an infinite solid will be briefly introduced. Progressively building up the background for waves in half space including liquid-solid boundaries. Special emphasis on theoretical and experimental work on leaky Rayleigh waves will be given. The thin layer substrate problem will be followed by the same structure loaded with water. Both Lamb waves and leaky Lamb waves will be introduced and application of these waves to fiber reinforced composites and welded plates will be given.

2. INFINITE ELASTIC SOLID

The wave Equation for a homogenous isotropic linear elastic solid may be written in a vector form for the displacement u

$$\mu\nabla^2 u + (\lambda+\mu)\nabla\nabla\cdot u = \rho\ddot{u} \qquad (1)$$

where λ and μ are the so called Lamé elastic constants, ρ is the density of the material. We may decouple the displacement vector in the form

$$u = \nabla\varphi + \nabla\wedge\psi \qquad (2)$$

where ϕ is the scalar potential and ψ is the vector potential and equation (1) decouples as

$$\nabla^2\varphi = \frac{1}{C_L^2}\ddot{\varphi} \qquad (3)$$

and

$$\nabla^2\psi = \frac{1}{C_T^2}\ddot{\psi} \qquad (4)$$

where

$$C_L^2 = \frac{\lambda + 2\mu}{\rho}, \text{ and } C_T^2 = \frac{\mu}{\rho}. \qquad (5a,b)$$

3. REFLECTION AND REFRACTION OF HARMONIC WAVES AT BOUNDARIES

For a joined half space of two homogenous isotropic and linear solids and incident plane wave in general produces two reflected and two refracted waves shown on figure 1. The material properties are λ μ ρ and λ^B μ^B ρ^B of the two solids.

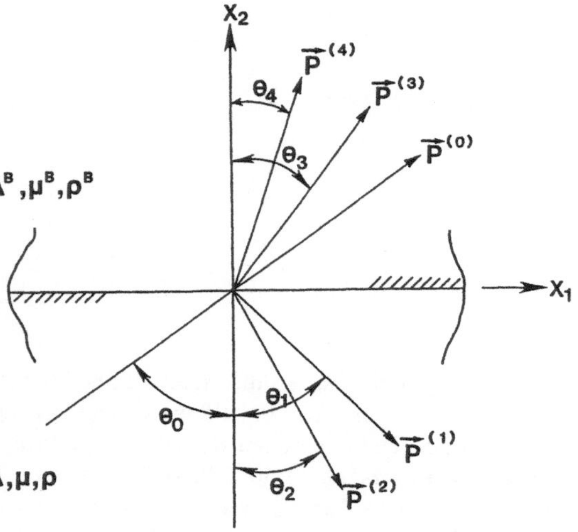

FIGURE 1 Incident reflected and refracted waves

The incident reflected and refracted waves may be represented as

$$U^{(n)} = A_n d^{(n)} \exp[i\, k_n (X_1 P_1{}^{(n)} + X_2 P_2{}^{(n)} - C_n t)] \tag{6}$$

where

n = 0 is the incident wave
n = 1 reflected L wave
n = 2 reflected S wave
n = 3 refracted L wave
n = 4 reflracted S wave

$\vec{P}$ is the propagation vector and $\vec{d}$ is the polarization vector for longitudinal waves: $\vec{d} = \pm\vec{P}$, $c_0 = C_L$ and for transverse wave $\vec{d} \cdot \vec{P} = 0$, $c_0 = C_T$. We distinguish between two types of shear waves for SV waves $d^{(0)} = i_3 \times P^{(0)}$, $c_0 = C_T$ and SH waves $d^{(0)} = i_3$, $c_0 = C_T$; for SH waves there is no mode conversion after reflection and refraction.

The reflection and transmission coefficient can be calculated by using boundary conditions. In general, the continuity of stresses and particles displacements are required at the interface.

4. RAYLEIGH SURFACE WAVES

A Rayleigh surface wave will propagate along the free surface of an elastic half space which is largely confined to the neighborhood of the boundary.

For a two dimensional case of plane wave the components of the particle displacements are given as

$$U_1 = Ae^{-bx_2} \exp[i\,k(x_1 - Ct)] \tag{7,a}$$

$$U_2 = Be^{-bx_2} \exp[i\,k(x_1 - Ct)] \tag{7.b}$$

$$U_3 = 0 \tag{7,c}$$

the particle motion is elliptically polarized (see figure 2). The Rayleigh velocity c can be obtained from the equation.

$$(2-\frac{C^2}{C_T^2})^2 - 4(1-\frac{C^2}{C_L^2})^{\frac{1}{2}}(1-\frac{C^2}{C_T^2}) = 0 \tag{8}$$

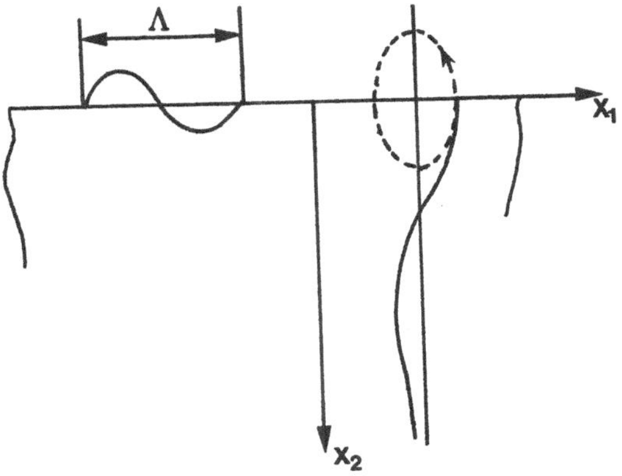

FIGURE 2 Rayleigh waves

and good approximation to the above equation is given for the Rayleigh velocity

$$C_R = \frac{0.862 + 1.14\nu}{1 + \nu}\, c_T \tag{9}$$

where ν is the so called Poisson ratio.

5. SURFACE WAVES AT LIQUID-SOLID INTERFACE

If the solid material is immersed into water, a Rayleigh like surface wave maybe generated at the interface. This wave is called leaky Rayleigh wave. The identification of this type of waves took place during the study of the displacement and distortion of reflected bounded beams at the Rayleigh angle. (2,3,4,5) The reflection coefficient has been examined at the Rayleigh angle to explain the distortion phenomenon, namely the shape of the reflected field distribution is quite different from that of the incident one. Calculations of the reflected field distribution at the Rayleigh angle includes material parameters of the interface expressed in terms of the so called Schoch displacement and parameters of the incident beam. The leaky Rayleigh velocity is slightly different than the solution given by equation (8) because of the presence of water. For this case equation (8) is replaced by equation (10):

$$4\left[\frac{C_T}{C}\right]^2\left[1-\left[\frac{C_T}{C}\right]^2\right]^{1/2}\left[\left[\frac{C_T}{C_L}\right]^2-\left[\frac{C_T}{C}\right]^2\right]^{1/2}+\left[1-2\left[\frac{C_T}{C}\right]^2\right]^2 = \tag{10}$$

$$= -\frac{\rho_w}{\rho}\left[\frac{(C_T/C_L)^2-(C_T/C)^2}{(C_T/C_w)^2-(C_T/C)^2}\right]^2$$

The leaky Rayleigh wave whose velocity can be obtained from equation (10) will radiate leaks into the water and combined together with specularly reflected wave produces the so called "null strip" (see figure 3).

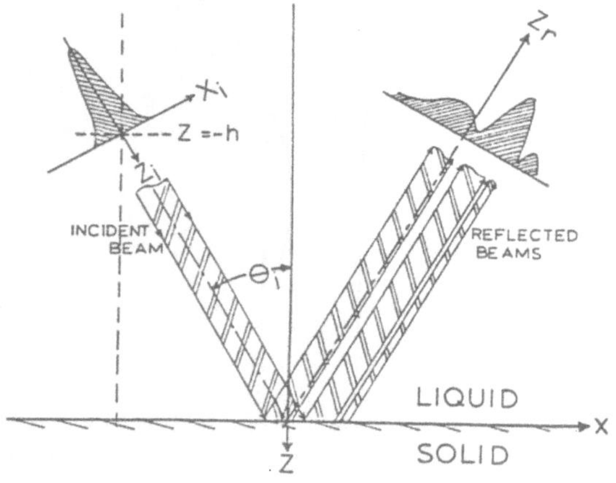

FIGURE 3 Reflection of finite beam from liquid-solid interface at the Rayleigh angle

The indentification of the null strip is one method to measure leaky Rayleigh velocity by using Snell's law.

On figure 4, which is a Schlieren photograph obtained for reflected ultrasonic beam at 3 different angles from water-stainless steel interface, the "null strip" is clearly identified.

Another phenomenon, which occurs at (6,7) the same angle where the null strip is produced and known as "backscattering" (see figure 5), may be used to measure the leaky Rayleigh velocity. In addition to the inhomogenous surface wave described above, equation(10) has a homogenous solution generally referred to as Scholte waves.

The velocity of the Scholte wave is close to the velocity of sound in the liquid and hence it cannot be generated the same way as the leaky Rayleigh waves. Recent work on the reflection of ultrasonic beam from periodically rough liquid-solid interface identified the presence of the Scholte wave (8,9). The reflection coefficient is plotted vs frequency (Figure 6) for a water-brass interface with a periodicity of 250μ. The minima occuring at 6 and 8 MHz correspond to the Scholte [velocity 1500 m/s] and leaky Rayleigh waves [velocity 2000 m/s].

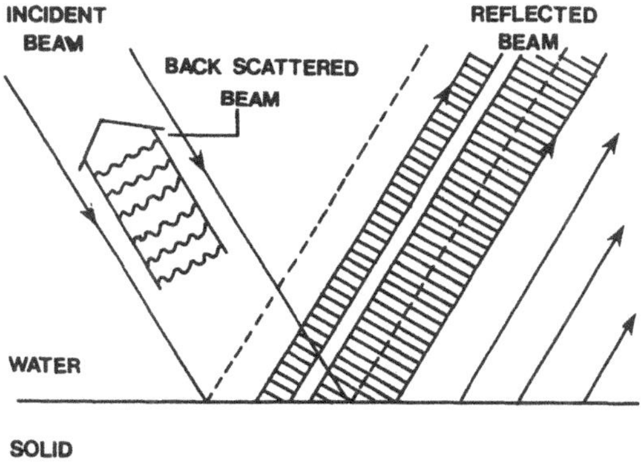

FIGURE 5 Illustration of finite beam ultrasonic phenomena at liquid-solid interface for Rayleigh angle of incidence.

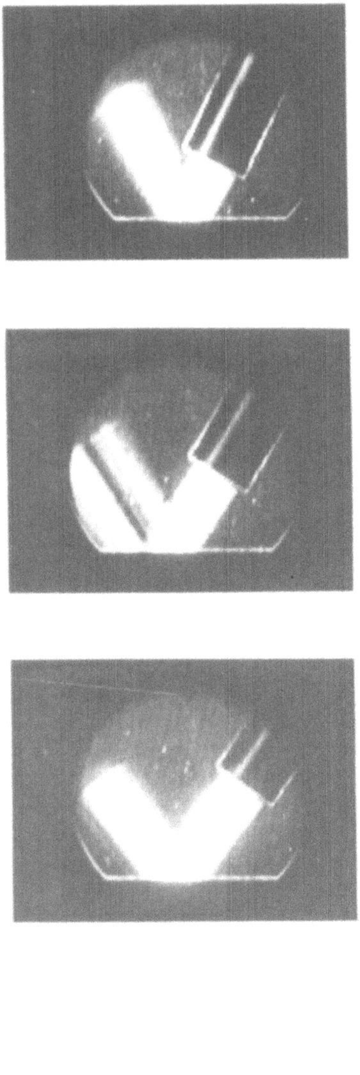

FIGURE 4 Relection of a Gaussian ultrasonic beam from a water-stainless steel surface for incident angles (a) 25°, (b) 30.5°, (Rayleigh angle), (c) 40°.

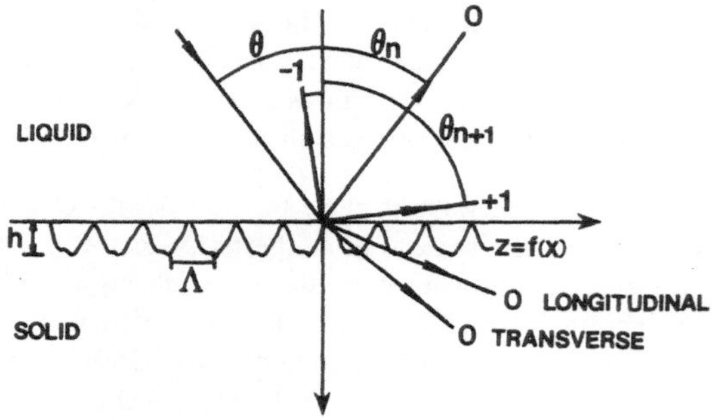

FIGURE 6 Coordinate system used to describe interaction of ultrasonic waves with periodic liquid-solid interface.

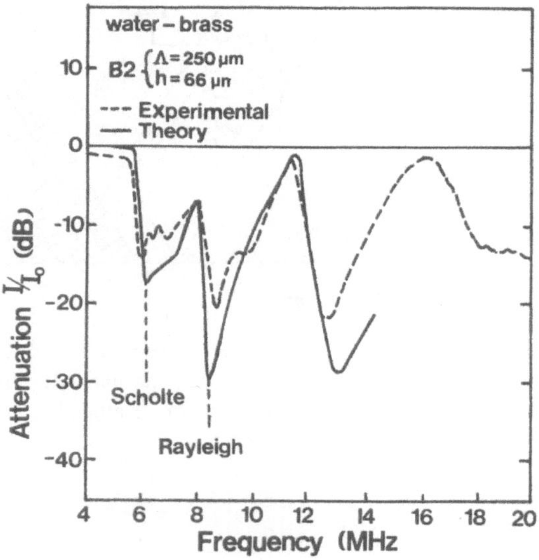

FIGURE 7 Comparison of theory ana experiment for the reflected ultrasonic spectrum for water-brass interface.

This phenomena using the same brass sample is shown on the Schlieren photographs (figure 7) taken for several discrete frequencies. Using normal incidence the zero order diffractions are shown for various frequencies. A diminished signal is observed near 6 and 8 MHz. Tone burst excitation and pulsed illumination was used. The grooves were parallel to the optical axis. The higher order difraction components will appear above 6 MHz when the grooves are perpendicular to the optical axis (figure 8).

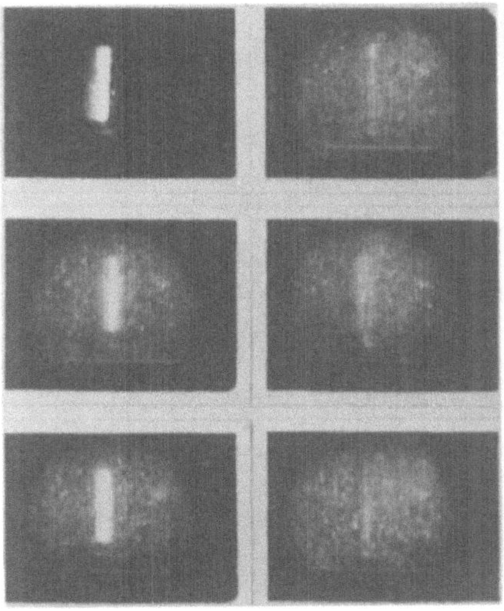

FIGURE 8 Schlieren photograph of reflected wave from periodic surface. Groove are parallel to the optical axis.

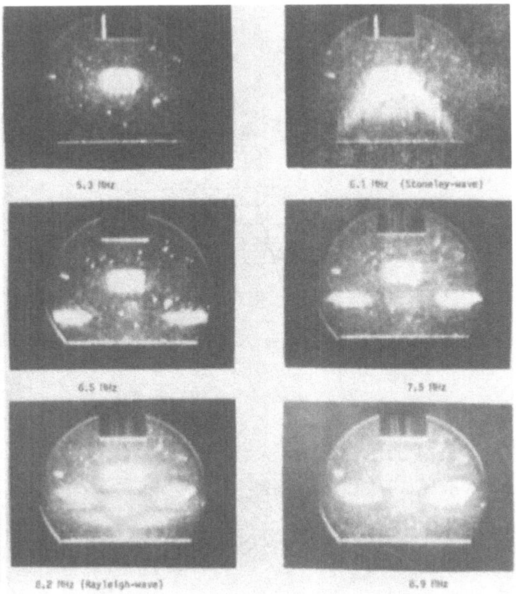

FIGURE 9 Schlieren photograph of reflected wave from periodic surface. Grooves are perpendicular to the optical axis.

6. WAVES IN THIN SOLID LAYER ON SOLID SUBSTRATE

Detailed treatments of the most general case of wave propagation in a thin solid layer welded to an infinite solid substrate is given in the literature (10) (figure 10). Both the layer and the substrate maybe anisotropic and piezoelectric. The layer thickness is usually assumed to be less then the wavelength. For isotropic nonpiezoelectric materials the displacements are completely uncoupled for the sagittal-plane displacements (Rayleigh-like waves) and transverse displacements (Love waves). Both type of waves are dispersive, i.e. their phase velocity is frequency dependent. For the Rayleigh-like modes the layer may "stiffen" the substrate if the Rayleigh velocity of the substrate is lower than the shear velocity of the layer. In the other extreme, the layer "loads" the substrate, because the velocity of the free surface Rayleigh mode on the substrate is decreased by the presence of the layer. Figure 11 shows the dispersion curves for these two cases.

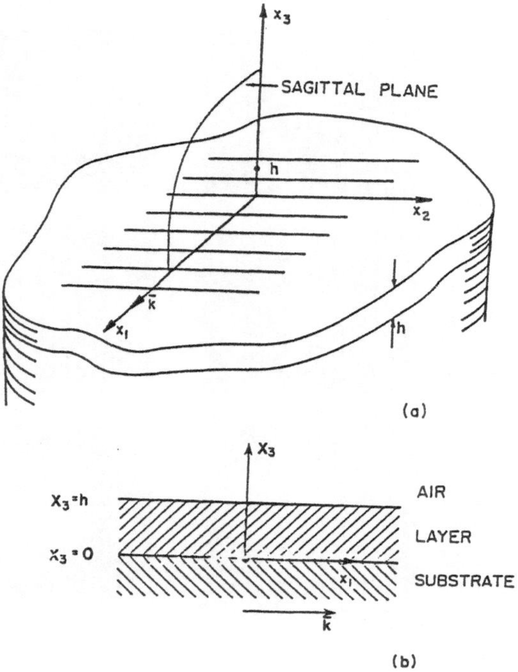

FIGURE 10 Schematics of the thin layer problem.

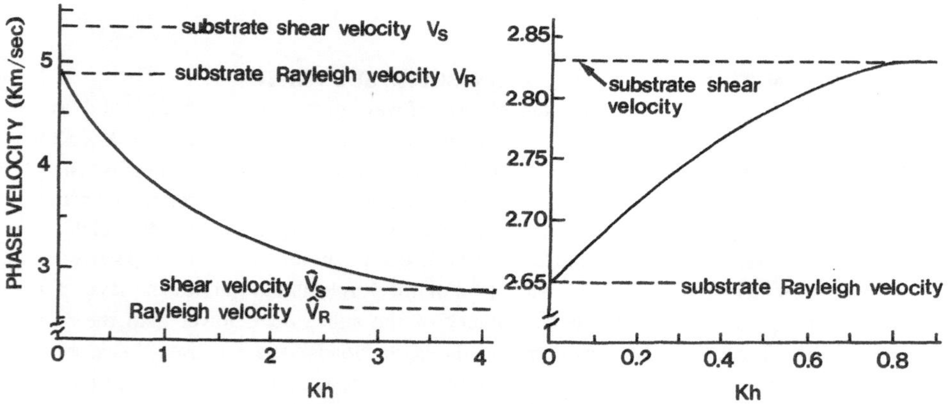

FIGURE 11a Dispersion curve for zinc oxide on silicon ($V_S < \hat{V}_S$). The layer loads the substrate.

FIGURE 11b Dispersion curve for silicon on zinc oxide ($V_S > \hat{V}_S$). The layer stiffens the substrate.

7. ULTRASONIC LEAKY RAYLEIGH WAVES IN THIN LAYER ON A SUBSTRATE (11,12)

Consider a thin elastic layer of thickness 2h bonded to a solid half space immersed in fluid (figure 12). Formal solution for the incident and reflected fields equations can be obtained by using scalar and vector potentials (ϕ and ψ) in each media and using the proper boundary conditions for each interface.

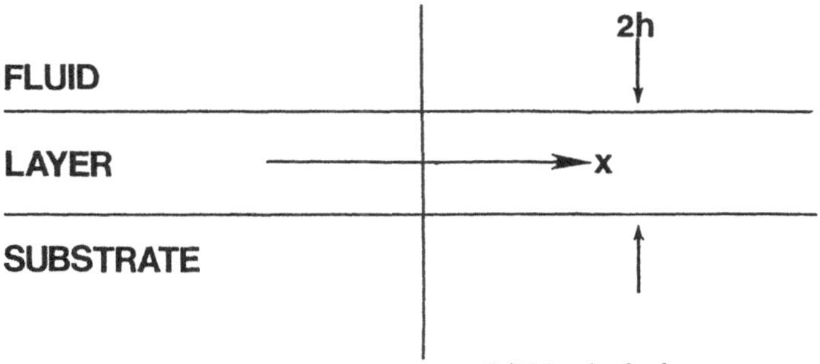

FIGURE 12 Schematics of the case where fluid loads the layer.

The reflection coefficient can be obtained for a longitudinal wave incident in the fluid as

$$R = \frac{\zeta_f\,(a_1 b_2 + a_2 b_1) + \rho_f \omega^2\,(\xi a_2 - \zeta_{1s} b_2)}{\zeta_f\,(a_1 b_2 + a_2 b_1) - \rho_f \omega^2\,(\xi a_2 \;\; \zeta_{1s} b_2)} \tag{11}$$

where

$$a_1 = \mu_s\,(2\xi^2 - K^2_{2s}) - 2ih\mu_o \zeta_{1s} \zeta^2_{2o} \tag{12}$$

$$b_1 = 2i\xi(i\mu_s \zeta_{2s} - h\mu_o \zeta^2_{2o}) \tag{13}$$

$$a_2 = 2i\xi[i\mu_s \zeta_{1s} - h(\lambda_o + 2\mu_o)\,\zeta^2_{1o}] \tag{14}$$

$$b_2 = \mu_s\,(2\xi^2 - K^2_{2s}) - 2ih\zeta_{2s}(\lambda_o + 2\mu_o)\,\zeta^2_{1o} \tag{15}$$

$$\zeta_{1,2} = (K_{1,2} - \xi^2)^{1/2}, \quad \zeta_f = (K_f^2 - \xi^2)^{1/2}$$

In the absence of the layer (h = 0) espression (11) reduces to the reflection coefficient of a solid medium submerged in a liquid.

The expression for the reflection coefficient contains the characteristic equation for the leaky Rayleigh wave which propagates along the interface of the fluid and the layer bonded to the substrate. The vanishing of the denominator in Eq. (11) gives the characteristic equation for such waves. The complex solution is given for

$$\xi = K_r + i\alpha, \tag{16}$$

where the leaky Rayleigh velocity is given as

$$C_R = \frac{\omega}{K_R} \tag{17}$$

and α is the attenuation coefficient related to the Schoch displacement Δ: $\alpha = 2/\Delta$. For the thin layer both C_R and α are dispersive. Figures 13 and 14 are plots of these behaviors for copper layer on stainless steel.

The experimental data shown on figure 15 are taken for various frequencies and identifing the Rayleigh angles from the appearence of the null strip (figure 16).

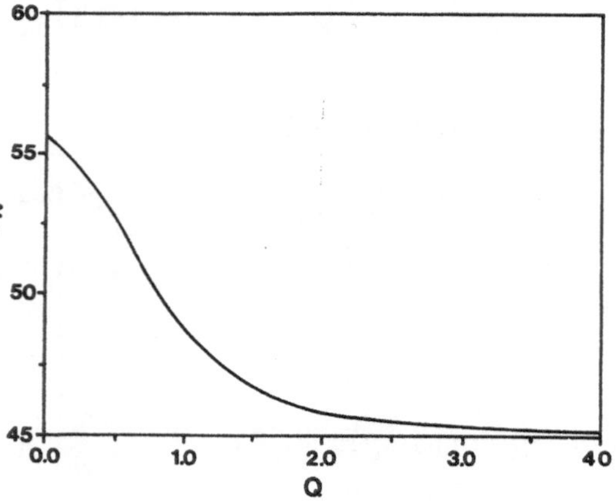

FIGURE 13 Displacement parameter over wavelength plotted as a function of the dimensionless frequency-thickness product, for the case of a copper layer on stainless steel.

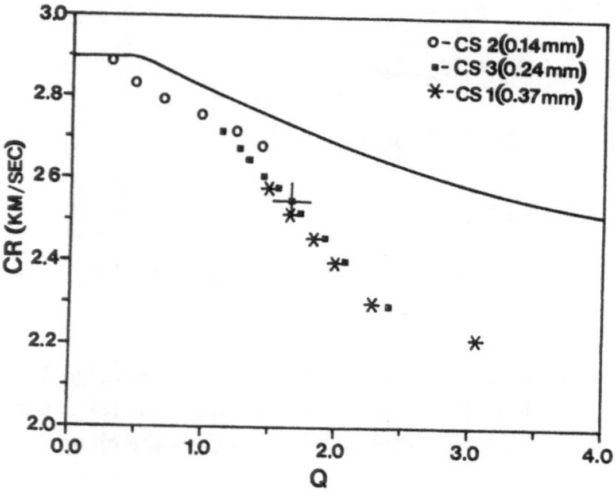

FIGURE 14 Surface wave phase velocity plotted versus frequency-thickness product. Solid line is approximate theory, while experimental data points for three samples correspond to symbols indicated on graph. Typical error bars are shown for one representative point.

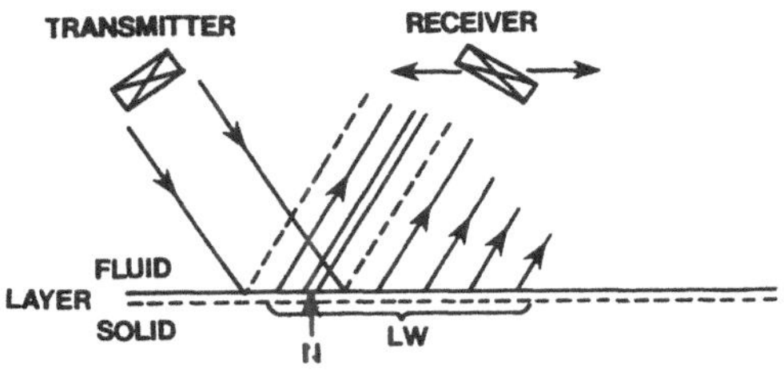

FIGURE 15 Schematic of leaky wave experiment. Transmitter is fixed, while receiver scans along x-axis. Dashed lines in reflected field indicate specular reflection. Null zone is denoted by N, and leaky wave reflected field by LW.

The behavior of the reflection coefficient from the same layer-substrate combination is shown on figure 16a, $Q = 0$ i.e. $h = 0$
16b, $Q = 2$
16c, $Q = 6$

The appearance of higher order modes are seen for $Q = 6$.

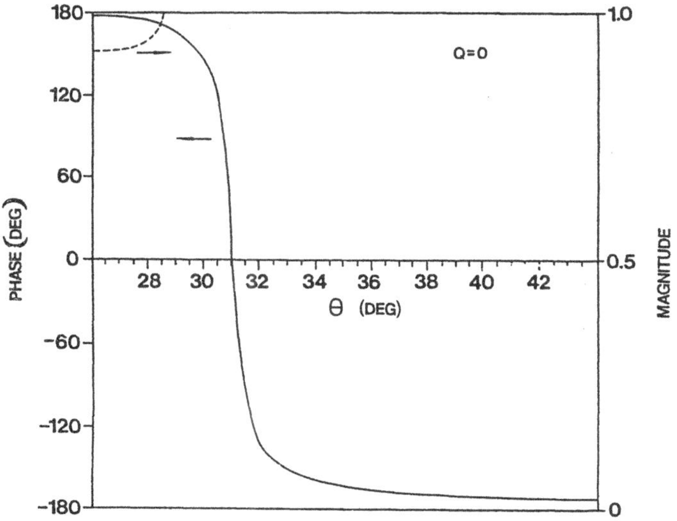

FIGURE 16a

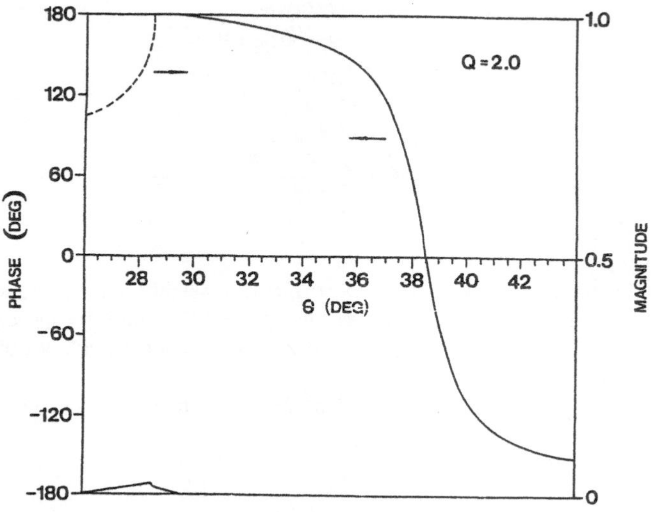

FIGURE 16b

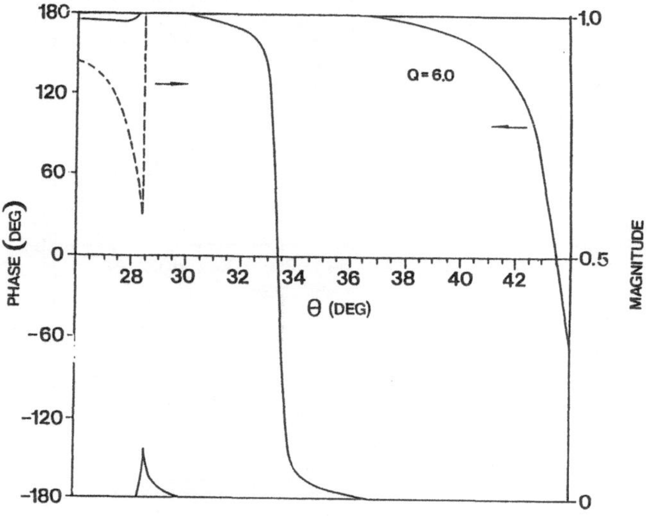

FIGURE 16c Reflection coefficient vs angle of incidence for the case of copper layer on a stainless steel a) Q = 0, b) Q = 2, c) Q = 6

8. WAVES IN ELASTIC LAYER (1)

In the previous section the layer was assumed to be thin compared to the wavelength. Now the case of harmonic waves in a layer of thickness 2h is considered without any assumption on the layer thickness (figure 11).

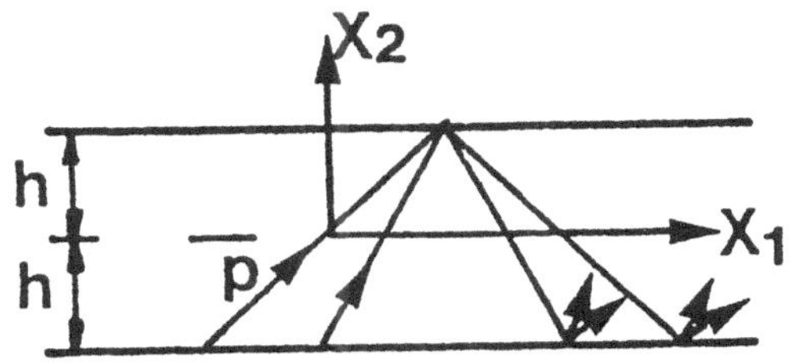

FIGURE 17 Waves in a layer

Harmonic waves can propagate between the two plane surfaces, reflected back and forth in a complicated pattern due to the mode conversion. The much simpler case of SH wave propagation can be easily discussed.

8a. SH WAVES IN THE LAYER

The particle displacement u_3 may be obtained from the wave equation:

$$\frac{\partial^2 U_3}{\partial x_1^2} + \frac{\partial^2 U}{\partial x_2^2} = \frac{1}{C_T^2}\frac{\partial^2 U_3}{\partial t^2} \qquad (18)$$

and written as:

$$U_3 = f(x_2)\,\exp i(Kx_1 - \omega t) \qquad (19)$$

Function $f(x_2)$ can be obtained from the boundary conditions at $\pm$ h:

$$\mu\frac{\partial U_3}{\partial x_2} = 0 \qquad (20)$$

giving:

$$f(x_2) = B_1 \sin(qx_2) + B_2 \cos(qx_2) \qquad (21)$$

where

$$q^2 = \frac{\omega^2}{C_T^2} - K^2 \qquad (22)$$

$$B_1 \cos(qh) \pm B_2 \sin(qh) = 0 \qquad (23)$$

$$B_1 = 0 \text{ and } \sin(qh) = 0 \qquad (24)$$

$$B_2 = 0 \quad \text{and} \quad \cos(qh) = 0 \tag{25}$$

The displacement is symmetric if $B_1 = 0$ and antisymmetric if $B_2 = 0$. In both cases the frequencies follow from

$$qh = \frac{n\pi}{2} \tag{26}$$

In figure 18 the displacement distribution for several modes are shown. Equation 22 can also be written for dimensionless frequency and wave mode as

$$\Omega^2 = h^2 + \xi^2 \tag{27}$$

where:

$$\Omega = \frac{2h\omega}{\pi C_T} \tag{28}$$

$$\xi = \frac{2Kh}{\pi} \tag{29}$$

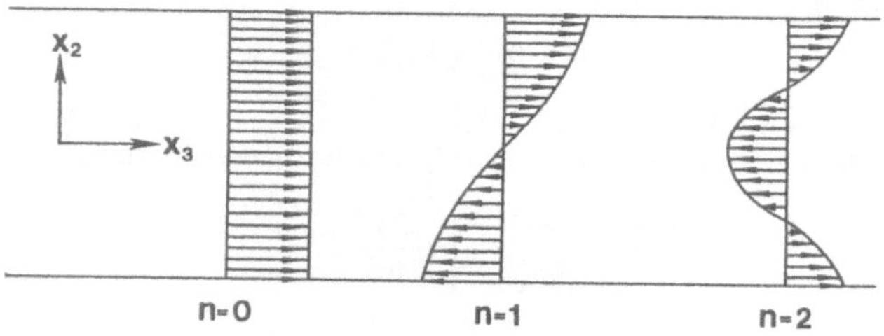

FIGURE 18 Displacement distributions in the $x_2 x_3$-plane

The velocity of the modes may be obtained as

$$\frac{C}{C_T} = \pm\left(1 - \frac{n^2}{\Omega}\right)^{-1/2} \qquad \frac{C}{C_T} = \pm\left(1 + \frac{n^2}{\xi^3}\right)^{1/2} \tag{30}$$

All modes, except $n = 0$, are dispersive.

8b. LAMB WAVES (13,14)

When both longitudinal and shear (SV) waves are present in the layer the analysis is much more complicated. It is convenient to decompose the displacement field into scalar and vector potentials.

$$U_3 = 0, \qquad \frac{\partial}{\partial x}(\,) = 0 \tag{31}$$

$$U_1 = \frac{\partial \phi}{\partial x_1} + \frac{\partial \psi}{\partial x_2} \tag{32}$$

$$U_2 = \frac{\partial \phi}{\partial x_1} - \frac{\partial \psi}{\partial x_2} \tag{33}$$

The uncoupled wave equations for the potentials are obtained as in Eqs. (3) and (4), and once again one may obtain symmetric and antisymmetric modes. The phase velocity of the various modes can be calculated from the characteristic equations. The relation between frequencies and wave numbers are expressed by the Rayleigh-Lamb frequency equations which take the following form for symmetric modes:

$$\frac{\tan[\frac{1}{2}\pi(\Omega^2 - \xi^2)^{\frac{1}{2}}]}{\tan[\frac{1}{2}\pi(\Omega^2/K^2 - \xi^2)^{\frac{1}{2}}]} = -\frac{4\xi^2(\Omega^2/K^2 - \xi^2)^{\frac{1}{2}}(\Omega^2 - \xi^2)^{\frac{1}{2}}}{(\Omega^2 - 2\xi^2)^2} \tag{34}$$

and the following one for the antisymmetric modes:

$$\frac{\tan[\frac{1}{2}\pi(\Omega^2 - \xi^2)^{\frac{1}{2}}]}{\tan[\frac{1}{2}\pi(\Omega^2/K^2 - \xi)^{\frac{1}{2}}]} = -\frac{(\Omega^2 - 2\xi^2)^2}{4\xi^2(\Omega^2/K^2 - \xi^2)^{\frac{1}{2}}(\Omega^2 - \xi^2)^{\frac{1}{2}}} \tag{35}$$

where:

$$\Omega = \frac{2h\omega}{\pi C_r}, \quad \xi = \frac{2Kh}{\pi}, \quad K = \left[\frac{2(1-\nu)}{1-2\nu}\right]^{\frac{1}{2}}$$

and ν is the Poisson's ratio.

The dispersion curves for the case $\nu = 0.34$ are given in figures 19 and 20 both for symmetric and antisymmetric modes for phase and group velocity respectively.

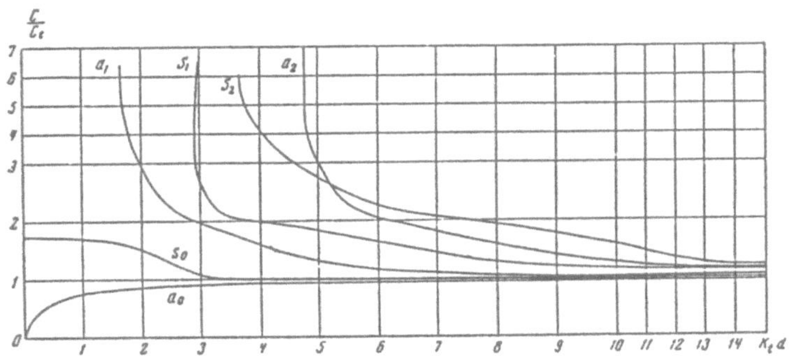

FIGURE 19 Dispersion curves for phase velocity (s = symmetric modes, a = antisymmetric modes)

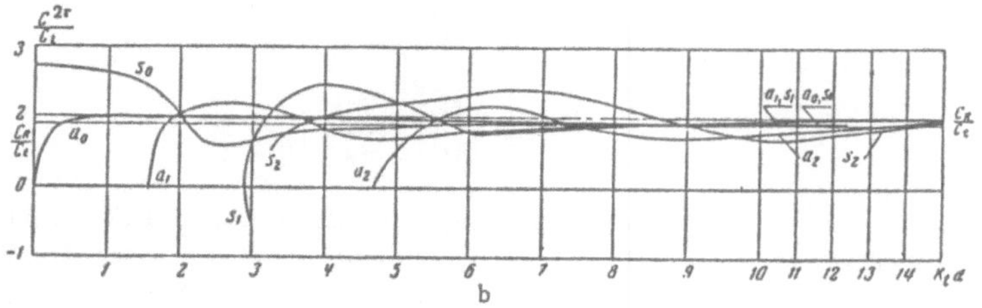

FIGURE 20 Dispersion curve for group velocity (s = symmetric modes, a = antisymmetric modes)

A special class of exact solutions, called Lame' modes, exist in plates. These modes result from a propagating train of SV waves at 45°. These modes have special significance in the applications to bonded plates. On figure 21 the formation of these waves are represented

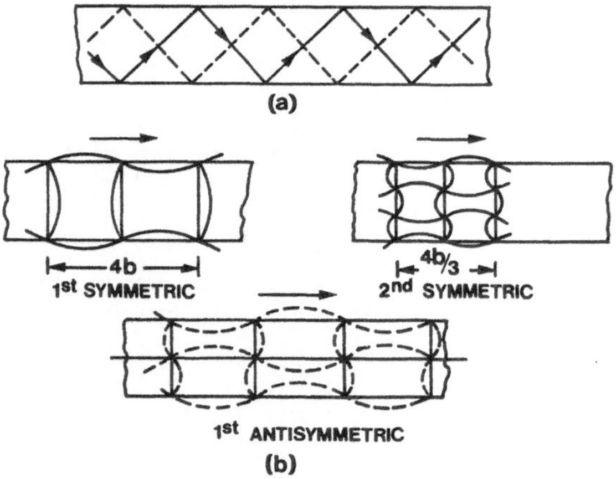

FIGURE 21 Formation of (a) symmetric Lame' modes by SV waves at 45°, and (b) the first few Lame' modes.

9. LEAKY LAMB WAVES (15)

When a plate is immersed in water, leaky Lamb waves may propagate in analogy to leaky Rayleigh waves propagating on surfaces. By considering the liquid-solid-liquid problem for an incident waves, the reflection coefficient can be calculated and from that the symmetric and antisymmetric modes can be obtained. These modes are shown for brass plate in water in figure (22).

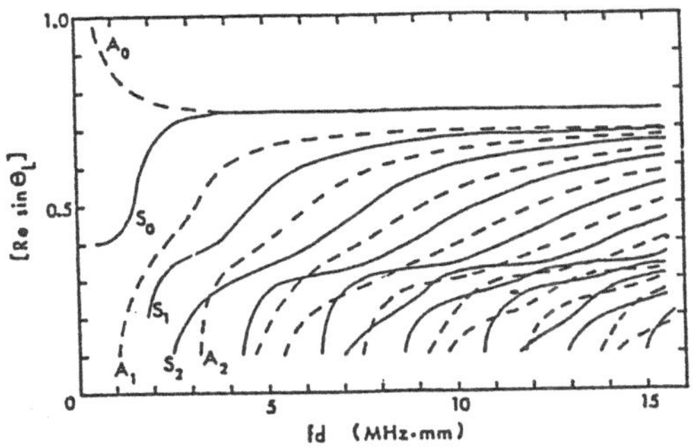

FIGURE 22 The real part of the pole locations in the complex plane as a function of the frequency width product (fd) for a brass plate in water. The solid lines are the symmetric poles and the dashed lines are the antisymmetric poles.

Let us now consider an acoustic wave incident on a fiber reinforced composite plate immersed in water in case the acoustic wavelength is much larger then the fibers. Figure (23) shows the parallel fibers in a host material. Figure 24 shows a bilayered structure as an intermediate step in the calculations. First the layered structure (figure 24) was analysed, deriving from this parallel stress model the properties of compound layer 1 in figure 25 as consisting of two dimensional "compound": layer 1 stacked in series with the (matrix) layer 2.

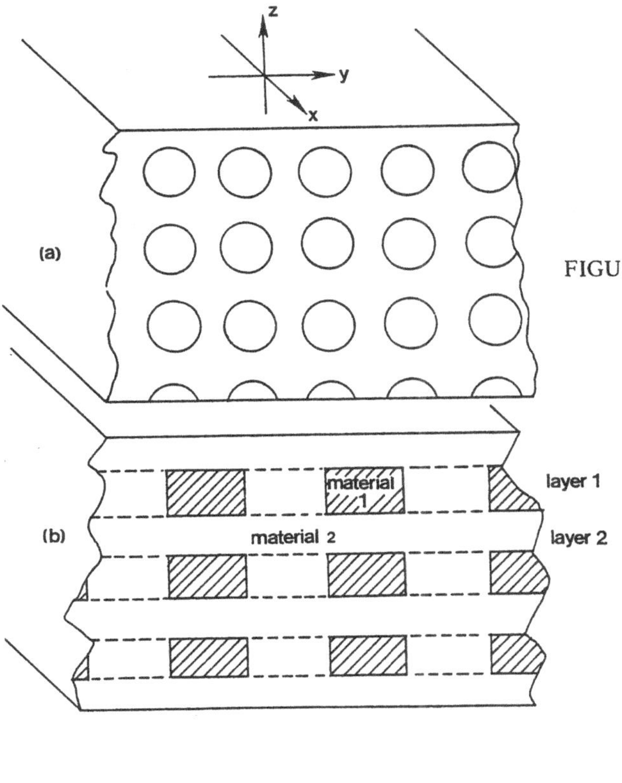

FIGURE 23 Fibrous Composite

a) Actual
b) Model

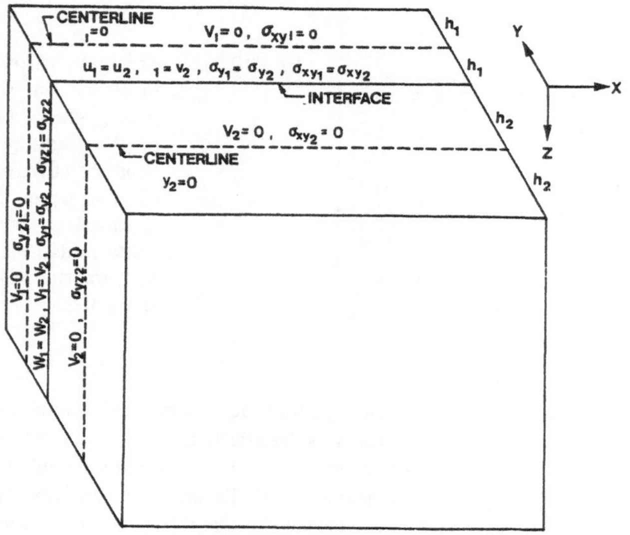

FIGURE 24 Bilayered structures with relevant boundary condition

The reflection coefficient (magnitude and phase) gives sharp minima in the magnitude and rapid reversal in the phase attesting the presence of propagating modes obviously called Lamb waves. From the angle of incidence the phase velocity of the Lamb waves can be obtained.

10. LAMB WAVE DIFFRACTION THROUGH AN APERTURE (17,18)

The problem is schematically illustrated in figure (26). An incident mode in region I onto the sheet juction (region III). The incident wave is partially reflected back into region I and diffracted at the tip into regions I and I'. The wave which passes into region III is transfering into Lamb modes, which are further reflected, diffracted and transmitted into region II. It has been shown that the reflection coefficient of a_1 mode becomes zero if

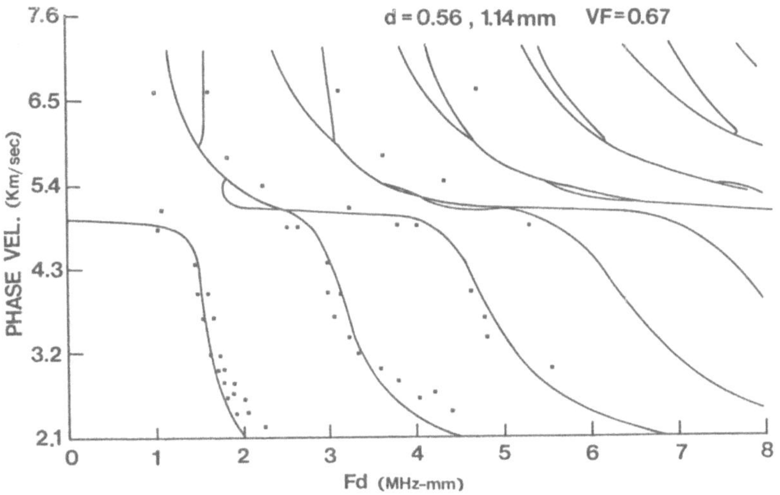

FIGURE 25 Velocity dispersion curves for glass-epoxy system with volume fraction of fiber = 0,67: theory (solid curve) and esperimental points.

$$K_t h = \sqrt{2}\,\pi \tag{36}$$

The Lamb mode at this point is the Lame mode which is a pure shear wave (SV) propagating at an angle 45°. To satisfy the Lame mode condition in region III it is necessary to excite an incident mode s_o at a given thickness so that Eq. (36) is satisfied. It may be calculated that for this mode the transmission coefficient for a circular aperture is given as

$$T \sim K_t h(d/h)^3 \tag{37}$$

Application of this concept to the measurement of the diameter of welded plates requires to generate an s_o mode in plate 1 and measure the transmission coefficient in plate 2. Figure (29) gives the schematic diagram for the method of measurements. Results for a large number of welded plates are plotted in figure (30). The theoretical prediction for transmission coefficient predicts well the spot size. Further developments indicate good correlation between spot size and weld strength (18).

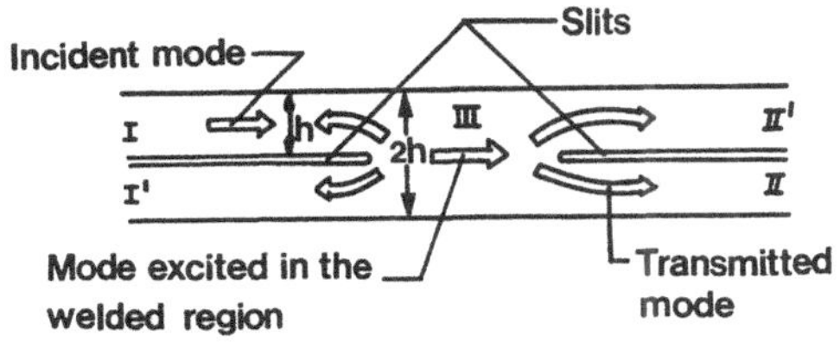

FIGURE 26 Schematic explanation of Lamb wave diffraction by on aperture.

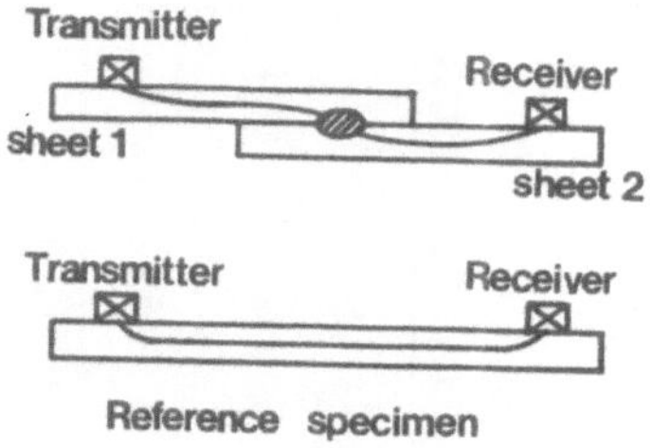

FIGURE 27 Schematic illustration of the method of measurement: (a) welded specimen; (b) reference.

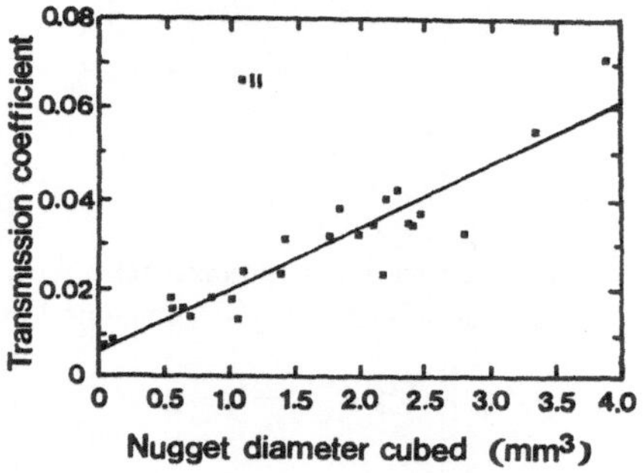

FIGURE 28 Transmission data vs. diameter cubed.

REFERENCES

1. J.D. Achenbach, "Wave Propagation in Elastic Solids", North-Holland Press (1973).
2. W.G. Neubauer, J. Appl. Phys., 44 48 (1973).
3. H.L. Bertoni and T. Tamır, Appl. Phys. 2 157 (1973).
4. M.A. Breazeale, L. Adler and G.W. Scott, J. Appl. Phys, 48 530 (1977).
5. O.I. Diachok and G.W. Mayer, J. Acoust, Soc. Amer. 47 155 (1970).
6. L. Adler, M. deBilly and G. Quentin, J. Appl. Phys. 53 8756 (1982).
7. M. deBilly, L. Adler and G. Quentin, J. Acoust. Soc. Amer. 72 1018 (1982).
8. A. Jungman, L. Adler and G. Quentin, J. Appl. Phys, 53 7 (1982).
9. M. Clayes, O. Leroy, A. Jungman and L. Adler, J. Appl. Phys. 54 10 (1983).
10. G.W. Farnell and E.L. Adler, Phys. Acous. edited by W.P. Mason and R.N. Thurston, Academic Press 9 35 (1972).
11. A.H. Nayfeh, D.E. Chimenti, L. Adler and R.L. Craine, J. Appl. Phys. 52 4985 (1981).
12. D.E. Chimenti, Appl. Phys. Lelt 43 46 (1983).
13. K. Graff: Wave Motion in Elastic Solids, Chanendon Press (1975).
14. I.A. Victorov: Rayleigh and Lamb Waves, Plenum Press (1967).
15. L.E. Pitts, T.J. Plona and G.W. Mayer I.E.E.E. Sonics and Ultrasonic SV 24 101 (1977).
16 D.E. Chimenti and A.H. Nayfeh, accepted for pubblication to J. Acoust. Soc. Amer. (Jan. 1986).
17. S.I. Rokhlin, J. Acoust. Soc. Amer. 69 922 (1979).
18. S.I. Rokhlin and L. Adler, J. Appl. Phys. 56 728 (1984).

DISCUSSION

Comment: Hutchins
How is that you have been able to observe a slow compressional wave in a porous rock, when other workers have failed?

Reply: Adler
In order to separate slow waves from fast waves in a Biot-material, our method can use a rather thin sample, and therefore the highly attenuated slow wave will show up. The opposite is true for the previously used methods.

Comment: Cheeke
In the reflectivity function as a function of the angle at a solid-liquid interface, there is an apparent difficulty at $\vartheta = \vartheta_{CL}$. The modulus of the reflection coefficient goes to unity and yet there appears to be a transverse wave transmitted into the solid. Do you have a simple explanation for this?

Reply: Adler
Considering the complex reflection coefficient $R(k_x)$ there are various square roots that produce some singularities. These singularities include (amongst others) branch points at $k_x = \pm k_l$, where k_l is the l-wavenumber. There is a relatively fast variation of the phase term ϕ (k_x) of the $R(k_x)$ near the longitudinal critical angle. This phase variation can be responsible for the transverse wave transmission even if the modulus is unity.

Comment: Busse
In the expression of the reflection coefficient, there is a difference between two complex numbers at the denominator. Is it obvious why the reflection coefficient cannot be larger than one?

Reply: Adler
Since we are dealing here with complex quantities, the expressions that are presented contain both amplitude and phase. Once these parameters are separated, the amplitude (modulus) of the reflection coefficient will never be larger than one.

DETERMINATION OF LINEAR AND NONLINEAR PARAMETERS OF THIN LAYERS

G. SOCINO

1. INTRODUCTION

Thin film layers play an important role in several applications in the field of acoustics. Piezoelectric films, such as for example ZnO films, are of great importance on the generation and detection of high frequency bulk and surface acoustic waves, in applications including acoustic microscopy and acousto-optic Bragg cells. Thin film overlays are widely used for improving the acoustic propagation characteristics and the transducer performance in surface acoustic wave passive devices, such as resonators and filters. Other applications rely on the use of thin films as guiding structures for surface acoustic wave delay lines, as well as for both acoustic and light waves in acousto-optic planar cells for integrated optics. There are materials, moreover, such as polymer polyvinylidene fluoride (PVF_2), which are stretched into the form of thin layers a few tens of micron thick for a number of piezoelectric acoustic wave transducer applications.

The mechanical and electrical properties of these films differ in a more or less pronounced way from those of the corresponding bulk materials and can be highly affected by the processes of their production. This implies that a direct characterization of the layers is often necessary. Acoustic investigation techniques have been developed for use in determining the elastic and piezoelectric second order constants of the layer materials. These techniques are based on the measurement of bulk thickness modes of the layers [1,2] or on the analysis of generalized Rayleigh modes propagating along the layer [3-7], this latter technique including the former one as a particular case. The potentiality of these techniques relies on the intrinsic characteristics of the acoustic modes considered. Each mode is, in fact, constituted by a combination of the longitudinal and shear acoustic bulk waves coupled by multiple reflection at the plane layer boundaries. These waves are usually called the partial waves of the acoustic mode. The procedure followed consists in analysing a spectrum of modes having different partial wave propagation directions into the layer. A comparison of the values of the phase velocities obtained experimentally with those calculated theoretically for a given set tentatively assigned of materials constants can lead, through a computer optimization process, to the determination of the second order elastic and piezoelectric constants.

Recently, the investigation of acoustic propagation in layers has been extended to nonlinear effects with the analysis of the interaction between acoustic modes propagating in a thin piezoelectric plate and a bias electric field [8,9]. It has been shown that the applied field affects the phase velocity of plate modes on a way which strongly depends on the mode considered as well as on the interaction geometry. This effect can be interpreted theoretically in terms of the dynamic equations and boundary conditions of nonlinear electroacoustics [10-12]. In the approximation of small amplitude waves superposed on bias, it turns out that the effect of a mechanical or electric bias field is that of modifying the values of the second order constants of the material by small amounts which are proportional to the intensity of the applied field [13,14]. The proportionality factors are combinations of second and higher order constants of the material, which depend on the crystallographic symmetry of the medium and on the geometry of the interaction. It will be shown that the values of such combinations can be experimentally determined by measurement of the variations in phase velocity produced by the bias field on different layer modes. The procedure is similar to the one used, in the case of bulk piezoelectric materials, for determining higher order material constants which are responsible for nonlinear effect [15-17]. Some theoretical background to these experimental methods will be briefly outlined in this paper, together with an analysis of their more interesting aspects.

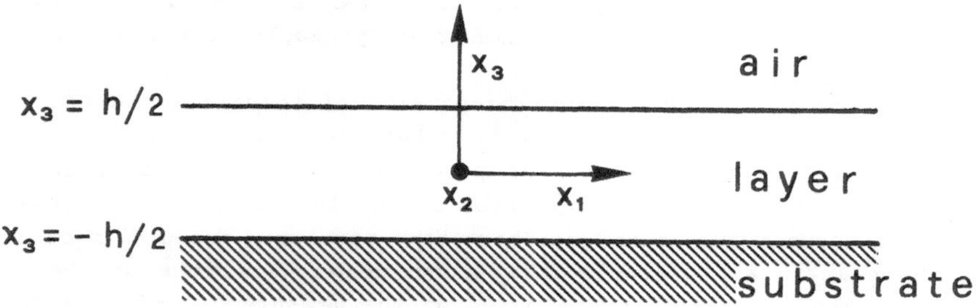

Fig.1. Geometry of the layer and coordinate system.

2. ACOUSTIC PROPAGATION IN THIN LAYERS

2.1. Dynamic equations and boundary conditions

The configuration of the material system under investigation is shown in Fig.1. A thin layer of thickness h is rigidly bonded to a substrate having different mechanical and electrical characteristics. Both media are assumed to be anisotropic and piezoelectric, so that the mechanical displacement in the layer and in the substrate are coupled to the electric field by the piezoelectric constants. A limiting case of this configuration is represented by a layer with the plane boundaries abutting free space. The waves of interest are of the acoustic type, which implies that the "quasi electrostatic" approximation can be used, i.e. one neglects the magnetic induction field associated with the wave and represents the electric field as the gradient of a scalar quantity.

The laboratory coordinate system x_1 shown in Fig.1 is used, with x_3 normal to the layer and x_1 along the wavevector direction of the acoustic mode. All the acoustic field quantities are supposed to be independent of the spatial coordinate x_2. In the quasi electrostatic regime the acoustic modes propagating along the layer must satisfy Newton's second law and Maxwell's equation in both the media in contact:

$$\begin{cases} \rho \ddot{u}_i = T_{ji,j} \\ D_{h,h} = 0 , \end{cases} \qquad \begin{matrix}\text{dynamic equations} \\ \text{in a charge free dielectric}\end{matrix} \tag{1}$$

together with the mechanical and electrical boundary conditions at the free surface and at the interface between the two media:

boundary conditions at the interface, $x_3 = -h/2$

$$u_i - u'_i = 0 \qquad D_3 - D'_3 = 0$$
$$T_{3i} - T'_{3i} = 0 \qquad \varphi - \varphi' = 0 \tag{2a}$$

boundary conditions at the free surface, $x_3 = h/2$

$$T'_{3i} = 0 \tag{2b}$$
$$D'_3 + \varepsilon_0 \varphi^0_{,3} = 0$$

ρ being the mass density , u_i the components of the particle displacement vector, φ the electric potential and T_{ij}, D_h the components of the stress tensor and of the electrical displacement vector, respectively; primed quantities refer to the layer. φ^0 stands for the electric potential in the free space, satisfying Laplace's equation $\varphi^0_{,hh}=0$, and ε_0 is the electric permittivity in vacuo. Summation over repeated subscripts is assumed and a comma followed by a letter stands for a derivative with respect to the corresponding spatial coordinate.

In the linear approximation, T_{ij} and D_h can be expressed as a linear combination of the derivative of the displacement components u_j and of the electric potential φ with respect to the spatial coordinate x_n, according to:

$$T_{ij} = c_{ijmn} u_{m,n} + e_{nij} \varphi_{,n}$$
$$D_h = e_{hmn} u_{m,n} - \varepsilon_{hn} \varphi_{,n} \tag{3}$$

where c_{ijmn}, e_{nij} and ε_{hn} are the components of the elastic stiffness tensor, of the piezoelectric tensor and of the dielectric permittivity of the substrate, respectively. All of them are rotated material constants, referred to the laboratory coordinate system of Fig.1. Similar expressions can be written for the layer on using the corresponding material constants.

Substitution of Eqs.3 into Eqs.1, gives the following set of four differential equations in the substrate:

$$\begin{aligned} \varrho \ddot{u}_i &= c_{ijmn} u_{m,nj} + e_{nij} \varphi_{,nj} \\ 0 &= e_{hmn} u_{m,nh} - \varepsilon_{hn} \varphi_{,nh} \end{aligned} \tag{4}$$

and an analogous set of equations in the layer

2.2. Layer modes

Plate mode solutions are sought in the monocromatic travelling wave form:

$$\begin{aligned} u_i &= a_i \exp(j \beta b x_3) \exp j(\beta x_1 - \omega t) \\ \varphi &= a_4 \exp(j \beta b x_3) \exp j(\beta x_1 - \omega t) \end{aligned} \tag{5}$$

which represents a plane wave with propagation constants β along x_1 direction and βb along the normal to the layer. Amplitudes a_j and a_4 can be determined on substitution of Eqs.(5) into Eqs.(4), which gives a system of four linear homogeneous equations in the unknown a'_s in the substrate and a similar system in the layer, of the type:

$$\begin{vmatrix} \Gamma_{11}-\rho v^2 & \Gamma_{12} & \Gamma_{13} & \Gamma_{14} \\ \Gamma_{12} & \Gamma_{22}-\rho v^2 & \Gamma_{23} & \Gamma_{24} \\ \Gamma_{13} & \Gamma_{23} & \Gamma_{33}-\rho v^2 & \Gamma_{34} \\ \Gamma_{14} & \Gamma_{24} & \Gamma_{34} & \Gamma_{44} \end{vmatrix} \begin{vmatrix} a_1 \\ a_2 \\ a_3 \\ a_4 \end{vmatrix} = 0 \tag{6}$$

where $v = \omega/\beta$ is the phase velocity along the x_1 direction and :

$$\begin{aligned} \Gamma_{ij} &= c_{3ij3} b^2 + (c_{i31j} + c_{i13j}) b + c_{i11j} \\ \Gamma_{i4} &= e_{33i} b^2 + (e_{31j} + e_{31i}) b + e_{11i} \\ \Gamma_{44} &= -(\varepsilon_{33} b^2 + 2\varepsilon_{13} b + \varepsilon_{11}) \end{aligned} \tag{7}$$

In order to obtain non trivial solutions, the determinant of the coefficients must be set equal to zero, which gives an eight order algebraic equation in the unknown b, with the phase velocity v as an assigned parameter, for either medium. All the solutions are retained in the layer, while only those lying in the lower half of the complex plane are taken in the substrate, so that the acoustic field components decay to zero below the interface. Upon sobstituting the obtained solutions $b^{(p)}$ into Eqs. 6, the corresponding polarization $a_j^{(p)}$ and electrical potential amplitudes $a_4^{(p)}$ of the partial waves can be determined. Each mode will then consist of a linear combination of eight partial waves in the layer and of four partial waves in the substrate, of type:

$$\begin{cases} u'_i = \Sigma_q A'_q a'^{(q)}_i \exp(j\beta b'^{(q)} x_3) \exp j(\beta x_1 - \omega t) \\ \varphi' = \Sigma_q A'_q a'^{(q)}_4 \exp(j\beta b'^{(q)} x_3) \exp j(\beta x_1 - \omega t) \end{cases} \quad \text{layer}$$

$$\begin{cases} u_i = \Sigma_p A_p a_j^{(p)} \exp(j\beta b^{(p)} x_3) \exp j(\beta x_1 - \omega t) \\ \varphi = \Sigma_p A_p a_4^{(p)} \exp(j\beta b^{(p)} x_3) \exp j(\beta b_1 - \omega t) \end{cases} \quad \text{substrate} \tag{8}$$

The number of partial waves reported here refers to the more general case and is lower when the crystallographic symmetry and the propagation geometry uncouple some of the partial waves from the others.

2.3 Dispersion relations

The amplitudes of the partial waves and the wave numbers β corresponding to layer mode solutions can be obtained, for any given value of the phase velocity v, by inserting Eqs. (8) into the boundary conditions, Eqs. (2), which gives the following system of twelve homogeneous linear equations in the unknowns $A^{(p)}$ and $A'^{(p)}$:

<u>boundary conditions at the interface</u>, $x_3=-h/2$

continuity of the particle displacement components u_i, i=1,2,3

$$\Sigma_p A_p a_i^{(p)} \exp(-j\beta b^{(p)} h/2) - \Sigma_q A'_q a'^{(q)}_i \exp(-j\beta b'^{(q)} h/2) = 0 \tag{9a}$$

continuity of the stress tensor components T_{3i}, i=1,2,3

$$\Sigma_p A_p [(c_{i31j} + c_{i33j} b^{(p)}) a_j^{(p)} + (e_{13i} + e_{33i} b^{(p)}) a_4^{(p)}] \exp(-j\beta b^{(p)} h/2) -$$
$$-\Sigma_q A'_q [(c'_{i31j} + c'_{i33j} b'^{(q)}) a'^{(q)}_j + (e'_{13i} + e'_{33i} b'^{(q)}) a'_4] \exp(-j\beta b'^{(q)} h/2) = 0 \tag{9b}$$

continuity of the electric displacement normal component

$$\Sigma_p A_p [(e_{3j1}+e_{3j3}b^{(p)})a_j^{(p)} - (\varepsilon_{31}+\varepsilon_{33}b^{(p)})a_4^{(p)}]\exp(-j\beta b^{(p)}h/2) -$$

$$-\Sigma_q A'_q [(e'_{3j1}+e'_{3j3}b'^{(q)})a'^{(q)}_j - (\varepsilon'_{31}+\varepsilon'_{33}b'^{(q)})a'^{(q)}_4]\exp(-j\beta b'^{(q)}h/2=0 \qquad (9c)$$

continuity of the electric potential

$$\Sigma_p A_p a_4^{(p)}\exp(-j\beta b^{(p)}h/2) - \Sigma_q A'_p a'^{(q)}_4 \exp(-j\beta b'^{(q)}h/2) = 0 \qquad (9d)$$

<u>boundary conditions at the free surface</u>, $x_3=h/2$

vanishing of the traction stresses T_{3i}, i=1,2,3

$$\Sigma_q A'_q [(c'_{i31j}+c'_{i33j}b'^{(q)}) a'^{(q)}_j + (e'_{13i}+e'_{33i}b'^{(q)})a'^{(q)}_4]\exp(j\beta b'^{(q)}h/2)=0 \qquad (9e)$$

continuity of the electric potential

$$\Sigma_q A'_q [(e'_{31i}+e'_{33i}b'^{(q)})a'^{(q)}_i - (\varepsilon'_{31}+\varepsilon'_{33}b'^{(q)} - j\varepsilon_o)]\exp(j\beta b'^{(q)}h/2)=0 \qquad (9f)$$

In order to obtain non trivial solutions, the determinant of the coefficients must be set equal to zero. This last condition determines the values of the product between wavenumber β and layer thickness h satisfying the transverse resonance conditions for the specific value of the phase velocity assigned. On following the same procedure for different values of the phase velocity, one can calculate the dispersion relation $v(\beta h)$ for the whole set of layer modes of interest. Numerical computation techniques are necessary for determining the dispersion relation in anysotropic media, except for a few cases of higher symmetry, where an analytical expression of the dispersion relations can be obtained.

It is not the purpose of this paper to go into details of acoustic propagation in thin layers. A thorough analysis of the subject can be found in some more specific paper in this field [18-21]. We shall, rather, briefly show how these waves can be used to investigate the second order elastic and piezoelectric constants of thin layer materials.

2.4. Second order constant evaluation

On the basis of the theory briefly outlined, the dispersion relations $v(\beta h)$ of layer modes can be evaluated for a specific crystallographic orientation of both layer and substrate, once the mass density and the set of elastic, piezoelectric and dielelctric constants involved in the specific geometry of propagation are known. The partial waves coupled in the acoustic mode considered and the set of constants involved can be derived from Eqs.(6) and (7).

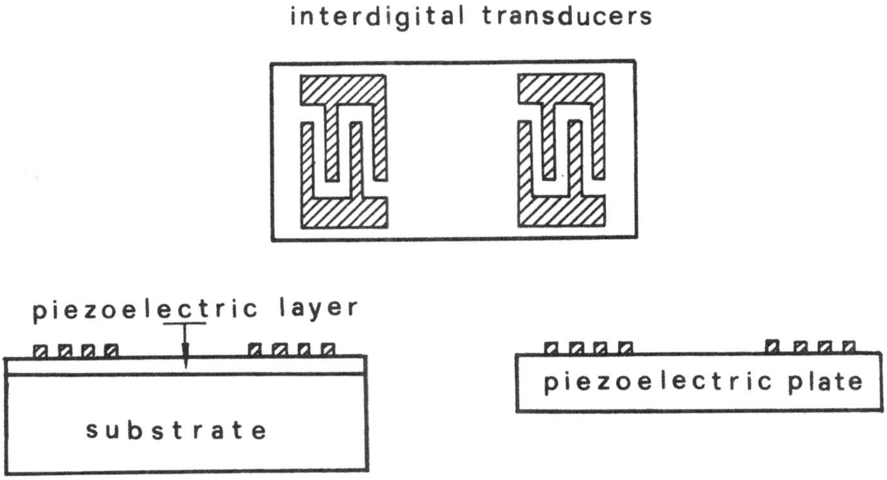

Fig.2 Acoustic delay line configuration in piezoelectric materials for phase velocity measurements.

The evaluation of second order constants from the analysis of measurements of the phase velocity of layer modes, is the inverse problem. In this case one tries to fit the experimental points with the theoretical dispersion curves through use of a computer optimization program, by tentatively assigning different sets of material constants.

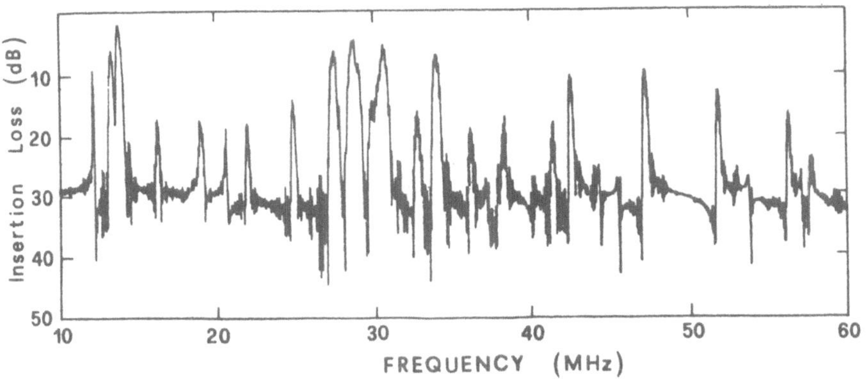

Fig.3. Transmission spectrum of a yz $LiNbO_3$ Lamb wave delay line (h=1.25mm β=10.864mm^{-1})

Experiments can be done either through the measure of the phase delay of the acoustic modes on using a delay line configuration, or by means of acoustooptic techniques, as for example through Brillouin scattering experiments. If either the layer or the substrate are piezoelectric, the acoustic modes can be excited and detected by means of interdigital transducers, as seen in the Fig.2.

In Fig.3 the transmission spectrum is shown, relative to Lamb modes propagating along the z direction of a y-cut $LiNbO_3$ plate with free boundaries. The electrode spacing of the interdigital transducers and the plate thickness h correspond to a value of βh equal 13.58; each transducer contains 25 electrode pairs.

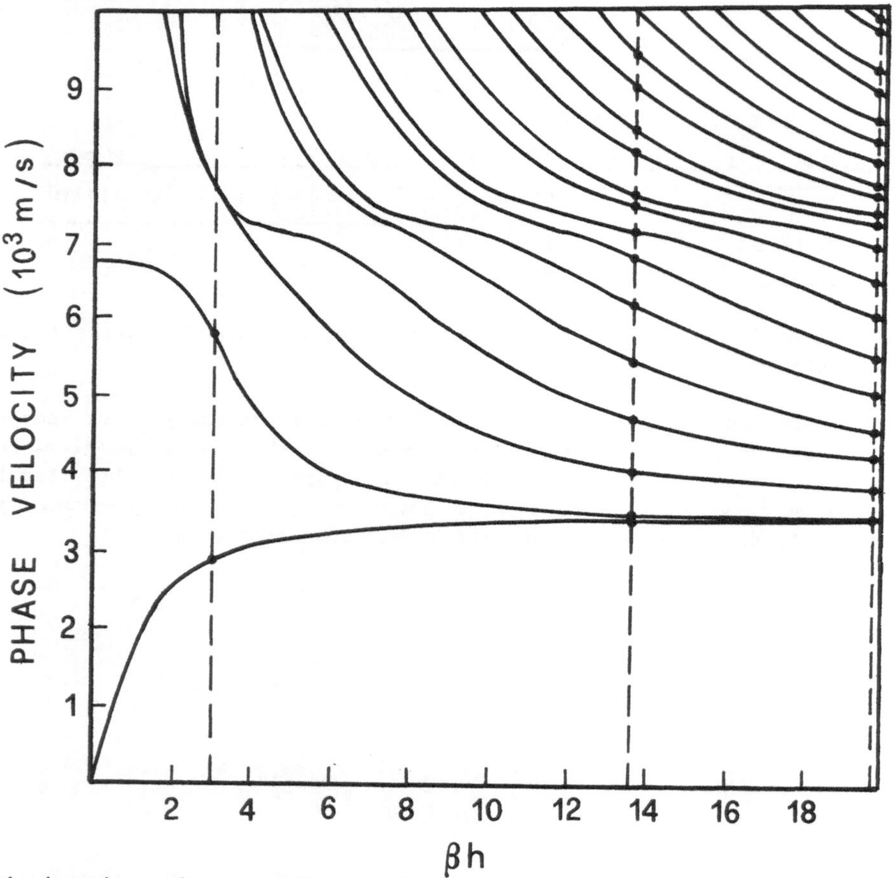

Fig.4. Dispersion relations and phase velocity measurements for Lamb waves propagating along the z direction of a y-cut $LiNbO_3$ plate.

The acoustic modes considered consist of partial waves polarized in the sagittal plane yz of the crystal and are coupled to the electric potential. No coupling exists with the SH shear waves. The expression of the coefficients Γ's involved in this acoustic propagation are:

$$\Gamma_{11} = c_{44}b^2+c_{33} \qquad \Gamma_{14}=e_{24}b^2+e_{33}$$

$$\Gamma_{33}=c_{11}b^2-2c_{24}b+c_{44} \qquad \Gamma_{34}=-e_{22}b^2+(e_{24}+e_{32})b$$

$$\Gamma_{13}=-c_{24}b^2+(c_{23}+c_{44})b \qquad \Gamma_{44}=-(\varepsilon_{11}b^2+\varepsilon_{33})$$

where the material constants are expressed in contracted notation and referred to the crystallographic axes.

The theoretical dispersion curves were computed by assigning to this set of constants the values given on the literature and are plotted in Fig.4 together with the experimental points. The measurements, relative to three values of parameter βh, well agree with the theoretical curves.

Fig.5 and 6 show the measured and calculated values of the dispersion curves relative to a ZnO thin film overlay on an Al_2O_3 ceramic substrate, and to a gold layer on a glass substrate, respectively. The dashed lines correspond to the Rayleigh wave velocity on the layer and to the shear wave velocity in the substrate which limit the range of layer mode velocities.

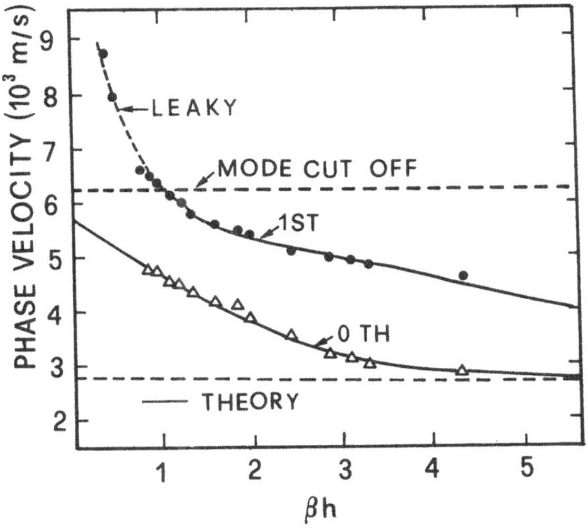

Fig.5. Experimental and theoretical values of plate mode phase velocity for ZnO films on Al_2O_3 ceramics (Ref.5).

This technique has also been successfully used for determining elastic constants of PVF_2 layers[3,4] and for the analysis of their dependence on temperature, in the range -50°C to -175°C[3].

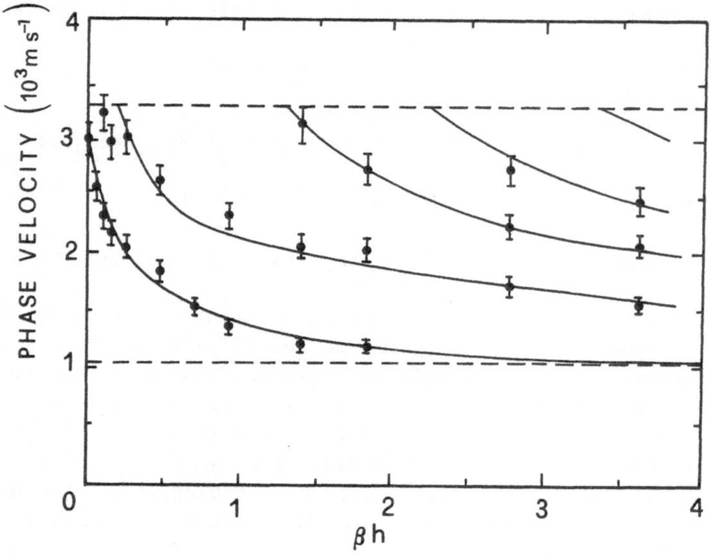

Fig.6. Dispersion curves and phase velocity measurements performed by Brillouin scattering experiments, for gold films on glass (Ref.7).

3. NONLINEAR PARAMETER DETERMINATION

3.1. Nonlinear dynamic equations

The nonlinear effects in acoustic wave propagation along the dielectric layer can be interpreted in terms of the dynamic equations and boundary conditions of the nonlinear electroacoustics of dielectric media. Higher order terms in the field quantities have to be taken into account, for the nonlinear case, in the constitutive relations. On referring to the material frame of reference x_1, the complete set of bilinearized electroelastic equations take the form:

$$\varrho \ddot{u}_i = P_{ji,j}$$
$$D_{m,m} = 0 \; , \qquad (10$$

where P_{ji} and D_h are the Piola - Kirchhoff stress tensor and the nonlinear electric displacement, respectively:

$$P_{ij} = c_{ijkl} u_{k,l} + e_{mij} \varphi_{,m} +$$
$$+(1/2)(c_{ijklmn} + \delta_{km} c_{ijnl} + 2\delta_{jm} c_{inkl}) u_{k,l} u_{m,n} + \qquad (11$$

$$+(e_{mijkl}+\delta_{kj}e_{mil})u_{k,l}\varphi_{,m}-$$

$$-1/2\,[e_{mnij}-\varepsilon_\circ(\delta_{im}\delta_{jn}+\delta_{in}\delta_{jm}-\delta_{ij}\delta_{mn})]\,\varphi_{,m}\varphi_{,n}+\ldots$$

$$D_m=-\varepsilon_{mn}\varphi_{,n}+e_{mij}u_{i,j}+$$

$$+1/2\,\varepsilon_{mnp}\varphi_{,n}\varphi_{,p}+1/2(\delta_{ik}e_{mlj}+e_{mijkl})u_{i,j}u_{k,l}-$$

$$-[e_{mnij}-\varepsilon_\circ(\delta_{mi}\delta_{nj}+\delta_{ni}\delta_{mj}-\delta_{ij}\delta_{mn})]\,\varphi_{,n}u_{i,j}+\ldots$$

where c_{ijklmn}, e_{mil} and ε_{mnp} are the third order stiffness, piezoelectric or electroelastic and dielectric constants, respectively, and e_{mnij} are the components of the elctrostrictive tensor.

Nonlinear acoustic propagation in a layer can be properly described by using the dynamic equations Eq.(10) and the constitutive nonlinear relations, Eq.(11), together with boundary conditions of the type:

$$\begin{cases} u_i-u_i'=0 \\ P_{3i}-P'_{3i}=0 \end{cases}\qquad \begin{cases} D_3-D'_3=0 \\ \varphi-\varphi'=0 \end{cases}\qquad \text{at } x_3=-h/2 \tag{12}$$

$$P'_{3i}=0 \qquad D'_3+\varepsilon_\circ\varphi^\theta_{,3}=0 \qquad \text{at } x_3=h/2$$

We shall briefly analyse in the following the specific case of nonlinear interaction between layer modes and a bias mechanical or electric field, under the symplifying assumption of small amplitude waves superposed on a bias.

3.2 Nonlinear interaction between a bias field and layer modes.

Let us consider a uniform mechanical field $u^\circ_{p,q}$ and a uniform electric field $E^\circ_h=-\varphi^\circ_{,h}$ to be applied to a layer. The components of the displacement vector and of the electric potential in Eq.(11) can then be written as the sum of a motional term, depending on time and position, and the uniform and constant contribution due to the biasing field. In the approximation of small amplitude acoustic waves superposed on a bias, we can neglect terms in the products of the dynamics field quantities. The constitutive relations result, then, linear in both $u_{k,l}$ and $\varphi_{,m}$, and take the form:

$$P_{ij}=(c_{ijkl}+\Delta c_{ijkl})u_{k,l}+(e_{mij}+\Delta e_{mij})\varphi_{,m}$$

$$D_m=(e_{mij}+\Delta e_{mij})u_{i,j}-(\varepsilon_{mn}+\Delta\varepsilon_{mn})\varphi_{,n} \tag{13}$$

with

$$\Delta c_{ijkl} = (c_{ilmn}\delta_{kj}+c_{ijnl}\delta_{mk}+c_{inkl}\delta_{mj}+c_{ijklmn})u^{\circ}_{m,n} +(e_{mil}\delta_{kj}+e_{mijkl})\varphi^{\circ}_{,m}$$

$$\Delta e_{mij} = (e_{mil}\delta_{kj}+e_{mijkl})u^{\circ}_{k,l} + [-e_{mnij}+\varepsilon_{\circ}(\delta_{mi}\delta_{nj}+\delta_{ni}\delta_{mj}-\delta_{mn}\delta_{ij})]\varphi^{\circ}_{,n} \tag{14}$$

$$\Delta\varepsilon_{mn} = [e_{mnij}+\varepsilon_{\circ}(\delta_{mn}\delta_{ij}-\delta_{mi}\delta_{nj}-\delta_{ni}\delta_{mj})]u^{\circ}_{i,j} - \varepsilon_{mnp}\varphi^{\circ}_{,p}$$

The effect of the bias field is simply that of modifying the second order material constants by perturbation terms which are linked to the bias field components through linear combinations of second and third order constant. On substituting Eq.12 into the dynamic equations, one obtains a set of linear differential equations of the type:

$$\varrho\ddot{u}_i=(c_{ijkl}+\Delta c_{ijkl})u_{k,lj}+(e_{lji}+\Delta e_{lji})\varphi_{,lj}$$
$$0 = (e_{jkl}+\Delta e_{jkl})u_{k,lj}-(\varepsilon_{jl}+\Delta\varepsilon_{jl})\varphi_{,lj} \tag{15}$$

which is identical to the one relative to the linear case, except for the perturbation terms adding to the material constants. This implies that layer mode solution can be sought by following the same procedure as in the linear case.

The combinations of material constants involved in the perturbation terms given in Eq.(14), depend on the crystallographic symmetry of the material and on the layer orientation, as well as on the characteristics of the bias field. An analysis of the dependence of the phase velocity of the acoustic layer modes on the bias field intensity can, thus, lead to the determinations of these combinations and give important informations about the third order material constants.

3.3. Experimental procedure

Experiments can be done by applying a uniform mechanical or electric bias field to the layer and by measuring the changes in phase velocity of the acoustic modes as a function of the bias field intensity. In Fig.7 the experimental arrangement is shown, used for analysing the effect of a bias electric field on Lamb modes propagating along a thin $LiNbO_3$ plate [7,8]. The acoustic line is connected to an amplifier in the configuration of an acoustically stabilized oscillator, with a tunable filter in the feedback loop in order to select the plate mode to be investigated. The oscillation frequency f of the oscillator is related to the phase velocity of the Lamb mode as follows:

$$f = nv/L_o \tag{16}$$

where L_o is the distance between the transducer in the unbiased plate and n the integer number of acoustic wavelengths contained in L_o. An almost uniform bias electric field is applied across the sample by means of two thin film electrodes deposited on the two large paces of the plate. The bias field produces a fractional change in the oscillation frequency which, according to Eq.(16), can be written as:

$$\Delta f/f = \Delta v/v \tag{17}$$

It has to be pointed out that, when referring to the material frame of reference x_i, the fractional change in phase velocity given in Eq.17 includes the effects of change in lenght and thickness of the plate produced by the bias field.

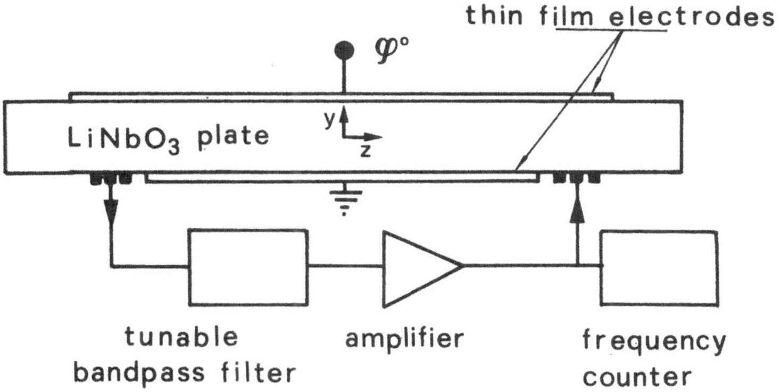

Fig.7. Experimental arrangement for measurements of changes in phase velocity produced by an electric field on Lamb modes propagating along a piezoelectric plate.

The fractional phase velocity variations of four Lamb modes propagating along the z direction of a y-cut $LiNbO_3$ plate are plotted in Fig.8 versus the bias electric field intensity. In order to obtain from the measurements the needed information relative to higher order material constants it is convenient to express $\Delta v/v$ through a series expansion in the applied electric field intensity E_o:

$$\Delta v/v = aE^\circ + bE^{\circ 2} + \text{higher order terms} \tag{18}$$

where the coefficients

$$a = [\delta(\Delta v/v)/\delta E^\circ]_{E^\circ=0} \quad \text{and} \quad b = [\delta^2(\Delta v/v)/\delta E^{\circ 2}]_{E^\circ=0} \tag{19}$$

give the linear and quadratic contribution to the electroelastic effect, respectively.

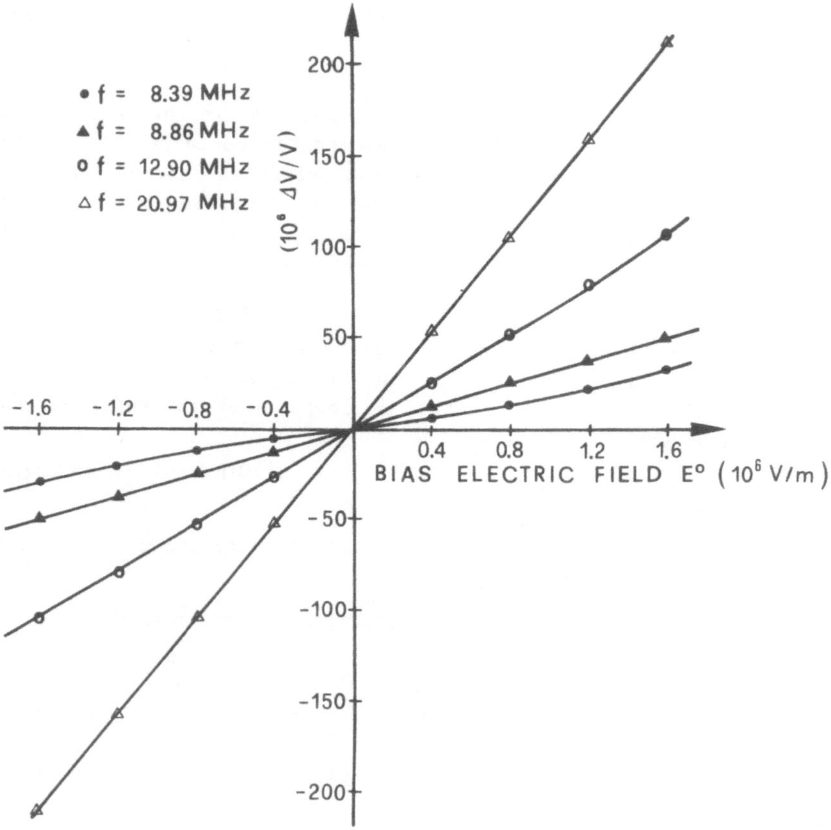

Fig.8. Fractional velocity change vs. bias electric field intensity for four Lamb modes propagating along a yz-$LiNbO_3$ plate ($\beta h=19.83$).

For a given layer orientation and propagation geometry, the values of coefficients a and b strongly depend on the acoustic mode considered and on the parameter βh. This feature is evident in Fig. 9 , where the values of the coefficient a are plotted for different modes propagating in a yz-$LiNbO_3$ plate, for two values of parameter βh.

3.4 Nonlinear parameter determination.

According to the theory previously outlined, the experimental values of coefficients a and b of Eq.(18) can be interpreted in terms of the modifications produced by the bias field on the second order material constants involved in the propagation geometry under investigation. These modifications are given by a linear combination of the second and third order constants of Eq.(14). In order to determine their values, the change in phase velocity of several acoustic modes with respect to the unbiased

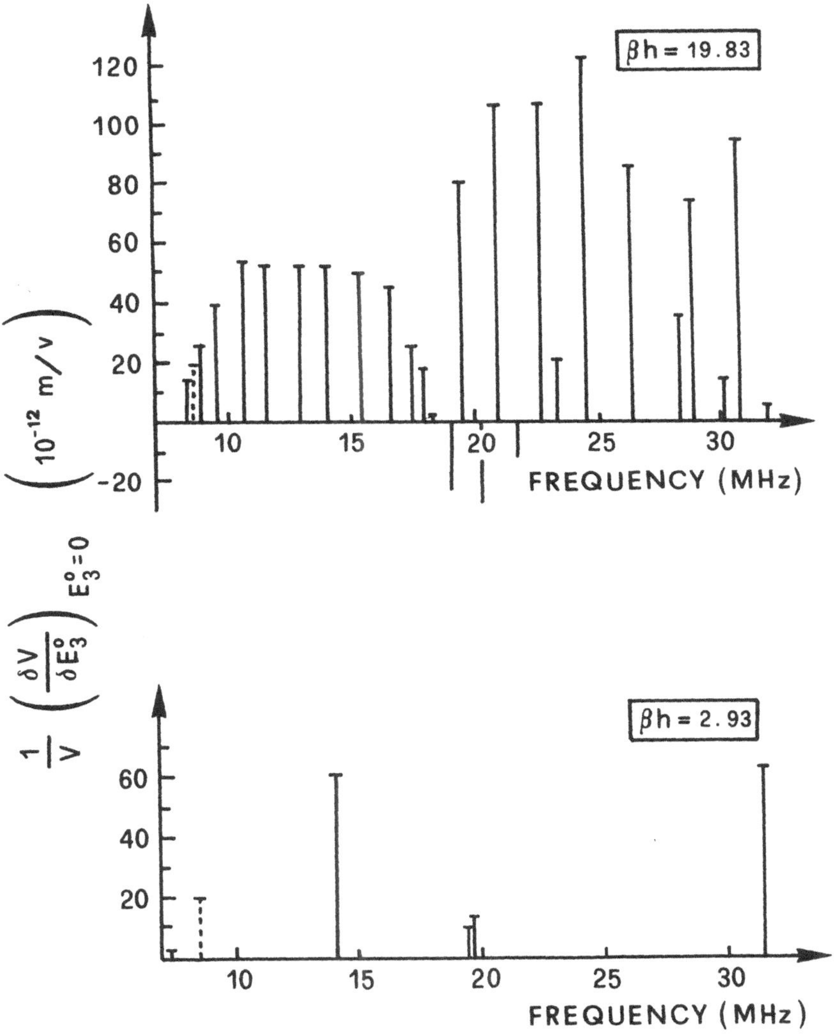

Fig. 9 . Measurements of the nonlinear electroacoustic parameter for Lamb modes propagating along a yz-$LiNbO_3$ plate.

condition is evaluated by tentatively assigning a numerical value to the set of perturbation terms Δc_{ijkl}, Δe_{mij} and $\Delta \varepsilon_{mn}$ of Eq.(14), involved in the experimental case considered. A comparison of the theoretical results with the experimental ones leads, through a numerical optimization technique, to the determination of these nonlinear parameters. The number of modes analysed has to be at least equal to the number of independent material constant combinations relative to the geometry of interaction investigated. In addition, since some of the perturbation terms are effective and others inconsequential in modifying the phase velocity of a given mode, one has to make an appropriate selection of the modes to be analysed in order to obtain more accurate and complete information.

An investigation of different interaction geometries and the use of different types of bias field allows one to evaluate several third order constant combinations. Provided that the number of independent combinations so evaluated is at least equal to the number of unknown constants, the procedure followed leads to the determination of the third order material constants.

The experimental method outlined is similar to the one used for determining the electroelastic coefficients of piezoelectric crystals through an analysis of the fractional frequency changes produced by a bias electric field on bulk wave resonators [15-17]. In the present case, the acoustic modes consist of combinations of longitudinal and shear partial waves propagating in different directions into the layer, so that the information obtained for a given interaction geometry is larger than in the case of bulk waves. The potentiality of the method lies mainly in this intrinsic characteristic of layer modes, which permits an evaluation of the elastic and piezoelectric nonlinear properties of the layer material through the analysis of a limited number of interaction conditions. This last property is of great importance in the case of thin film layer characterization, since the possibilities of investigation are limited by the configuration and by the crystallographic orientation of the layer.

REFERENCES

1. Ohigashi H.: Elecromechanical Properties of Polarized Polyvinylidene Fluoride Films as studied by the Piezoelectric Resonance Technique, J.Appl. Phys. 47, 949 (1976).
2. Auld B. and Gagnepain J.: Shear Polarized PVF_2 film studied by the Piezoelectric Resonance Method, J.Appl. Phys. 50, 5511 (1979).
3. Wagers R.S.: Low Temperature Lamb Wave Propagation in PVF_2 J.Appl. Phys. 51, 5797 (1981).
4. Horvat P., Gagnepain J.J. and Auld B.: Propagation of fundamental SH and Lamb in PVF_2 Films IEEE Ultrason. Symp. Proc. 511 (1979).
5. Shiosaki T., Takeda F. and Kawabatia A: High coupling and high velocity surface acoustic waves using c-axis oriented ZnO films on translucent Al_2O_3 ceramics, IEEE Ultrason. Symp. Proc. 83 CH 1947-1, 323 (1983).
6. Jelks E.C. and Wagers R.S.: Elastic Constants of Electron-beam Deposited Thin Films of Molybdenum and Aluminum on $LiNbO_3$, IEEE Ultrason. Symp. Proc. 83 CH 1947-1, 319 (1983).
7. Hillebrands B., Baumgart P., Mock R., Güntherodt G., Bechtmold P.S.: Dispersion of Localized Elastic Modes in Thin Supported Gold Layers

Measured by Brillouin Scattering, J.Appl.Phys. 58, 3166 (1985).
8. Palma A., Palmieri L., Socino G. and Verona E.: Acoustic Lamb Wave-electric Field Nonlinear Interaction in yz-$LiNbO_3$ Plates, Appl.Phys. Lett.46, 25 (1985).
9. Palma A., Palmieri L., Socino G., Verona E.: Nonlinear electro-acoustic Effect in Lamb Wave Propagation in $LiNbO_3$ Plates, IEEE Ultrason.Symp. Proc. 84 CH 2112-1, 958 (1984).
10. Tiersten H.F.: On the Nonlinear Equations for Thermoelectroelasticity, Int. J. Engin. Sci, 9, 587 (1971).
11. Nelson D.F.: Theory of Nonlinar Electroacoustics of Dielectric, Piezoelectricc and pyroelectric Crystals, J. Acoust. Soc. Am. 63, 1738 (1978).
12. Nelson D.F.: Electric, optic and acoustic interactions in dielectrics, J. Wiley, New York, 1979.
13. Baumhauer J.C., Tiersten H.F.: Nonlinear Electroelastic equations for small fields Superposed on a bias, J.Acoust. Soc. Am. 54, 1017 (1973).
14. Tiersten H.F.: Perturbation Theory for Linear electroelastic Equations for Small Fields Superposed on a bias, J.Acoust. Soc. Am. 64, 832 (1978).
15. Gagnepain J.J. and Besson R.: Nonlinear Effects in Piezoelectric Quartz Crystals, Physical Acoustics, vol. 11, W.P. Mason and R.N. Thurston Eds., Academic Press. (1975). p. 245-288.
16. Hruska K.: Polarizing Effect with Piezoelectric Plates and Second-order Effects, IEEE Trans.Sonics and Ultrason. SU-18, 1 (1971).
17. Brendel R.: Material Nonlinear Piezoelectric Coefficients for Quarz, J.Appl.Phys. 54, 5339 (1983).
18. Auld B.A.: Acoustic Field and Waves in Solids, J.Wiley and Sons, New York, 1973.
19. Solie L.P.: Piezoelectric Waves on Layered Substrates, J.Appl. Phys. 44, 610 (1973).
20. Farnell G.W. and Adler E.L.: Elastic wave Propagation in Thin layers, Physical Acoustics Vol. 9, W.P. Mason and R.N. Thurston Eds. Academic Press, New York (1973).
21. Wagers R.S.: Plate modes in Surface Acoustic Wave Devices, Physical Acoustics, Vol. 13, W.P.Mason and R.N. Thurston Eds. Academic Press, New York (1977) p. 49-78.

MEASUREMENT OF THE COHERENT AND INCOHERENT CONTRIBUTIONS OF ULTRASONIC SCATTERING IN INHOMOGENEOUS MEDIA

R. C. Chivers

1. INTRODUCTION

Inhomogeneity in the intrinsic parameters of a medium which affect acoustic or elastic wave propagation (i.e. density, compressibility or the elastic constants, and the local attenuation coefficient) will, in general, cause a propagating acoustic or elastic wave to be scattered. This results in changes in the amplitude, frequency, phase velocity or direction of propagation of the incident wave. The literature is rich with analytical approaches to the propagation problem. Most assume a regular geometry for the incident wave (i.e. planar or spherical). The effects that then provide the direction of the analytical development are essentially those on which experimental measurements may be made. Specifically, these are the waves that can be measured as having been scattered in different directions (excluding, for the moment, the forward direction), the amplitude and phase of the waves emerging in the same direction as the incident wave (which are related to measurements of velocity and attenuation), and the fluctuations in amplitude and phase which are observed in the waves propagating in a particular direction.

One of the major areas that remains under development is the utilization of these theories in experimental situations where the transducers used do not have ideal characteristics. The discussion will be based largely on the problems that arise in connection with materials' investigation in the low megahertz region of ultrasonic frequencies.

In the first place, the transducers used have complex field patterns in general. Point sources are available in the form of breaking capillaries or laser generated pulses for solids. For fluid materials, it appears hard to make a source of adequate strength. At the other extreme, the planarity of the waves produced is limited by the dimensions of the transducers available, and their ability to be manufactured in such a way as to produce pistonlike behaviour (indeed the assessment of the vibrational behaviour of the source has only been feasible in recent years and is not yet within the means of most workers in the field). Single element ceramic transducers are susceptible to mechanical problems due to their comparative thinness even if the problems of manufacturing them in large sheets is overcome. Multi-element ceramic transducers require great care in mounting to achieve planarity and uniformity. The possibility of large area $PVDF_2$ transducers does not appear to have been explored. The major practical disadvantage of large transducers, even if they were readily available, would be the large experimental volume required. Thus for transmission, the tendency is to use transducers with elements ranging from a large fraction to a few centimeters in diameter.

In reception, the main complication in the use of finite receivers arises from the fact that fluctuations in phase across the face of the transducer caused by inhomogeneities in the propagation medium can lead to errors of interpretation [1]. Assuming that, in spite of the phase fluctuations, the amplitude of the wave is constant across the face of the transducer, the amplitude of the voltage output will be less than that produced by an incident plane wave of the same pressure amplitude due to 'phase cancellation' effects. In practice, there is theoretical and experimental evidence [2] that phase fluctuations are usually accompanied by amplitude fluctuations so the effect may be more correctly termed a 'wave fluctuation' artefact.

Before proceeding to discuss the effect of these practical constraints on the measurements of scattering, of attenuation and velocity, and of fluctuations, mentioned above, it is useful to revise the concepts of coherent and incoherent scattering. The distinction made between these components of the scattering of waves in an inhomogeneous medium - which has been closely paralleled in the definitions used by other workers - is clearly seen in the analysis of Foldy [3] for multiple scattering by discrete scatterers.

If the distribution of scatterers is irradiated by an incident wave ψ_0, and the total scattered wave is ψ_s, then the total radiation density as a result of the interference of these two waves will be given by:

$$\langle|\psi_0 + \psi_s|^2\rangle = |\psi_0 + \langle\psi_s\rangle|^2 + \langle|\psi_s|^2\rangle - |\langle\psi_s\rangle|^2 \qquad (1)$$

Thus the effect of the inhomogeneities is to produce a term $\langle\psi_s\rangle$ which is added to the incident wave as a complex amplitude - the coherent part of the scattering - and two terms which add to the square of the modulus of the sum of ψ_0 and $\langle\psi_s\rangle$. The combination of these last two terms is called the incoherent part of the scattering, adding as intensities rather than as complex amplitudes. Thus if a plane wave propagating through a homogeneous medium (which may include absorption) produces the complex but planar pressure amplitude ψ_0, the effect of adding inhomogeneities to the medium will produce an emergent wave which, in general, varies in amplitude and phase across a plane parallel to that of the original wavefronts. The mean values of the amplitude and phase of the fluctuations give the amplitude and the phase of $\psi_0 + \langle\psi_s\rangle$, i.e. of the total new wave, that propagates in a manner similar to ψ_0 but with a new overall amplitude and phase. Comparison of the new values of the amplitude and phase with those of ψ_0, permit assessment of the contribution of the (coherent) scattering to attenuation of the wave and to the change in phase velocity effected. The fluctuations about the mean value constitute the incoherent part of the scattering and contribute as the last two terms in equation (1).

It can be seen that the coherent part of the scattering will become significant if there is regularity in the spatial ordering of the scattering centres from the extreme limit if they form a boundary), or if they are sufficiently closely packed [4], or if they are strong enough scatterers for multiple scattering to be significant.

2. EFFECT OF REALISTIC TRANSDUCERS ON MEASUREMENTS

2.1 Scattering measurements

The need to define angles or directions in connection with scattering leads to the use of finite transducers [5]. The incorporation of the radiation characteristics of these probes into the analysis by which differential scattering cross-sections are calculated from the actual measurements made is a relatively complex process. Most of the developments appear to have been made in connection with measurement of the scattering from biological tissue. Attention has been concentrated on backscattering measurements. Different authors have used the -3dB [6,7] or the -6db measured beam profiles [8], others use an ideal piston source model [9,10] or a Gaussian fit to the measured beam patterns [11]. In general, different approaches will lead to different numerical results for the differential scattering cross-sections. More recently, an analysis has been presented in which arbitrary spatial distributions for the transmitting and receiving transducers can be incorporated [12] provided that facilities are available to measure them.

The penalty usually paid for the incorporation of more complex approximations to reality as far as the transducers are concerned is a simplification of the scattering situations that are covered. Specifically, all the authors mentioned but Madsen, et.al. [10], explicitly limit their analyses to uncorrelated scattering by very weak scatterers. Thus the Born approximation is used and the scattering considered is incoherent. Madsen, et.al., discuss the coherent part of the scattering that may be backscattered and then utilize a regime for measurement in which the only contributions to the coherent scattering arise from the edges of the gate used to select the volume. They emphasise that their work is an extension of the analysis of Glotov [13]. Machado, et.al., [14] have produced experimental results suggesting that the multiple scattering becomes significant at 5MHz for concentrations of 0.6mm polystyrene spheres in a weak sugar solution above 100 cm^{-3}. The interpretation of their results is based on the formalism of Ishimaru [15].

Probably as a result of considering only the incoherent scattering, the emphasis on the measurements is to take the mean square pressure amplitude as the appropriate parameter to measure. For some authors [e.g. 6,11] there is an explicit requirement in the analytical formulation for the transducers to be far from the scattering volume. Others [e.g. 7] introduce this as a means of reducing 'phase cancellation' artefacts. Otherwise, the potential fluctuations in the scattered wavefronts and the resultant artefacts appear to be ignored, except for a model calculation by Shung and Drzerzianowski [16].

2.2 Attenuation and velocity measurements

The principle approach to attenuation and velocity measurements in the laboratory is the replacement method in which a specimen of known thickness is interposed between two aligned transducers. Measurements of the phase and amplitude of the received wave are compared to those obtained with a known fluid between the transducers. Sound velocity and attenuation for the specimen are calculated from the respective changes. To a first approximation, planar wave propagation is assumed. To a second approximation, diffraction corrections are used to calculate the effects of the finite nature of the transducers. These almost always rely on the assumption that the transducers are ideal in their radiation behaviour. A method for defining an effective radius for such transducer corrections has been proposed [17] and investigated in some detail [18], but its value in experimental systems has not yet been proven. Furthermore, no analysis appears to exist for corrections appropriate to the most common case of the medium between the transducers being piecewise discontinuous.

The presence of inhomogeneities in the specimen under investigation, which cause the emergent wavefront to exhibit fluctuations will lead to additional difficulties as mentioned in the previous section. There are two main types of problems. The first concerns the effect of the addition of (discrete) scatterers to a homogeneous matrix whose properties are either known or can be measured relatively easily. In this case, the interest is in assessing the extent to which the scatterers have changed the bulk (plane wave) attenuation and sound speed (the coherent contribution), and the extent to which fluctuations are present (the incoherent contribution). The second type of problem is one in which the medium is known to be inhomogeneous and to produce scattering, but in which the identity of the scatterers and the matrix (and thus the appropriate scattering requires) are a priori unknown.

In both of these cases the use of one type of large aperture transducers provides only limited information. The conventional piezoelectric transducers which are phase sensitive and can thus be used for measuring sound speed give the value of $\psi_0 + \langle\psi_s\rangle$ for an incident plane wave ψ_0. The phase insensitive receivers [e.g. 18,19] appear to measure the total 'intensity' , i.e. $\langle|\psi_0 + \psi_s|^2\rangle$ and thus cannot measure sound speed. However, knowledge of ψ_0 and use of each type of receiver should permit separation of the coherent and incoherent parts of the scattering. The major practical problem then remains one of ensuring that the apertures of the receivers are adequate for statistical sampling.

It is relevant here to point to a question of definitions. For a plane wave, the amplitude attenuation coefficient, α, defined by

$p(x) = p(o)\ e^{-\alpha x}\ e^{i(\omega t-kx)}$ where $p(x)$ is the acoustic pressure in the plane wave travelling in the x direction, can be converted to the intensity attenuation coefficient μ by the simple relation $\mu = 2\alpha$. However, if the 'intensity' attenuation coefficient for an inhomogeneous material is measured it will incorporate both coherent and incoherent parts of the scattering. The related 'amplitude' attenuation coefficient will incorporate only the coherent part of the scattering and there will not be a simple relationship between the two, unless the incoherent part of the scattering is in fact negligible.

The present work describes the use of a small piezoelectric receiver for measuring one or both of the scattering contributions. The first is limited to penetrable media and, at the present time, to measuring the coherent part of the scattering. The second actually measures the fluctuations in a plane perpendicular to the axis of the incident wave, and uses this data together with equation 1 to estimate the two contributions. The principles of the calculations and some preliminary results are presented.

3. COHERENT SCATTERING IN PENETRABLE MEDIA

The use of miniature hydrophone probes for field measurements in penetrable media has been demonstrated by Lewin [20] for human tissue. The basis of this work is the analytical formulation of the axial field from a finite (weakly) focussed pistonlike bowl transducer of O'Neill [21]. It can be shown [21] to have the form:

$$|p(z)| = \left[\frac{2\rho c u_0}{1 - \frac{z}{A}}\right] e^{-\alpha z} \sin k\delta \qquad (2)$$

where ρ, c, α are the density, sound speed and amplitude absorption coefficient in the fluid; $k = 2\pi/d$; z is the axial distance; u_0 is the particle velocity at the surface of the source; A is the radius of curvature; and δ is given by:

$$2\delta = [(z-h)^2 + a^2]^{1/2} - z \tag{3}$$

where h and a are defined as in figure 1.

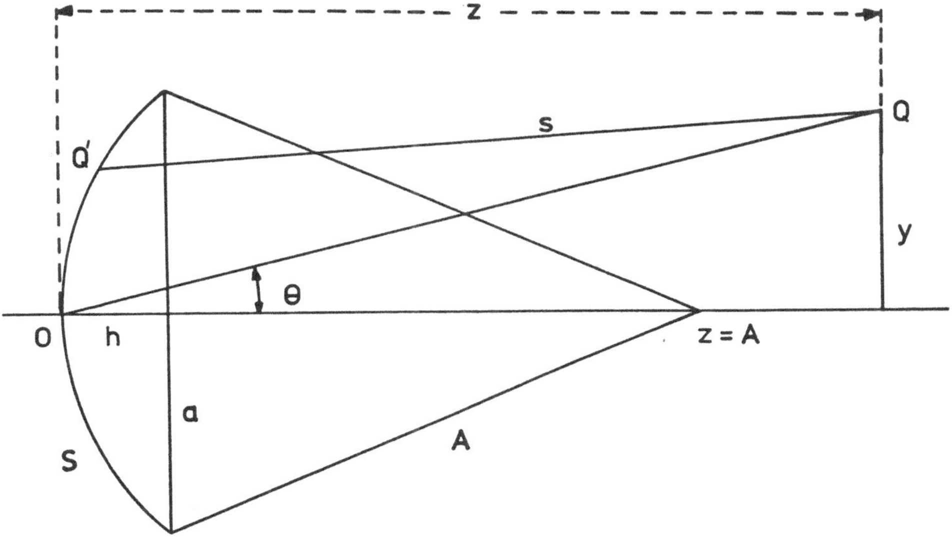

FIGURE 1. Geometry of the focussed bowl radiator.

Taking measurements in two media denoted by subscripts $_1$ and $_2$ respectively, we find:

$$\ln \frac{|p_1|}{|p_2|} = \ln \left[\frac{\sin k_1\delta}{\sin k_2\delta} \cdot \frac{\rho_1 c_1}{\rho_2 c_2} \right] - (\alpha_2 - \alpha_1)z \tag{4}$$

Thus if the first term on the right-hand side is independent of z, we can obtain a straight line plotting $\ln(|p_1|/|p_2|)$ against z. The gradient will be $\alpha_1 - \alpha_2$. The conditions under which the first term is independent of z are discussed elsewhere [22]. The main dependence comes in δ and this can be removed if $k_1 \simeq k_2$. If α_1 is known or measured (or is effectively zero as in the case of water), α_2 can be determined.

The preliminary tests were made on castor oil using a bowl of nominal diameter 1.5cm, focal length 10cm and resonant frequency of 2MHz. The receiver was a needle-type probe with a ceramic (PZT) element of 1.0mm diameter. The measurement system is shown schematically in figure 2.

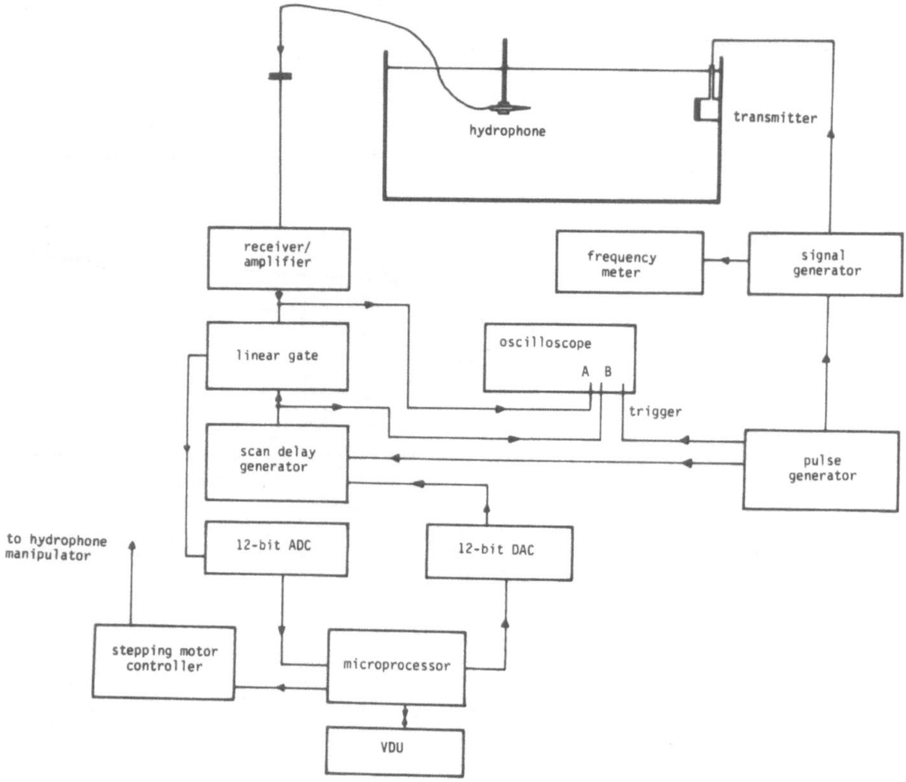

FIGURE 2. Electronic measurement system.

A typical plot of the axial distribution in castor oil is shown in figure 3. Comparison of 5 sets of axial measurements in the range 1.8 - 2.0MHz with equivalent axial plots in water yielded values of absorption for castor oil (23.9 neper m^{-1}) in reasonable agreement with the spread of values found in the literature.

Suspension of a 1% volume concentration of polystyrene spheres (diameter 300-600μm) in the castor oil was followed by repetition of the axial plots. In this case, it is assumed that the effect of the scatterers on the phase of the coherent part of the scattering is negligible (so that k_1 remains equal to k_2 in equation (4)). As may have been expected, the log ratio of the curves with and without the scatterers showed considerable fluctuation. Figure 4 shows a typical example. Fitting a least squares line between 4 and 9cm gave a value for $\alpha_2 - \alpha_1$ at the frequency of the measurement, i.e. of the coherent contribution of the scatterers to the amplitude attenuation.

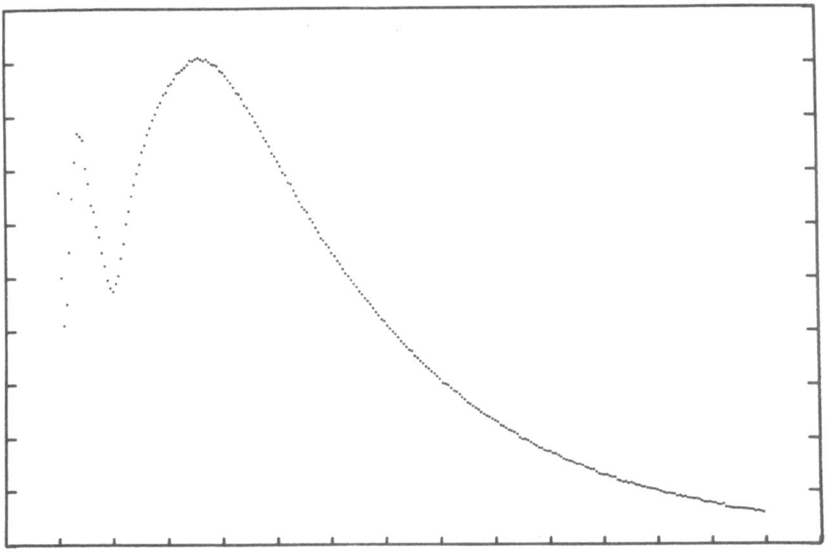

FIGURE 3. Axial pressure amplitude distribution in castor oil at 25°C. and 1.87MHz. Horizontal axis: axial distance 0-15cm. Vertical axis: arbitrary linear scale.

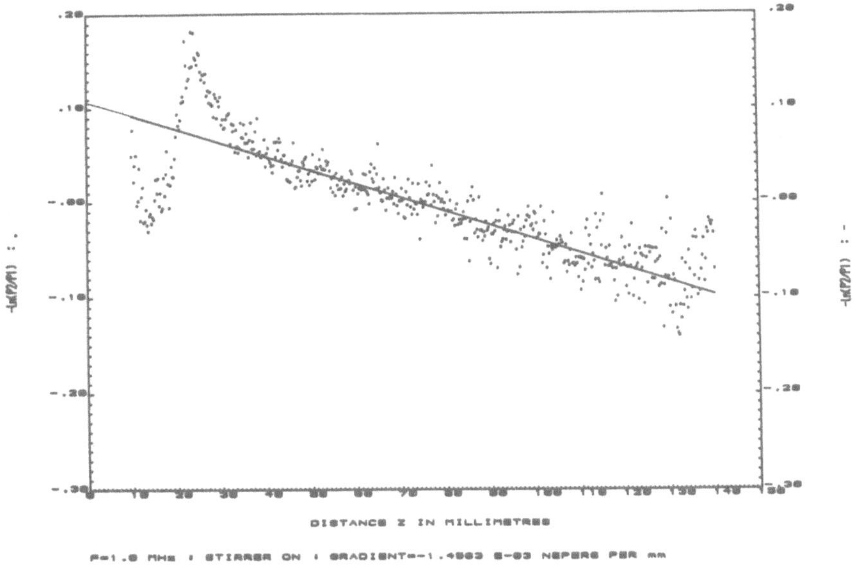

FIGURE 4. Logarithm of the ratio of pressure amplitude distributions in castor oil and castor oil with 1% polysytrene beads at 20°C. and 1.0MHz.

It was felt necessary to check that this was in reality the coherent scattering contribution to the attenuation. The choice of polysytrene was made largely on considerations of an appropriate density to maintain a reasonably stable suspension in castor oil. Thus in equation 4 the first term is closely equal to zero. Measurements may thus be made at a known value of z in the two media and provided adequate averaging is performed $\alpha_2 - \alpha_1$ may be calculated. This facilitates measurement as a function of frequency. Accordingly $\alpha_2 - \alpha_1$ was determined as a function of frequency at 0.25MHz steps between 1.0 and 2.5MHz at values of z of 5cm and 6cm. The results of the values of $\alpha_2 - \alpha_1$ were compared to those predicted by the theory of Waterman and Truell [23]. The results are shown in figure 5 where it can be seen that the level of agreement is good. Deviations may be expected due to the uncertainties associated with the particle size distribution and with the value of the shear modulus used for the polystyrene.

It is tempting to try to identify the fluctuations about the mean value observed in figure 4 with the incoherent part of the scattering. This is presently under investigation.

4. WAVE FLUCTUATION MEASUREMENTS

Some of the reasons for making wave fluctuation measurements that are not related specifically to separating the coherent and incoherent parts of the scattering are given in the companion paper to this one [2] in which the method of making the measurements and some typical results are given. It can be seen from the tables of fluctuations presented in reference [2] that the phase fluctuation of the glass bead models is only significant at the larger concentrations of the larger bead sizes. From the values for tissue, 40° is a typical variational range, thus calculations have been performed for the 7.5% concentration of 500-600μm scatterers in silicone rubber. Using equation (1) and assuming an incident plane wave, the ratio of the coherent to incoherent scattering intensities is approximately 3:2. This implies a high degree of coherent scattering which will be important if a similar ratio obtains for tissue.

5. CONCLUSION

The separation of the coherent and incoherent contributions to the scattering of ultrasound by an inhomogeneous medium is of importance because of the fact that most of the theoretical bases for the quantitative measurement of differential scattering cross-sections explicitly assume that the scattering is incoherent. Three approaches to the measurement of one or both of the components are presented in the present paper: using finite receivers, using a point receiver to measure propagated fluctuations, and using a point receiver in a focussed field. The last can only be used in a penetrable medium and is at its most effective when assessing the effect of adding discrete scatterers to an existing (fluid) matrix. To date it has only been shown that it is capable of measuring the coherent part of the scattering.

The other two methods may both be used with solid or semisolid materials. However, both require either plane wave irradiation (or a known close approximation to it) and the use of parallel-sided specimens (or some procedure for avoiding phase differences due solely to thickness variations). One of the methods uses finite receivers and exploits the difference in the receiving characteristics of piezoelectric and acousto-opto-electric receivers, which gives directly the incoherent scattered intensity. Knowledge of the incident beam permits the coherent component

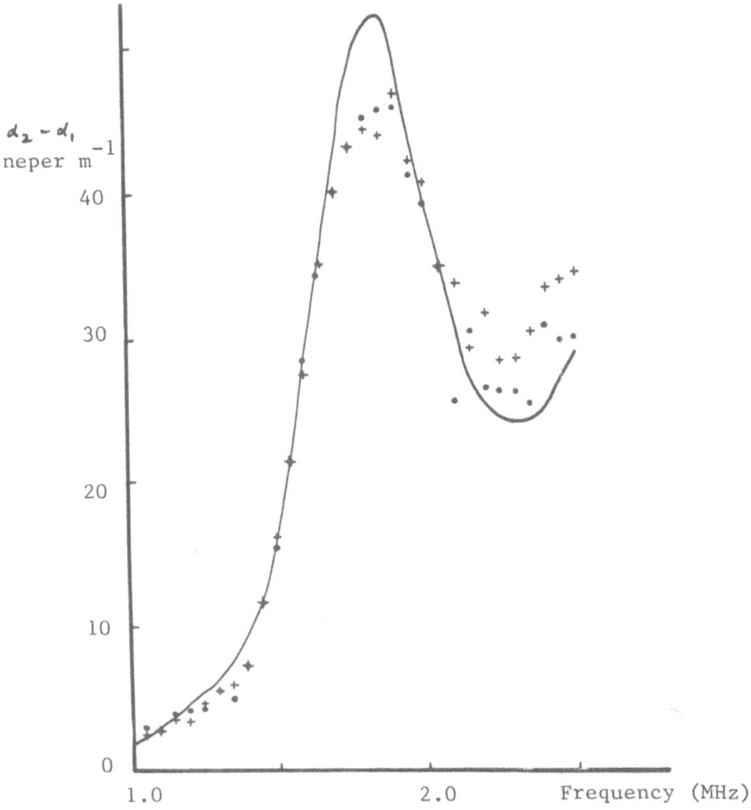

FIGURE 5. Excess attenuation (α_2 - α_1) as a function of frequency measured at 5cm (+) and 6cm (•) compared to the theory of Waterman and Truell (--) (temperature between 19.0 and 21.6°C.).

to be assessed. The last, most tedious, method measures the actual fluctuations observed in a plane parallel to that of the incident wavefronts. From these both coherent and incoherent components can be evaluated. Calculations are in progress to evaluate measurements obtained on suspension systems and fresh beef liver. One of the major outstanding questions concerns the need to define an adequate aperture for the finite receivers or for the area over which fluctuation measurements are taken.

ACKNOWLEDGEMENTS

The author is grateful to a number of coworkers who have contributed significantly to the work presented here: Drs. J.D. Aindow, L.W. Anson, Anna Markiewicz, and Messrs. J. Adach and P.R. Filmore. The invaluable assistance of Messrs. F.G. Bristow and E.A. Worpe and their colleagues who built the instrumentation is also acknowledged.

REFERENCES

1. Marcus PN and Carstensen EL: J. Acoust. Soc. Am. 58, 1334-1335, 1975.
2. Chivers RC: Elsewhere in this volume, 1985.
3. Foldy LL: Phys. Rev. 67, 107-119, 1945.
4. Kur'yanov BF: Soviet Phys.-Acoustics 10, 160-164, 1964.
5. Zverev VA: Soviet Phys.-Acoustics 3, 348-359, 1957.
6. Sigelmann RA and Reid JM: J. Acoust. Soc. Am. 53, 135-1355, 1973.
7. O'Donnell M and Miller JG: J. App. Phys. 52, 1056-1065, 1981.
8. Nicholas D, Hill CR, and Nassiri DK: Ultrasound Med. Biol. 8, 7-15, 1982.
9. Lizzi FL, Greenebaum M, Felippa EJ and Gilbaum M: J. Acoust. Soc. Am. 73, 1366-1373, 1983.
10. Madsen EL, Insana MF and Zagzebski JA: J. Acoust. Soc. Am. 76, 913-923, 1984.
11. Campbell JA and Waag RC: J. Acoust. Soc. Am. 74, 393-399.
12. Waag RC: Private communication, 1985.
13. Glotov VP: Soviet Phys.-Acoustics 8, 220-222, 1963.
14. Machado JC, Sigelmann RA and Ishimaru A: J. Acoust. Soc. Am. 74, 1529-1534, 1983.
15. Ishimaru A: Wave propagation and scattering in random media (Vols. I and II). New York: Academic Press Inc., 1978.
16. Shung KK and Drzerzianowski JM: Ultrasonic Imaging 4, 56, 1982.
17. Filmore PR, Bossellaar L, and Chivers RC: J. Acoust. Soc. Am. 68, 80-84, 1980.
18. Busse LJ, Miller JG, Yuhas DE, Mimbs JW, Weiss AN and Sobel BE: Ultrasound in Medicine, Vol. 3, pp. 1519-1535. New York: Plenum Press, 1977.
19. Heyman JS: J. Acoust. Soc. Am. 64, 243-249, 1978.
20. Lewin PA: Ultrasonic International 81, pp. 434-439. Guildford: IPC Press, 1981.
21. O'Neill HT: J. Acoust. Soc. Am. 21, 516-529, 1949.
22. Adach J, Anson LW and Chivers RC: J. Phys. E. (submitted for publication).
23. Waterman PC and Truell RJ: J. Math. Phys. 2, 512, 1962.

AMPLITUDE AND PHASE FLUCTUATIONS IN ULTRASONIC WAVES PROPAGATING IN INHOMOGENEOUS MATERIALS

R. C. Chivers

1. INTRODUCTION

With a significant area of uncertain overlap, the ultrasonic methods of investigating materials can be grouped into those that are concerned with essentially homogeneous materials, those that are concerned with discrete, relatively prominent variations, and those concerned with a more widespread inhomogeneity. While the foremost is of great value in chemical and material micro-structural applications, it is beyond the scope of the present discussion. The division between the last two types of applications is more complex. Obvious examples of the use of discrete, relatively prominent inhomogeneities range from sizeable flaws to acoustic microscopy. These are often the situations in which the dimension of the inhomogeneity is small compared to the wavelength, (but this is neither an infallible rule, nor a complete description). In each of these cases, there is a condition that simplifies the wave analysis. In the former, it is the fact that the flaw is usually the isolated scattering structure. In the latter, simplification comes from the use of thin specimens or focal zones which thus essentially reduce the problem to a two-dimensional one. The concern in the present work is with the third group in which neither of these is true, namely where the inhomogeneity is distributed throughout the scattering volume. Our interest is in the fluctuations that may be observed in the amplitude and phase of the waves emerging from the scattering medium (in a plane perpendicular to the direction of the incident plane wave). Attention is further confined to the low megahertz frequency range, with wavelengths typically in the range 75μm-1.5mm.

There are two motivations for this work. The first concerns the validity of the assumption that is often made of plane wave (or straight line ray) propagation in a material upon which scattering or attenuation measurements are to be made. The second concerns the potential use of fluctuation measurements in the characterization of the material. The present contribution consists of a brief theoretical review followed by a discussion of the technical problems of making measurements of this type, together with some experimental results. Before proceeding to the theoretical background, it is worth discussing in some detail the significance of the first motivation of the work.

2. THE SIGNIFICANCE OF THE PLANE WAVE ASSUMPTION

Although it is immediately extendable to other materials, the discussion here will be centered upon soft biological tissues. The primary ultrasonic propagation parameters that are measured are velocity and attenuation. The methods of measurement all assume plane wave propagation, i.e. that a normally incident plane wave will emerge, after traversing a parallel sided specimen as, essentially, a plane-wave. If it does not, the amplitude attenuation coefficient and the sound velocity will be dependent on the position in the plane in which they are measured, and a more careful interpretation of the material propagation parameters will be needed.

In addition to this, the analyses of scattering (e.g. [1,2]) require the input of values for the sound speed and amplitude attenuation in the scattering medium. Indeed, one of the major contributions to the uncertainty in the values of the scattering measurements arises from the uncertainty in the value of the attenuation used [3]. However, explicit in the scattering analysis is the assumption of plane wave propagation. The importance of this can be seen in the consideration of backscattering measurements (Figure 1). The region from which scattering is to be measured is selected by a time gate on the signals reflected back to the (pulsed) emitter. Even if a plane wave was emitted, the traversal through the medium from its front surface to the gated scattering region will, in general, produce fluctuations across the wavefront that is incident on the scattering volume, after scattering a wavefront with a new set of fluctuations sets off on its journey back to the transducer. On the way back it will suffer further fluctuation. If there were no limitation on the way in which a receiving transducer might process this returned wavefront it would still be almost impossible to account for the fluctuations induced by travel across the tissue between the transducer and the sample volume and thus to identify, and measure, the scattering produced.

In fact, we do not have a great range of choice in the way in which receiving transducers convert incident pressure waves into electrical signals. The power sensitive transducers which are thus phase insensitive cannot be used to effect the time-gating required. Thus we are left with piezoelectric devices which integrate the complex pressure received over their aperture. This may lead to errors due to the "phase-cancellation" artefact [4]. It will occur both for attenuation and scattering measurements.

It is clearly important to know if significant fluctuations do occur in the waves propagating through soft tissue if meaningful attenuation measurements are to be made. There is evidence that the wavefronts incident on the receiver were, in some experiments, distorted [5], but it is not clear [6] whether this was due to the non-planar incident wavefronts, to variations in specimen thickness or to intrinsic inhomogeneities in the tissue. If it was the latter, then the attenuation values obtained with phase insensitive detectors need careful interpretation [4]. In addition, it may imply [7], from the argument above, that scattering measurements are in general impossible to make in a meaningful way.

Certainly the apparent observation of significant wave fluctuations in attenuation measurements on tissues [5], and the fact that reported measurements of the ultrasonic backscattering cross-section per unit volume per unit solid angle vary by over an order of magnitude for the same (fresh liver) tissue [8], provides circumstantial evidence in favour of the significant disruption of the wavefronts by the tissue inhomogeneities.

The implications of such a conclusion are sufficiently significant to make a more careful assessment of the evidence. It then becomes clear that as far as the reported effects on attenuation measurements are concerned, there is little doubt that the wavefronts reaching the receiver were perturbed. However, there are three potential sources for the perturbation. Firstly, the incident waves are seldom planar, secondly variations in the specimen thickness may have significantly contributed to the

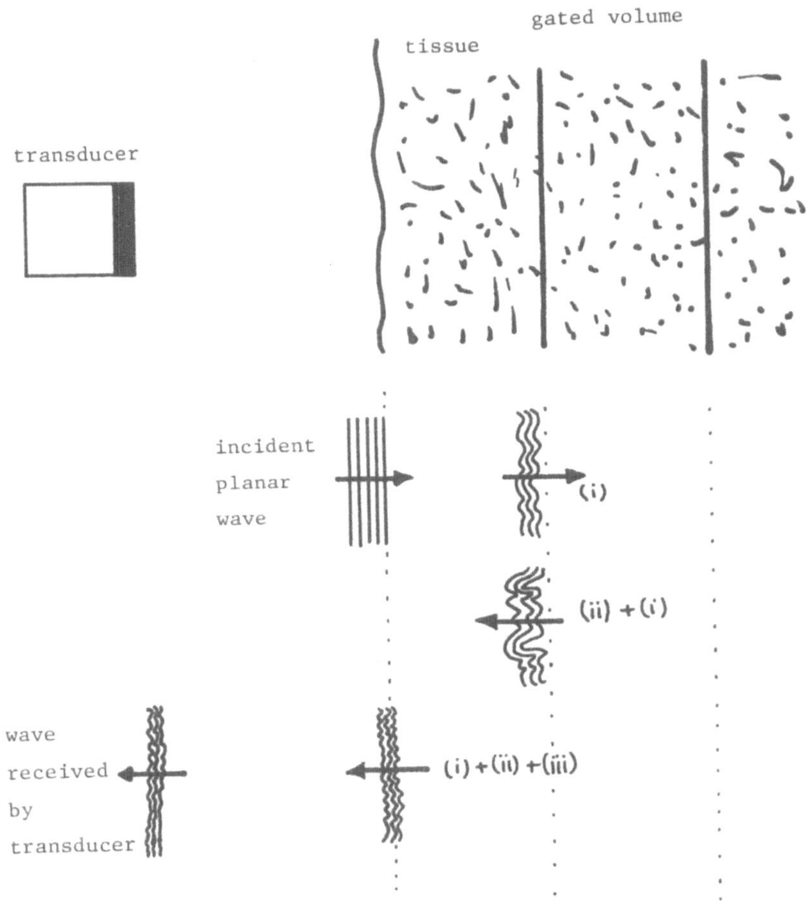

FIGURE 1. The compounding of fluctuations in the presence of significant propagation fluctuations. (i) represents the fluctuations resulting from travel between the tissue boundary and the gated scattering region, (ii) represents the fluctuations superimposed on (i) by the scattering process, (iii) represents the fluctuations superimposed on (i) and (ii) resulting from travel back to the tissue boundary. These are propagated to and received by a phase sensitive receiver. In general, the scattering (ii) cannot be distinguished from the propagation effects (i) and (iii).

fluctuations, and thirdly the fluctuations may have arisen from the intrinsic inhomogeneities of the tissue. Thus in order to identify the third of these, the first two need to be controlled and eliminated by careful experimental procedure. A primary need for this is the - technically difficult - measurement of phase distributions discussed in section 4.

The interpretation of the scattering measurement variation also requires care. Zverev [9] has shown that under certain conditions the mean square scattered pressure can be expressed as an integral of the product of a term which is the Fourier conjugate of the correlation function of the medium, and a second term which depends only on the experimental configuration, especially of the transducers. More generally, Waag [10] has shown that the mean square pressure can be expressed as

$$\langle |p|^2 \rangle = \int |\Lambda(\underline{K})|^2 \, k^4 \, S(\underline{K}) d\underline{K}$$

where $k = 2\pi/\lambda$, $\underline{K}$ is the scattering vector, and $S(\underline{K}) = F_{3\text{-}D} \{B(\underline{r})\}$ where

$$B(\underline{r}) = \left(B_{mm}(\underline{r}) + 2B_{m\rho}(\underline{r}) \cos\theta + B_{\rho\rho}(\underline{r}) \cos^2 \theta\right),$$

B_{ij} are correlation functions, m = compressibility, ρ = density.

The function $|\Lambda(\underline{K})|^2$ is essentially the 'normalization factor' that defines the volume and weighting of the measured scattering in terms of the transmitter and receiver sensitivities (both spatially and in frequency), and the time gates used to identify the scattering region.

Some discussion of the practical approaches to the calculation of Λ in a given experimental situation appear elsewhere [4]. However, there appears to be a lack of agreement on the correct procedure which may stem from the fact that a general analysis of the problem has only just been developed [10].

3. THEORETICAL BACKGROUND

The main analysis that appears available at the present time is that of Chernow [11] which has been extended by the present author [12]. Writing the interrogating wave as $p_0 = A_0 \exp[-i(\omega t - kx)]$, the resulting wave at $\underline{r}$ may be written as

$$p = A(\underline{r}) \exp[-i(\omega t - S'(\underline{r})]$$

$$= A_0 \exp[-i(\omega t - \psi(\underline{r}))]$$

$$\text{where } \psi(\underline{r}) = S'(\underline{r}) - i \ln \frac{A(\underline{r})}{A_0}$$

For the mean square fluctuations of amplitude $\langle B^2 \rangle$ and phase $\langle S^2 \rangle$ we can write

$$S(\underline{r}) = S'(\underline{r}) - kx$$

$$\text{and } B(\underline{r}) = \ln \frac{A(\underline{r})}{A_0}$$

Under the assumptions of largescale inhomogeneities (i.e. ka >> 1 where a is a correlation distance), and a region of integration whose dimension L >> a we find:

$$\frac{\langle S^2\rangle}{\langle B^2\rangle} = \frac{\langle(\Delta m)^2\rangle}{4m_0^2} kL \int_0^\infty [N_{mm}(\xi,0,0) \mp \int_0^\infty si\nu N_{mm}(r^1)d\nu]\, d\xi$$

$$+ \frac{\langle(\Delta\rho)^2\rangle}{4\rho_0} kL \int_0^\infty [N_{\rho\rho}(\xi,0,0) \mp \int_0^\infty si\nu N_{\rho\rho}(r^1)d\nu]\, d\xi$$

$$+ \frac{[\langle(\Delta m)^2\rangle\langle(\Delta\rho)^2\rangle]^{1/2}}{4\rho_0 m_0} 2kL \int_0^\infty [N_{m\rho}(\xi,0,0) \mp \int si\nu N_{m\rho}(r')d\nu]d\xi$$

using the notation of Chernow. $N_{\infty\infty}$ are the relevant correlation coefficients, ρ is the density and m the compressibility.

This expression can be simplified for extreme values of the wave parameter $D = 4L/ka^2$. Inserting correlation coefficients of the form $\exp{-r^2/a^2}$, i.e. assuming isotropy and that the correlation distances for density and for compressibility are identical while the cross correlation is zero, we obtain, for large values of the wave parameter:

$$\frac{\langle S^2\rangle}{\langle B^2\rangle} = \left[\frac{\langle(\Delta m)^2\rangle}{4m_0{}^2} \pm \frac{\langle(\Delta\rho)^2\rangle}{4\rho_0{}^2}\right] k^2La \frac{\sqrt{\pi}}{2}$$

while for small values of the wave parameter we have:

$$\langle S^2\rangle = \left[\frac{\langle(\Delta m)^2\rangle}{4m_0{}^2} + \frac{\langle(\Delta\rho)^2\rangle}{4\rho_0{}^2}\right] k^2La\sqrt{\pi}$$

$$\langle B^2\rangle = -\left[\frac{\langle(\Delta m)^2\rangle}{4m_0{}^2} + \frac{\langle(\Delta\rho)^2\rangle}{4\rho_0{}^2}\right] \frac{8}{3}\,\frac{L^3}{a^3}\sqrt{\pi}$$

These results differ slightly from those of reference [12]. Leeman and Fiddy [13] pointed out that some of the terms previously included could be shown to be zero, provided the derivative of the correlation functions are zero at the origin. It can, however, be further shown that the behavior of the correlation functions at the origin is incidental, the terms will be zero due to the low angle assumption introduced as a result of the long wavelength regime assumed [14].

There is much excellent work on fluctuation analysis in both the fields of optics [e.g. 15,16] and acoustics [17,18] but for application to many of the materials of current interest it is limited by being based on the long wavelength approximation and by the lack of inclusion of absorption [19]. The analysis above may be of value in indicating some features that may be of experimental interest. The frequency dependence and the potential dependance on L are measured. L, in fact, defines the length of material through which the interrogating wave has passed before detection. For inhomogeneities that are not large compared to the wavelength, two effects may be anticipated. The first is that the region of significant effect in terms of contribution to the fluctuations may only extend to a few times the near-field diffraction distance of the inhomogeneities. The second is that the presence of absorption may reduce this distance considerably, depending on the strengths of the scatterers and of the absorption. The situation for small scatterers in a lossless medium has recently been analysed by Soczkiewicz [20].

4. EXPERIMENTAL METHOD

The details of the experimental method used have been described elsewhere [6]. The principle is that of detecting tonebursts (of typically 20 cycles) after they have propagated through the medium under investigation using a miniature hydrophone of active element diameter typically of 1mm. The peak to peak amplitude of the wave is measured on a calibrated oscilloscope. The relative phase of the wave is measured from zero crossing which is a predetermined number from the beginning of the pulse (to ensure it is well into the steady state part of the pulse), and which lies between the peak and the trough from which the amplitude measurements are made. The zero crossing detector developed was slightly amplitude dependent so variable attenuation was introduced to ensure that measurements were always made with approximately the same amplitude of signal going into the detector.

The measurement of the relative phase was the most demanding of the experimental problems. Precise measurements require excellent scanning mechanics, a time delay measurement system with high stability and a good degree of temperature control. Detailed discussion of the numerical considerations are available [21]. The order of magnitude of the constraints may be seen by the fact that one degree of phase at 2MHz corresponds approximately to a mechanical movement of 2 microns, to a timing capability of 2ns and to a temperature stability (for typical transmitter: receiver spacings) of 0.02°C.

Mechanical precision was achieved by careful design and construction (and indeed checked by a technique involving ultrasonic phase measurements [22]). Temperature stability was achieved by the use of a temperature controlled tank outside the inner measurement tank [6] and was constantly monitored. The timing methods available of sufficient precision were essentially those of a narrow-band sing-around technique [23] or of time interval averaging. Theoretical and experimental analysis provided a criterion based on the experimental conditions (specifically transmitter: receiver travel time and the duration of the experiment) [24]. For the present experiments, time interval averaging was the method chosen.

5. EXPERIMENTAL RESULTS

Measurements have been reported on two types of materials: suspensions of glass beads of different diameters in a clear silicone rubber [19] and fresh beef liver [6,25]. The primary requirement was to define a region of the irradiating field that was a good approximation to a plane wave. Extensive field measurements compared with a modified theoretical model provided security in establishing this condition [26]. With the silicone rubber blocks - which were moulded with parallel sides, the problem of specimen thickness variation does not arise. For the tissue, the effect of specimen thickness variation was investigated by measuring the fluctuations with the specimen in water and then with the specimen immersed in saline with the same speed of sound as the tissue. Measurements were made at one position for the silicone rubber blocks (thickness 3cm) and at 5 positions on each of the tissue specimens at a frequency of 2MHz.

The results are summarized in Tables 1 and 2, where the peak to peak values of the phase and amplitude fluctuations are given. A more sophisticated anlaysis of the data obtained is discussed in the companion paper to this one [4]. It is interesting to make some comments in passing. Firstly, as may be expected, the fluctuations in phase and in amplitude tend to increase together. Secondly, the fluctuations increase with both increasing scatterer size and increasing concentration. With the largest bead size and concentration they are typically ±30° of phase and ±60% of amplitude. The results on tissue are inconclusive. One of the samples in water exhibited fluctuations as great as the greatest from the models.

Removal of the components due to variations in specimen thickness reduced these by a half, but they are still significant. The second specimen showed almost no fluctuations when immersed in velocity matched saline. Sectioning of the specimens provided no features that could be correlated in a positive way with these fluctuations.

TABLE 1. Extreme variations in amplitude and phase observed as model systems (amplitude normalised to a nominal value of 1.0).

Bead size (mm)	500-600			
Vol. Concentration	0	2%	7.5%	17.2%
Amplitude variation	0.16	0.34	0.90	1.22
Phase variation (°)	14	17	38	58
Bead size (mm)	250-300			
Vol. Concentration	0.8%	2%	8%	
Amplitude variations	0.20	0.48	0.60	
Phase variations (°)	7	16	17	
Bead size (mm)	90-106			
Vol. Concentration	0.1%	0.4%	5%	
Amplitude variations	0.32	0.16	0.16	
Phase variations (°)	12	10	12	

TABLE 2. Extreme variations in amplitude and phase observed on fresh beef liver in velocity matched saline (amplitude normalised to a peak of 1.0).

	Position				
	1	2	3	4	5
Specimen A					
Amplitude variation	0.20	0.36	0.36	0.57	0.43
Phase variation (°)	13.1	24.6	71.5	40.0	40.2
Specimen B					
Amplitude variation	0.13	0.47	0.11	0.19	0.26
Phase variation (°)	19.5	35.9	25.2	43.7	41.3

6. CONCLUSION

Analysis of the fluctuations of ultrasonic waves propagating through inhomogeneous media requires both theoretical and experimental attention. It has applicability to the practical problems of making measurements of both attenuation and scattering. Measurements of the fluctuations is technically demanding but quite within current technological limits and is an area that deserves greater attention. The potential of measurements such as these to be used as the basis of wave field extrapolations is limited essentially only by the finite area of the receiver and of the field that is sampled. It is nevertheless important to consider the directionality of the scattering expected as well as that of the (finite) receiver [9].

REFERENCES

1. Chivers RC and Hill CR: Phys. Med. Biol. 20, 799-815, 1975.
2. Campbell JA and Waag RC: J. Acoust. Soc. Am. 74, 393-399, 1983.
3. Bamber JC: Ph.D. Thesis. London: University, 1980.
4. Chivers RC: Elsewhere in this volume, 1985.
5. Busse LJ, Miller JG, Yuhas, DE, Mimbs JW, Weiss AN and Sobel BE: Ultrasound in Medicine, Vol. 3, pp. 1519-1535. New York: Plenum Press, 1977.
6. Aindow JD: Ph.D. Thesis. Surrey: University, 1983.
7. Chivers RC: Ultrasound Med. Biol. 7, 1-20, 1981.
8. Chivers RC: In Imaging with Non-Ionizing Radiations (Ed. D.F. Jackson) pp. 50-84, Glasgow: Blackie and Sons, 1983.
9. Zverev VA: Soviet Phys.-Acoustics 3, 348-359, 1957.
10. Waag RC: Private communication, 1985.
11. Chernow LA: Wave propagation in a random medium. New York: Dover, 1960.
12. Chivers RC: J. Phys. D. 13, 1997-2003, 1980.
13. Leeman S and Fiddy MA: J. Phys. D. 14, L89-90, 1981.
14. Ross G and Chivers RC: J. Acoust. Soc. Am. (submitted for publication).
15. Radcliffe JA: Rep. Prog. Phys. 19, 188-267, 1956.
16. Strohbehn JW, Wang T and Speck JP: Radio Science 10, 59-70, 1975.
17. Komissarov VM: Soviet Phys.-Acoustics 10, 143-152, 1964.
18. Shirokova TA: Soviet Phys.-Acoustics 9, 78-81, 1963.
19. Chivers RC and Aindow JD: J. Phys. D. 16, 2093-2102, 1983.
20. Soczkiewicz E: Elsewhere in this volume, 1985.
21. Aindow JD and Chivers RC: J. Phys. E. 15, 83-86, 1982.
22. Chivers RC and Aindow JD: Ultrasonics 21, 70-72, 1982.
23. Aindow JD and Chivers RC: J. Phys. E. 15, 1027-1030, 1982.
24. Aindow JD and Chivers RC: J. Acoust. Soc. Am. 73, 1833-1837, 1983.
25. Chivers RC and Aindow JD: Ultrasonics International, 83 pp. 37-42, Guildford: Butterworths Scientific Ltd, 1983.
26. Aindow JD and Chivers RC: J. Acoust. Soc. Am. (in press) 1985.

DISCUSSION

Comment: Zequiri
On one of your diagrams you showed that the detail or texture was due essentially to the scatter processes occurring within the tissue. To what extent do scanner operation characteristics affect the final image?

Reply: Chivers
This really depends on which of the extreme scattering models is most representative of tissue. If the tissue consists of a large population of weak, small scatterers then the evidence is that the pictures obtained are primarily characteristic of the pulse shape (the scatterers sample the pulse). If the scatterers are strong and sparse and there is a significant amount of multiple scattering the information is largely on the scatterers but there may be difficulties with quantitative approaches. The real situation is probably between these extremes as evidenced by the experience of the clinical operators.

Comment: Cheeke
Given the difficulty in measuring the attenuation due to scattering, would it be sensible to adopt an approach to measure the absorbed ultrasonic energy, and then deduce that scattered from a knowledge of the incident field?

Reply: Chivers
Certainly the magnitude of the scattering contribution has been estimated from measurements of attenuation and absorption. Unfortunately, they can seldom be made on the same specimen. The results lie approximately in the middle of the range of values that I quoted, i.e. that scattering contributes 5-10 percent of the attenuation.

ACOUSTOOPTICAL INTERACTION IN OPTICALLY ACTIVE MEDIA

A. ŚLIWIŃSKI

1. INTRODUCTION

Recently, in the literature some papers have appeared on acoustooptical interaction of waves progressing in anisotropic, acoustically [1,2] and optically [3] active media; in the case, however, when both waves, the acoustical and optical ones are progressing simultaneously in parallel [4-6] in the same direction. The aim of this paper is to present a consideration of the interacting process in a special uniaxial system in which two push-pull ultrasonic waves are generated as shear rotational waves propagating in a rod (corresponding to a nonlinear medium) of circular cross-section while optical beams are impinging side walls of the rod. The optical beams are coherent and linearly polarized and can be thought as the sum of left-hand and right-hand polarized components (Fig. 1).

2. INTERFERENCE OF ROTATIONAL ACOUSTICAL WAVES GENERATED IN PUSH-PULL CONDITIONS IN ACOUSTOOPTICALLY ACTIVE LAYER

Let us consider a transparent uniaxial crystal rod (or an uniaxial layer of a liquid crystal texture) of the length 2d (Fig. 1) at the end of which two transducers of shear vibrations [7] P_1 and P_2 of the same resonance frequency Ω are situated. They are supplied from the same signal generator to create proper phases of radiated ultrasonic waves at both ends of the rod. One can treat the shear wave of linear polarization excited in the rod (the polarization plane of which rotates due to the acoustical activity of the medium) as consisting of two components of opposite circular polarization and consider the behaviour of these two circular components in both waves (from both sides) when interacting at $z = d$. The phases at $z = d$ should be matched to obtain a constructive interference of the contra-

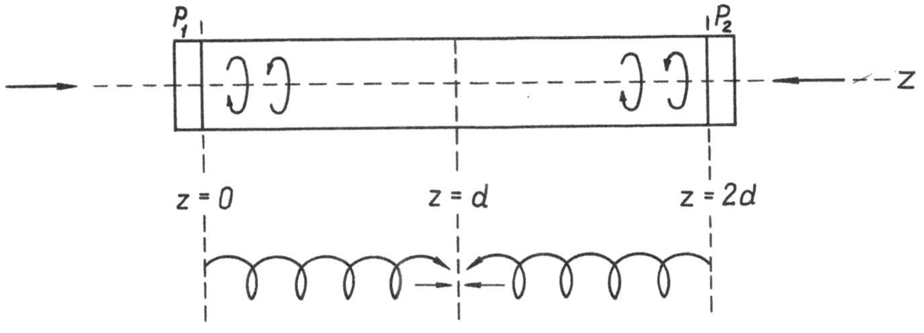

FIGURE 1

circular waves i.e. to create a standing wave (circular standing wave) and not to cancel the wave. That will take place when both transducers are supplied in phase coincidence. Then, they will travel opposite each other, either two waves of left-hand circular polarization or ones of left-hand circular polarization. Supplying the transducers with the signals opposite in phase one obtains waves shifted in phase by 180° and with opposite polarization components, for instance left-hand circular from one side and right-hand circular from the other, that will cancel in the interaction.

Let us consider the situation when two ultrasonic pulses are generated (Fig. 2) from both sides and the shear wave signals are both of the same circular polarization and are in phase coincidence. There are two possibilities for constructive interference (Fig. 2a,b):

a) both are of right-hand circular polarization (D) or

b) both are of left-hand circular polarization (L).

Let us assume we change the state of polarization one after the other (controlling the transducers electronically) with a given frequency. In a particular case the frequency may be equal to the frequency Ω. Then, widths of the pulses will be equal during one half of a period and the situation in time at the plane z = d is represented in the Fig. 2c. One can notice that the amplitude of shear deformation goes to zero with the frequency $\frac{\Omega}{2}$.

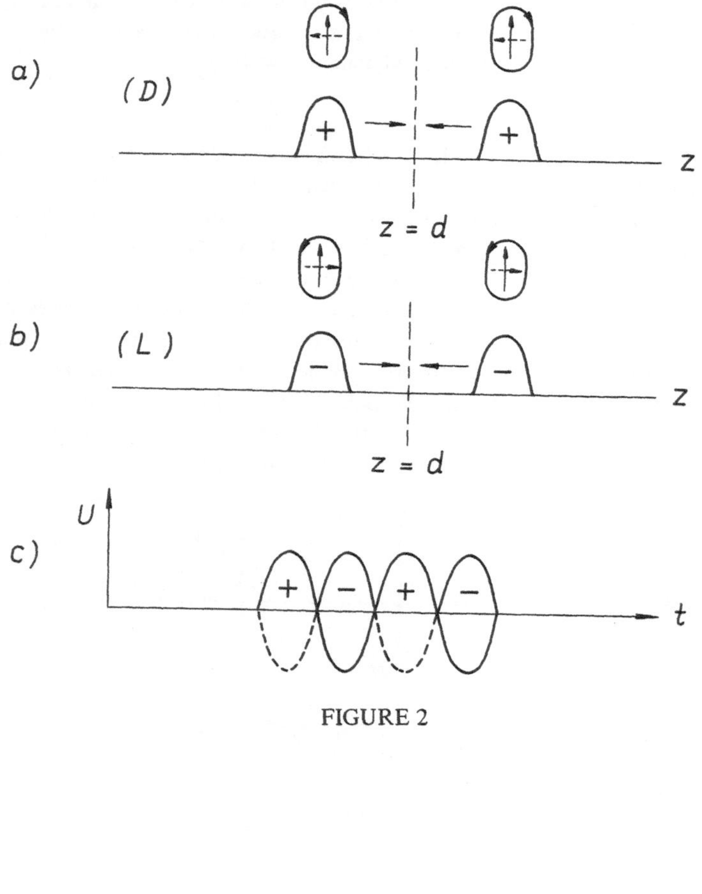

FIGURE 2

Let in the first case (D) (first half of a period) the transducers generate right-hand circular components:

$$U^{(D\rightarrow)} = \frac{U_o}{2} (\vec{a} + i\vec{b}) \exp [i (K_+ z - \Omega t)]$$

and

$$U^{(D\leftarrow)} = \frac{U_o}{2} (\vec{a} + i\vec{b}) \exp [-i (K_+ z + \Omega t)]$$

where the arrow symbolizes the sense on the propagation direction z of shear right-hand circular deformation U^D, in a general case of elliptical polarization and $\vec{a}$ and $\vec{b}$ are versors representing directions of half axis of the ellipse, U_o is the wave amplitude starting at the inputs, i.e. for $z = 0$ and $z = 2d$ (Fig. 1), K_+ stands for wave number of right-hand circular component. With the overlapping of two waves traveling one against the other, one obtains at $z = d$

$$U_d^{(D)} = U_d^{(D\rightarrow)} + U_d^{(D\leftarrow)} = U_o (\vec{a} + i\vec{b}) [\exp i K_+ d + \exp(-i K_+ d)] \exp (-i\Omega t),$$

or

$$U_d^{(D)} = 2 U_o (\vec{a} + i\vec{b}) \exp (-i \Omega t) \cdot \cos K_+ d . \qquad (1)$$

In the second case (L) (second half of a period) the transducers are producing left-hand circular components:

$$U^{(L\rightarrow)} = \frac{U_o}{2} (\vec{a} - i\vec{b}) \exp [i (K_- z - \Omega t)]$$

and

$$U^{(L\leftarrow)} = \frac{U_o}{2} (\vec{a} - i\vec{b}) \exp [-i (K_- z + \Omega t)]$$

where K_- is the wave number of the left-hand circular component.
After overlapping of left-hand circular components at $z = d$ one obtains

$$U_d^{(L)} = U_d^{(L\rightarrow)} + U_d^{(L\leftarrow)} = U_o (\vec{a} - i\vec{b}) [\exp i K_- d + \exp (-i K_- d)] \exp (-i\Omega t)$$

or

$$U_d^{(L)} = 2 U_o (\vec{a} - i\vec{b}) \exp (-i \Omega t) \cos K_- d . \qquad (2)$$

An acoustically active medium is characterized by the fact that the right-hand left-hand components have different propagation velocities (different wave numbers K_+ and K_-, respectively). For media of uniaxial symmetry the wave numbers can be expressed as follows [1,2,6]:

$$K_+ = \frac{\Omega}{\sqrt{C_{44}}} + \frac{1}{2}\,\Omega^2 \rho\, r_{54,3} \qquad \text{and} \qquad K_- = \frac{\Omega}{\sqrt{C_{44}}} - \frac{1}{2}\,\Omega^2 \rho\, r_{54,3} \tag{3}$$

where C_{44} - elasticity modulus of shear deformations,
- medium density,

$\rho r_{54,3}$ - coefficient of acoustical rotation - a coresponding matrix component of such coefficients for an uniaxial system; the coefficient is related to the piezo-electric effect when the medium is a piezo-eletctric.

It is worth to remark that the velocity difference of left-hand and right-hand circular components in an acoustically active medium (in analogy to the optically active medium) leads to the rotation of the polarization plane of the shear wave. The rotation angle is given by [1,2,6]:

$$\phi = \frac{1}{2}\,\Omega \ell \left[\frac{1}{v_+} - \frac{1}{v_-}\right] = \frac{1}{2}\,(K_+ - K_-)\,\ell = \frac{1}{2}\,\Omega^2 \rho\, r_{54,3} \cdot \ell\,,$$

where $v_\pm$ - phase velocity of components, respectively,

ℓ - distance traveled by the wave.

The specific rotation of a medium (the characteristic quantity determined on the unit distance) in the field of acoustic wave is equal to

$$\phi = \frac{1}{2}\,(K_+ - K_-) = \frac{1}{2}\,\Omega^2 \rho\, r_{54,3}\,. \tag{4}$$

3. INTERFERENCE OF CONTRA-CIRCULARLY POLARIZED LIGHT BEAMS IN ACOUSTOOPTICALLY ACTIVE LAYERS IN PRESENCE OF CONTRA-CIRCULARLY POLARIZED ULTRASONIC WAVES

Now, let us assume that in the system (Fig. 1) from both sides equal coherent and linearly polarized light waves of angular frequancy ω are incident in the presence of the acoustical field described by formulae (1) and (2) (Fig. 3). Both of these beams can be treated in the medium as a sum of two waves elliptically polarized. Using the results obtained by W.W. Gvozdiyev at al. [6] for the light wave amplitude traveling in an acoustooptically active medium one can write:

$$E^{(\rightarrow)} = \sum_{\tau=1,2} U(z) * E_\tau \frac{\vec{a}' + i\,\gamma_\tau \vec{b}'}{\sqrt{1+\gamma_\tau^2}} \exp\left[i\,(K_\tau z - \omega t)\right]$$

and

$$E^{(\leftarrow)} = \sum_{\tau=1,2} U(z) * E_\tau \frac{\vec{a}' + i\gamma_\tau \vec{b}'}{\sqrt{1+\gamma_\tau^2}} \exp[-i(k_\tau z + \omega t)],$$

where the arrow denotes the sense of the wave in direction z, $E\tau$ amplitude of the elliptically polarized light waves, right-hand ($\tau = 1$) or left-hand circular ($\tau = 2$), respectively, $U(z) = \exp[\phi_s \cdot z \cdot n']$, in a general case denotes the matrix of rotation of the angle $\phi_s \cdot z$ and n' antisymmetric tensor, dual to the versor $\vec{n}$ parallel to the z - axis, γ_τ ellipticity, i.e. the ratio of electric field components along the versors $\vec{a}'$ and $\vec{b}'$, k_τ wave numbers of respective beams, * stands for convolution.

After overlaping of the light beams at z = d we have

$$E_d = E_d^{(\rightarrow)} + E_d^{(\leftarrow)} = \sum_{\tau=1,2} E_\tau U(d) \frac{\vec{a}' + \gamma_\tau \vec{b}'}{\sqrt{1+\gamma_\tau^2}} [\exp(i k_\tau d) - \exp(-i k_\tau d)] \exp(-i\omega t)$$

or

$$E_d = 2 \sum_{\tau=1,2} U(d) E_\tau \frac{\vec{a}' + \gamma_\tau \vec{b}'}{\sqrt{1 = \gamma_\tau^2}} \exp(-i\omega t) \cos k_\tau d. \tag{5}$$

According to [6] the wave numbers can be expressed as

$$K_{1/2} = \frac{\omega}{C} \sqrt{\bar{\epsilon} + \eta^2 \pm \sqrt{(\Delta\epsilon)^2 + 4\eta\bar{\epsilon}}} \tag{6}$$

and the ellipticity

$$\gamma_{1/2} = \frac{2\eta^2 - \Delta\epsilon \pm \sqrt{(\Delta\epsilon)^2 + \sqrt{(\Delta\epsilon)^2 + 4\eta\bar{\epsilon}}}}{2\eta\sqrt{\bar{\epsilon} + \eta^2 \pm \sqrt{(\Delta\epsilon)^2 + 4\eta\bar{\epsilon}}}}, \tag{7}$$

where

$$\bar{\epsilon} = \frac{\epsilon_0 + \epsilon_\tau}{2} + U_0^2 (K_+^2 + K_-^2)\ (B_1 + B_2),$$

$$\Delta\epsilon = \frac{\epsilon_\tau - \epsilon_0}{2} + 2 B_2 U_0^2 K_+ K_-,$$

$$\eta = \frac{C}{\omega} (\phi_s - \phi_{em}), \qquad \phi_{em} = -\alpha_{11} \frac{\omega}{C}.$$

In the formula ϵ denotes the electric permeability of the medium, $\Delta\epsilon$ variation induced by the ultrasonic wave of the amplitude U_0; the variation is nonlinear one and it is proportional to the square of amplitude and to the product of $K_+ \cdot K_-$ (comp. [2]), so it depends on the frequency and on the acoustical activity of the medium,

$\bar{\epsilon}$ - depends on these quantities in similar way,

B_1, B_2 - piezoelectric constants of the second rank - corresponding combinations of the sixth rank tensor [6],

α_{11} - component of the optical rotation tensor of the medium,

ϕ_{em} - specific optical rotation of the medium,

η - resultant acoustooptical rotation of the medium.

The expression [5] is quite general. One can obtain from it (at given conditions of the system examined) expression for separate vibration modes as a result of interaction with a given distribution of an ultrasonic field. Let us consider the following cases:

1) two contra-circular ultrasonic waves having right-hand circular polarization and starting with coincident phases.
In this case, in the system contra-circular light waves of right-hand circular polarization can propagate and the left-hand circular components will be cancelled. So, one has

$$E^{1)}_{d} = 2U^{(D)} E_1 \frac{\vec{a}' + i\gamma_1 \vec{b}'}{\sqrt{1+\gamma_1^2}} \exp(-i\omega t) \cos k_1 d; \qquad (8)$$

2) two contra-circular ultrasonic waves having left-hand circular polarization and starting with coincident phases.
In this case the right-hand circular components of light beams will be cancelled and the left-hand circular beams will propagate and overlap. So, one has

$$E^{2)}_{d} = 2\,U^{(L)}\, E_2 \frac{\vec{a}' - i\gamma_2 \vec{b}'}{\sqrt{1+\gamma_2^2}} \exp(-i\omega t) \cos k_2 d. \qquad (9)$$

In the total cycle of vibrations of the ultrasonic transducers the process of successive cancelling and appearing of the light (alternating with once right-hand, once left-hand circular polarization) occurs. Variations of the resultant amplitude will be 50-100% of the light amplitude incident on the system. It concerns to the one meeting act of the waves.

It is difficult to say if the modulator of light which could be designed based on this phenomenon, would be better than the existing ones; however, the concept is, as far as the author knows, a new one and rather exciting if we concider the system situated between two parallel mirrors (in Fabry-Perot interferometer) - Fig. 3.

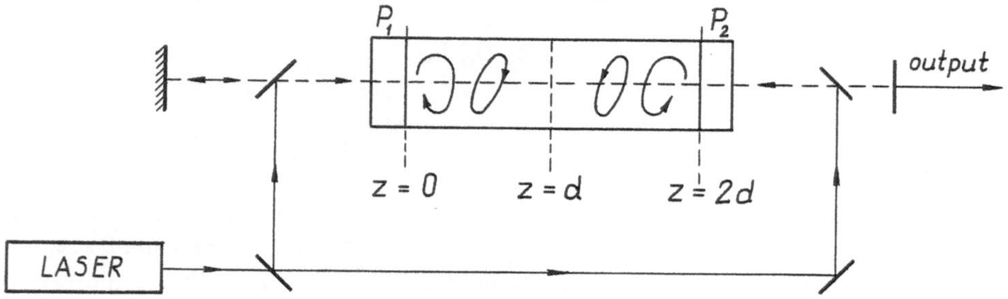

FIGURE 3

Assuming that we are able to fulfill proper phase matching conditions between light and ultrasonic field to allow the beams to pass through the system several times, we have the possibility to pump the light with an ultrasonic wave. In this case, we could fulfill the requirements with mirrors and repeat coherently the interacting process in the acoustical systems many times collecting the small contributions of the energy transfered from the shear waves into the light. The process is analogous to maintaining laser action in anomalous dispersive medium in a laser cavity. In our case the acoustooptically active device of two opposite propagating shear waves represents such anomalous dispersive medium for light. Because of the modulation with successive cancelling and appearing of the light, the system will be pulsating with the frequency $\Omega/2$. Every component of two opposite state of polatiration will appear with frequency Ω.

Coming back to the formulas (8) and (9) and using definitions of $U^{(D)}_{(z)}$ and $U^{(L)}_{(z)}$, as well as formula (4) one obtains

$$E^{1)}_{d} = 2E_1 \exp \frac{1}{2} (K_+ - K_-) \cdot 2d \, \frac{\vec{a}' + i\gamma_1 \vec{b}'}{\sqrt{1+\gamma_1^2}} \exp(-i\omega t) \cos K_1 d \tag{8a}$$

and

$$E^{2)}_{d} = 2E_2 \exp \frac{1}{2} (K_- - K_+) \cdot 2d \, \frac{\vec{a}' - i\gamma_2 \vec{b}'}{\sqrt{1+\gamma_2^2}} \exp(-i\omega t) \cos k_2 d \tag{9a}$$

Next, taking into account the fact that we have a pure optical effect connected with optical circulation

$\phi_{em} = -\alpha_{11} \frac{\omega}{C} = \frac{1}{2} (K_1 - K_2)$, we obtain

$$E^{1)}_{d} = 2 \frac{\vec{a}' + i\gamma_1 \vec{b}'}{\sqrt{1+\gamma_1^2}} E_1 \exp [(K_+ - K_-) - (K_1 - K_2)] d \cdot \exp(-i\omega t) \cos K_1 \cdot d \tag{8b}$$

$$E^{2)}_{d} = 2 \frac{\vec{a}' - i\gamma_2 \vec{b}'}{\sqrt{1+\gamma_2^2}} E_2 \exp [(K_- - K_+) - (K_2 - K_1)] d \cdot \exp(-i\omega t) \cos K_2 \cdot d \tag{9b}$$

Calculating, respectively, intensities as proportional to E^2, we have

$$I^{1)}_{d} = \frac{4E_1^2}{1+\gamma_1^2} \exp \left| 2d \, [(K_+ - K_- - (K_1 - K_2)] \right| \cdot \cos^2 K_1 d \tag{10}$$

$$I^{2)}_{d} = \frac{4E_2^2}{1+\gamma_2^2} \exp \left| 2d \, [(K_- + K_+) - (K_1 - K_2)] \right| \cdot \cos^2 K_2 d \tag{11}$$

If the matching condition

$$\frac{\omega}{C} (K_1 - K_2) = \frac{\Omega}{v} (K_+ - K_-)$$

are satisfied (it means that optical rotation is equal to the acoustooptical one) both the effects will overlap. Then, after using (3), (4) and (6) one obtains

$$\frac{\omega}{C}\left[\sqrt{\bar{\epsilon}+\eta^2+\sqrt{(\Delta\epsilon)^2+4\bar{\epsilon}\eta}}-\sqrt{\bar{\epsilon}+\eta^2-\sqrt{(\Delta\epsilon)^2+4\bar{\epsilon}\eta}}\right]=\frac{\Omega}{v}\rho r_{54,3}\Omega^2. \quad (12)$$

Introduction of $\bar{\epsilon}$, η and $\Delta\epsilon$ to this expression leads to rather algebraically complicated relationships not allowing to get a transparent analytic expression. We will not give them here. However, one can calculate the ratio between two intensities in two phases of illumination corresponding to the first and the second half of the acoustical wave period. From expressions (10) and (11):

$$\frac{I^{1)}_d}{I^{2)}_d}=\frac{E_1^2\,(1+\gamma_2^2)}{E_2^2\,(1+\gamma_1^2)}\,\sin h\left\{2d\,[(K_+-K_-)-(k_1-k_2)]\right\}\frac{\cos k_1 d}{\cos k_2 d} \quad (13)$$

4. FINAL REMARKS AND CONCLUSIONS

The mechanism of acoustooptical interaction in the system described above stimulated us to construct the device for light modulation by shear ultrasonic waves according to the idea considered. The efficiency of the pulsator in our first experiments is very poor. Further experiments are being continued. The calculations presented in this paper are not finisched, too. It is required to take into account, in the further calculations, optical losses for compensation of them by ultrasonic energy. For doing this a simpler analytical formula should be found.

REFERENCES

1. Portigal D.L., Burstein E., Acoustical Activity and Other First Order Spatial Dispersion Effects in Crystals, Phys. Rev. 170 (1968), 673-678.
2. Serdynkov A.N., Krugovoj dichroizm w akistikie krystalov z prostaanstviennoj dispersiey, Kristalografia 22 (1977), 459-462; Sov. Cristallography 22 (1977) 262-4.
3. Bokut B.W., Serdynkov A.N., Szepielewicz W.W., K fenomenologiczeskoj teorii poglo-shchayushchikh opticheski aktivnych sried, Optika i Spektroskopia 37 (1974), 120-124; Sov. Optics and Spectroscopy 37 (1974), 65-67.
4. Guliayev Yu.W., Loshchenkova E.P., Shkierdin G.N., Akustodielektricheskij effekt w kristalach, Fiz.Tverd.Tiela 22 1980, 150-155; Sov. Solid State Phys. 22 (1980) 1691-5.
5. Pienyas W.A., Siemchenko I.W., Serdynkov A.N., Wysokochastotnyj effekt kerra w dispergirujushchoj sriedie, Zurn. Prikl. Spektr. 25 1981, 363-366; Sov. Appl. Spectroscopy J. 25 (1981), ...
6. Gvozdyev W.W., Siemchenko I.W., Sierdynkov A.N., Spiralnaya modulacya dielektricheskoj pronichayemosti hiperzvukom i izmiereniye akusticheskoj wrashchatelnoy aposobnosti. Akust. Zurn, 29 1983, 326-328; Sov. Acoust. J. 29 (1983), ...
7. Bergmann L., Der Ultraschall, Zürich, 1954.

DISCUSSION

Comment: Zarembowitch
I have a comment and a question.
I find your idea very elegant but I have some concern regarding the order of magnitude of the proposed effect: the acoustical activity, indeed, is very small in the MHz region; it becomes larger in the GHz region but then the attenuation becomes prohibitively large. Do you expect to find a reasonable compromise?
Secondly, what is the acoustical active medium you use in your experiments?

Reply: Sliwinski
Thank you for your comment. What you said is true: I believe that a compromise will be possible between the two competitive effects, i.e. the gain of the optical signal (small effect) and the attenuation. I expect that in case of proper matching conditions inside of the Fabry-Perot interferometer a laser-like action will be possible. Till now, as I said, we haven't been successful with our device. We have used quartz rods combined with torsional transducers of frequency 50-80 kHz. After cutting such rods in their central position ($z = d$, in Fig. 1) we could put in between a thin layer of liquid crystal (either nematic or cholesteric). It is well known that the acoustical and optical activity of liquid crystals is much greater (few orders of magnitude) than that of solids.
The effect I am looking for in acoustooptics has a counterpart in mechanics, when a button is held with a cotton thread passing through its holes. Children do usually play with such a toy. Starting with small rotation and then stretching the thread periodically one can excite steady state rotational oscillations of the button, pumping energy through the spinning loop. This is a parametric system, a nonlinear one in which the elastic properties of the loop are dynamically oscillating, giving rise to some kind of self-excited generator.

II. METHODS

INHOMOGENEOUS MATERIALS STUDIED WITH BRILLOUIN SCATTERING

A. ZAREMBOWITCH, J. BERGER, M. FISCHER, F. MICHARD.

1. INTRODUCTION

The generic term inhomogeneous materials+ refers to a large variety of situations : composite materials, superlattices, liquid crystals, human tissues, materials with an homogeneous composition and a gradient of properties... This variety of situations calls for a conceptual effort to find out common features between these situations.

Indeed, an inhomogeneous medium is a medium which is not invariant under translation ; one consequence is that the phase velocity of elastic waves propagating in the medium depends on the spatial frequency. The importance of this spatial dispersion depends on $\lambda/_a$ (where λ is the acoustical wavelength and a is a characteristic length in the medium), which from the experimental point of view implies an extension of the frequency domain beyond that generally used for investigating homogeneous media.

Another consequence is the non local character of the elastic response of the medium which raises conceptual and experimental difficulties. In addition, diffusion phenomena are always large in inhomogeneous media and nonlinear effects are often enhanced.

Since the character of this school is predominantly experimental it appeared appropriate to illustrate these different characteristics of inhomogeneous media by using as an Ariane's thread an experimental technique well suited for their examination : BRILLOUIN scattering (B.S.).

The BRILLOUIN scattering of light by thermal elastic waves offers a local probe which extends towards high frequencies (in the order of 10^{10} Hertz) the information provided by conventional ultrasonic waves in the media in which they are propagating. This experimental tool, although more restricted than ultrasonic techniques (e.g., B.S is mainly limited to transparent media), is nevertheless able to replace ultrasonic waves in many experimental situations involving very small samples, very fragile samples, samples studied under high pressure, media ultrasonically opaque etc.

\+ Use of the term "inhomogeneous" can be misleading. From etymology "homogeneous" applies for one-component systems with no gradient. "Inhomogeneous" could thus be restricted to one-component systems for which there exists some gradient. On the opposite, when two or more components are involved the term "heterogeneous" seems more appropriate.

.../...

In this paper we propose first, to examine more in detail the concept of homogenity, then to discuss examples of BRILLOUIN scattering in order to show how elastic waves can be used to resolve some theoretical and experimental problems in inhomogeneous media.

2. THE CONCEPT OF INHOMOGENEITY IN THE CASE OF MATERIALS

Starting with homogeneous materials we can say that a medium is homogeneous when any point of the medium is representative of the properties of the whole medium. In other words, the properties of homogeneous media are invariant under spatial translation (isotropy is invariance under rotation).

If the relation between the stress tensor Tij and the strain tensor S_{kl} are assumed to be linear and if we refer to the principle of causality owing to which Tij (t) at the time t is determined only by the present strain S_{kl} (t) and the strain at the past moments t'< t **we have** :

$$Tij\ (\underset{\sim}{r},t) = \int_{-\infty}^{t} dt' \int Cijkl\ (t,t',\underset{\sim}{r},\underset{\sim}{r}').\ S_{kl}\ (\underset{\sim}{r}',t').d\underset{\sim}{r}'$$

where Cijkl is the elastic tensor and $\underset{\sim}{r}'$ a vector of the Euclidian space.

If the properties of the medium do not change with time (time is uniform) then, the kernel Cijkl depend only on the difference (t-t') ; furthermore, if the medium is homogeneous all its points are equivalent and Cijkl depend only on the difference $(\underset{\sim}{r}-\underset{\sim}{r}')$.
Under such conditions :

$$Tij\ (\underset{\sim}{r},t) = \int_{-\infty}^{t} dt' \int Cijkl\ (t-t',\underset{\sim}{r}-\underset{\sim}{r}').\ S_{kl}\ (r',t')\ dr' \qquad (1)$$

If we FOURIER transform equation (1) we have :

$$Tij\ (\omega,\underset{\sim}{k}) = Cijkl\ (\omega,\underset{\sim}{k}).\ S_{kl}\ (\omega,\underset{\sim}{k}) \qquad (2)$$

where $\omega,\underset{\sim}{k}$ are respectively the frequency and the wave vector.

Now if we assume that the medium is strictly continuous in space, there are no characteristic lengths l_c in the medium ; consequently the medium is invariant under translation but also under similarities operations ; as a consequence, C_{ijkl} do not depend on $\underset{\sim}{k}$. If there are no characteristic time τ_c for the medium, C_{ijkl} is invariant under dilatation of time and C_{ijkl} do not depend on ω.
Indeed, such hypothesis **are not physical** (all media have characteristic lengths and times) **and** often **is**

Indeed "heterogeneous" stresses the discontinuity of some physical quantity while "inhomogeneous" emphazises continuous changes. For example, acousticians use "inhomogeneous" to describe the ocean because of the gradient in pressure. On the other hand concrete is better represented by "heterogeneous".

$\omega\tau_c >> 1$ or $\omega\tau_c << 1$ and

$kl_c >> 1$ or $kl_c << 1$

In such cases the frequency dependence of C_{ijkl} (temporal dispersion) and the k dependence (spatial dispersion) can be neglected.

We will now examine the opposite situation (inhomogeneous media) where the spatial dispersion comes from the loss of translational invariance; here, we have obviously to distinguish the case of inhomogeneity at a microscopic scale from the case of inhomogeneity at a macroscopic scale.

2.1 Translational invariance at a microscopic scale

At a microscopic scale, homogeneous media do not exist because of the discontinuous structure of matter (atoms, vacuum...). A linear chaîn of atoms (the classical picture for a crystal - fig 1) is not in general invariant under translation, but only under translations of a, 2 a.., na ; a well known consequence is a spatial dispersion - a dependence of V on k (where V is the phase velocity and k the wave number).

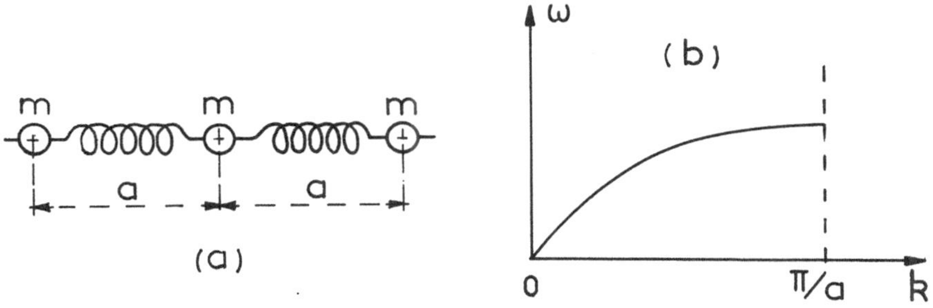

Figure 1 - (a) linear chaîn of atoms - (b) dispersion curve.

At that stage, a question arises : when we are dealing with materials and ultrasonic waves at a macroscopic scale, can we bypass the inhomogeneous structure of matter and thus ignore the spatial dispersion ?

A part of the answer can be found in accurate measurements of acoustic wave velocities that we have performed in good crystals of alkali halides (NaCl, KCl, KBr) in the ultrasonic frequency range (10 to 100 MHz) and in the hypersonic range through Brillouin scattering experiments (10 to 30 GHz) |1|. The result is that measured velocities are the same in both cases within an accuracy better than 10^{-3}. We can note that in such crystals there is no reason for temporal dispersion due to a relaxation mechanism.

Another approach for investigating possible dispersion is the following : in the absence of dispersion it can be shown from the KRAMERS - KRÖNIG relations |2| that the attenuation is proportional to the square of the frequency. We have checked such a dependence, again in the case of good crystals of alkali halides, as shown in fig 2.

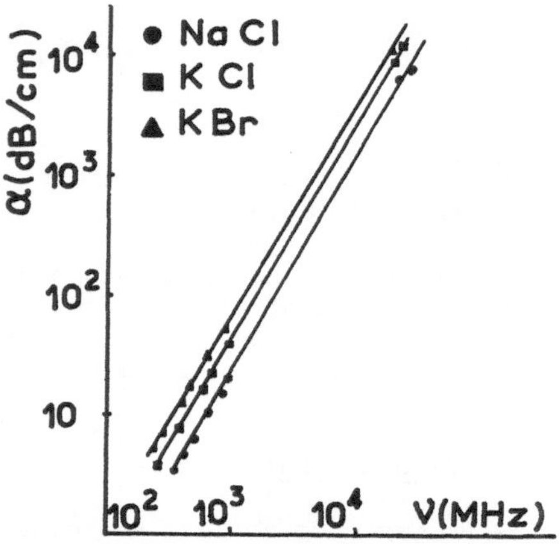

Figure 2 - Attenuation vs frequency for longitudinal waves propagating along |110| in some alkali halides. (Ultrasonic and B.S measurements).

It seems then that the microstructure of matter can be forgotten when measurements are performed at a macroscopic scale. Indeed, that statement is valid for media organized in a highly symmetrical way ; further examples will show that this statement is generally not confirmed for less symmetrical media.

Inhomogeneities at a microscopic scale can be "seen" at a macroscopic scale

2.1.1 . Acoustical activity

Acoustical activity is the rotation of the plane of polarization of a transverse elastic wave propagating in some media of low symmetry, (acoustically active media are also optically active |3|). Such media are not homogeneous since the elastic response is not local : the stress in a point depends not only on the local strain but also on the gradient of the strain. The elastic constants can then be written :

$$C_{ij}(\omega,\underset{\sim}{k}) = C_{ij}(\omega) + i\, d_{ij,l}\, k_l \ldots$$

This effect is known to be due to the lack of symmetry of the atoms in the molecules or of the atomic arrangement in the lattice.

In the case of a crystal which has a threefold axis of symmetry (quartz for instance), the square of the velocities for right (+) and left (-) circularly polarized waves are shown to be |4|:

$$V_{\pm}^2 = \frac{1}{\rho} \quad (C_{44} \pm d_{54,3}\, k_3).$$

As a result, the plane of polarization of the transverse waves rotates. A rotation of 130°/cm GHz has been observed in quartz crystals : a macroscopic effect due to inhomogeneities at a microscopic scale !

2.1.2 Light scattering by fluctuations

In the 16th century LEONARD DA VINCI suggested that the blueness of the sky could be due to the scattering of light by particles of air. Later, it was understood that fluctuation in density was an additionnal necessary requirement for the occurrence of scattered light. If the fluctuations are large, the scattering is important (critical opalescence near a phase transition) ; if the medium is perfectly ordered the scattering is small.

In fact density can fluctuate with pressure P or with entropy S. The pressure fluctuations of the density propagate through the material ; they scatter incoming photons at a displaced frequency and give rise to BRILLOUIN scattering. The entropy fluctuations will not propagate ; they will give rise to RAYLEIGH scattering without change of frequency. From thermodynamics one can easily show that |5|:

$$\frac{I_B}{I_R} = \frac{C_v}{C_p - C_v}$$

LANDAU - PLACZEK relation, where I_B and I_R are respectively the BRILLOUIN and RAYLEIGH intensities and C_p and C_v the specific heats at constant pressure and volume.

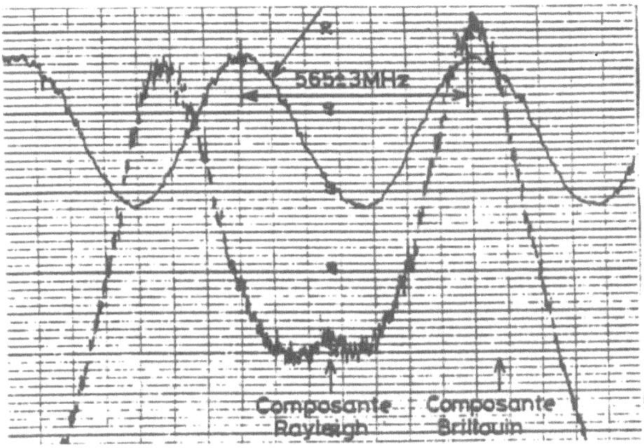

Figure 3 - B.S spectrum in a ferroelastic crystal La $P_5 O_{14}$. The upper arrow shows the Michelson signal used for calibration.

Indeed RAYLEIGH scattering is mainly due to impurities and defects ; fig. 3 shows a very unusual spectrum obtained from a crystal of exceptonally good quality and $I_B > I_R$. In general $I_R >> I_B$.
Nevertheless light scattering is a daily proof that inhomogeneities at a microscopic scale lead to macroscopic consequences.

2.1.3 Incommensurate materials

We know from crystals that atoms like to be arranged in a regular way. This trend is easily satisfied for atoms which obey a single logic ; the more complicated situation where they have to obey two (or more) conflicting logics leads to the concept of frustration (fig. 4).

In crystals, the compromise found by nature to overcome that conflict is illustrated in fig. 5. As a result the lattice periodicity and the magnetic periodicity measured respectively by a and b are such that $\frac{a}{b}$ is not a simple rational number. A phase of this type in a medium is called "incommensurate phase" ; it has lost all kinds of translationnal invariance since it is invariant neither under translation of m x a

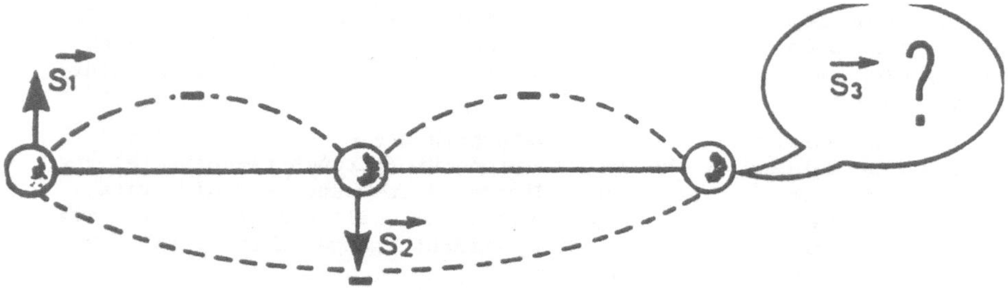

Fig. 4 - The interaction between atoms is antiferromagetic ; S_3 is frustrated

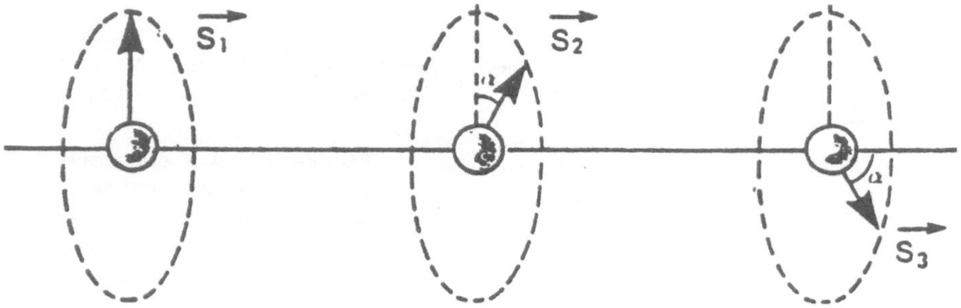

Fig. 5 - The interaction between first neighbourgs is ferromagnetic ; the interaction between second neighbourgs is antiferromagnetic.

nor n x b (where m and n are integer).

Various materials exhibit incommensurate structures ; the most familiar examples are helicoïdal or sinusoïdally modulated magnetic or ferroelectric crystals. The general behaviour when the temperature of such a crystal is varied is that the crystal undergoes with decreasing temperature several phase transitions from a commensurate to an incommensurate and again to a commensurate phase for a "lock-in temperature" corresponding to a critical change when the ratio $a/_b$ approaches a simple rational number.

It is remarkable that this change from a periodic organization (commensurate phase) to a modulated phase (inhomogeneous medium) can be observed easily when ultrasonic waves are propagated in the medium. Thus the macroscopic ultrasonic probe is a good tool for investigating tiny modifications at the microscopic level. Several examples illustrating that statement will be given in part 4.

2.2 Translationnal invariance at a macroscopic scale

When we examine solids at a macroscopic scale, the most homogeneous materials are to be found among amorphous media ; these media can reach a high degree of homógeneity and isotropy on account of the statistical distribution of their atoms ; for example, if we measure the refractive index of excellent samples of fused silica, the figures we find when probing the sample at different points are identical within an accuracy of 10^{-5}.

In order to check whether that translationnal invariance was satisfied also for the elastic properties of the medium, it was necessary to design an elastic probe able to perform local measurements ; such a device is described in detail in ref |6|.

An acousto-optical interaction in BRAGG's regime is performed in a small volume of the medium, allowing the measurement of an ultrasonic wavelength in terms of optical wavelengths ; this technique provides either the absolute value of the ultrasonic phase velocity with an error of 10^{-2} if the measurement is performed over a short distance (about 10 microns) or the absolute value with a high accuracy ($>3.10^{-4}$) if the measurements are averaged over a large distance .

The statistical treatment of a large number of measurements leads to an unexpected result : the main limit in the accuracy when we measure the velocity of ultrasonic waves propagating in the medium is the inhomogeneity of the medium.

The best sample of fused silica that we have got had not a sufficient homogeneity to allow absolute measurements of the velocity with an accuracy better than 10^{-4}. It is interesting to note that, from that point of view, elastic waves are a better probe for smooth inhomogeneity than the optical probe.

So we conclude that even at a macroscopic scale, the possibility of achieving perfectly homogeneous materials is very questionable.

3. BRILLOUIN SCATTERING TECHNIQUE

BRILLOUIN scattering is the result of an interaction in a material between elastic waves and light. The elastic waves considered in BRILLOUIN scattering are the thermal waves that propagate spontaneously in a solid. This interaction can be described either in classical or in quantum mechanical terms.

In classical terms, an elastic wave propagating through a material, produces in space a perodic variation of the refractive index ; this three-dimensional phase grating moves through the medium with the velocity of sound, V. An incident light beam is selectively reflected when the BRAGG condition is satisfied namely :

$$\lambda = 2\Lambda \sin \frac{\theta}{2} \qquad (3)$$

Where λ and Λ are respectively the wavelength of the light and elastic wave ; θ the angle between the incident and the reflected beam.

Since the grating is moving with the velocity V, the scattered light is shifted in frequency (DOPPLER effect)

$$F = \Delta f = \pm \frac{V}{C} \frac{\lambda}{\Lambda} f \qquad (4)$$

Where f is the frequency of the incident light and C the velocity of light.

In quantum mechanical terms, BRILLOUIN scattering is the scattering of an incident photon with wavevector $\underset{\sim}{k}_1$ and frequency f_1, by a phonon with wavevector $\underset{\sim}{K}$ and frequency F. The scattered photon is characterized by a wavevector $\underset{\sim}{k}_2$ and a frequency f_2.

The conservation of momentum and energy implies

$$\underset{\sim}{k}_1 - \underset{\sim}{k}_2 = \underset{\sim}{K} \qquad (5)$$

$$hf_1 - hf_2 = \pm hF \qquad (6)$$

The relations (5) and (6) are equivalent to (3) and (4). As the velocity of light C is usually 10^5 times greater than the velocity of sound,

$f_1 \simeq f_2$ and

$|\underset{\sim}{k}_1| \simeq |\underset{\sim}{k}_2|$

It follows that

$$\theta < |\underset{\sim}{k}| < 2|\underset{\sim}{k}_1|,$$

so, for visible light,

$$\theta < F < 100 \text{ GHz}$$

The frequency of the selected phonons depends on the geometry of the interaction. The most classical geometries are such that $\theta = 90°$ or $\theta = 180°$ (backward scattering).
From θ we deduce Λ - equation (3)
from F, f, λ, Λ we deduce V - equation (4)
from the width of the BRILLOUIN line we obtain the attenuation coefficient of the sound wave |7|.
The general characteristics of a BRILLOUIN scattering apparatus have been discussed by several authors |8, 9, 10, 11, 12, 13...|.
A typical arrangement is shown in fig (6).

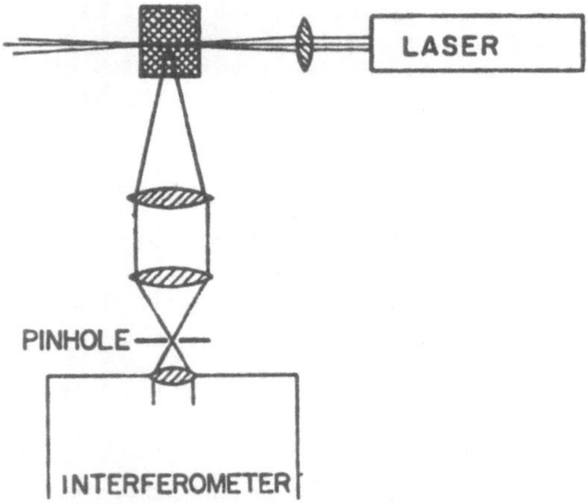

Figure 6 - BRILLOUIN scattering at 90°.

The main experimental difficulties come from :

- the very weak intensity of scattered light (typically 100 photons/sec for a 100 mw exciting beam, the collection of scattered phonons being performed over a solid angle of 10^{-2} steradian)

- The very small frequency shift of BRILLOUIN lines
- The high intensity of the RAYLEIGH line which is often 10^3 to 10^5 times greater than the BRILLOUIN line.

These experimental difficulties can be overcome with a high luminosity, high resolution, high contrast spectrometer. Such requirements can only be satisfied by using a FABRY-PERROT interferometer.

A plane FABRY-PERROT (F.P) interferometer consists of two plane mirrors coated in order to obtain a high reflectivity R, parallel to one another with a spacing e. The transmission function of a F.P under normal incidence is the AIRY function :

$$\frac{I}{I_o} = \frac{\left(1 - \frac{A}{1-R}\right)^2}{1 + \frac{4R}{(1-R)^2} \sin^2\left(\frac{2\pi ne}{\lambda_o}\right)}$$

Where I is the transmitted intensity

I_o and λ_o the intensity and the wavelength of the incident beam

n and e the refractive index and the thickness of the optical cavity

R and A the reflection and absorption factors of each face.

The transmission function is shown in fig. (7) . The maxima are obtained when the matching condition is satisfied :

$2\,ne = k\,\lambda_o$ where k is an integer

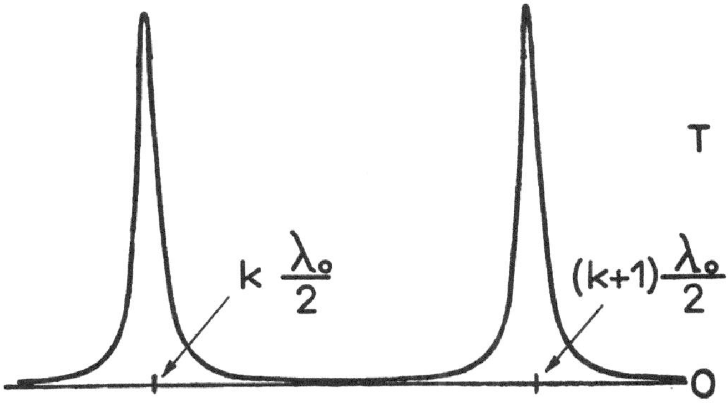

Figure 7 - Transmission function.

The main characteristics of a F.P interferometer are the free spectral range, the resolution, the finesse, the contrast - Typical figures for these parameters are given in Table I.

To obtain the BRILLOUIN lines (B.S. spectroscopy) the spectrometer is scanned either by moving a mirror by means of piezoelectric stacks, or by changing the refractive index of the air which depends on pressure. In order to minimize the absorption of light, a dielectric reflection coating is performed. Soft dielectric coating of $ZnS - MgF_2$ layers do not affect the mirror flatness which can remain better than $\lambda/200$.

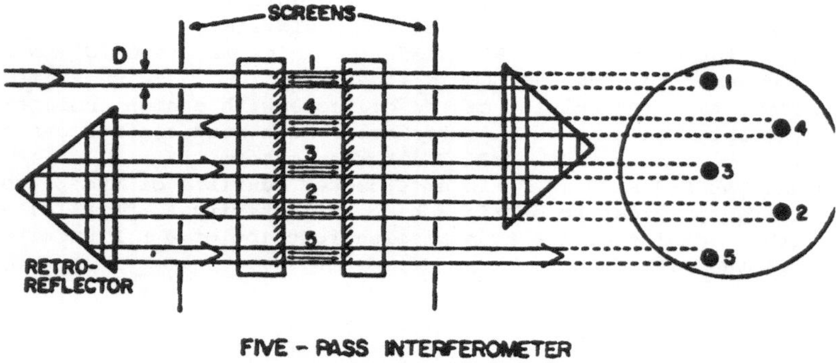

Figure. 8 - Multipass interferometer

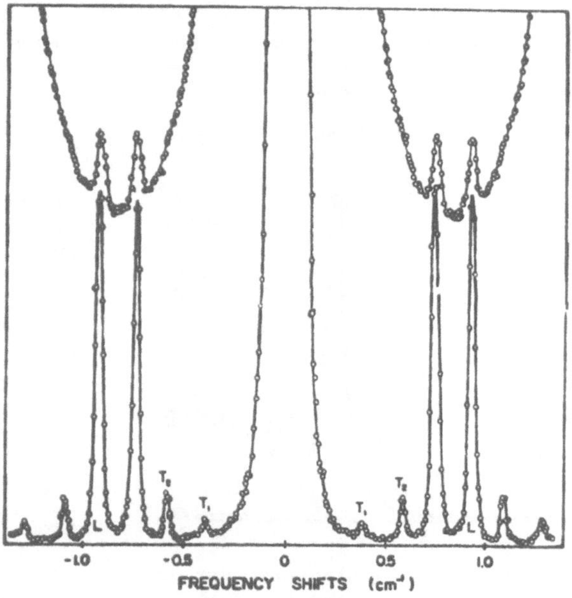

Figure. 9 - BRILLOUIN spectra of Sb SI using a one (upper curve) and two pass interferometer (lower curve).
From SANDERCOCK. Opt. Comm. 2, 73, (1970).

		F.P (R = 0.93, e = 3.5mm)	F.P (R = 0.97, e = 15mm)
Free spectral range	$(F.S.R) = \frac{1}{2ne}$	43 GHz	10 GHz
Finesse	$F = \frac{\pi\sqrt{R}}{1-R}$	60 (2 pass) 72 (3 pass)	70 (1 pass)
Contrast	$C = \frac{4R}{(1-R)^2}$	10^5 (2 pass)	2.10^3 (1 pass)

Table I

During the last 30 years the F.P interferometer has known several beautiful improvements. In such a short review we can only mention some of them, among the most significant.

The multipass interferometer

A contrast of 10^3 to 10^4 is often unsufficient to eliminate the effect of the elastic peak which is sometimes 10^5 times greater than the BRILLOUIN lines. These lines are then observed by the side of the RAYLEIGH line. Increasing the thickness of the interferometer improves the resolution but decreases the free spectral range[+]. A solution to this problem has been achieved by passing several times through the same interferometer ; this idea suggested by DUFOUR |14| has been realized the first time by HARIHARAN |15| with a double pass and later by SANDERCOCK with a five pass interferometer.

This arrangement shown in fig.(8) is equivalent to having several identical interferometers in series except that the separate passages are automatically synchronized. The theoretical improvement is considerable ; for n passages

[+] The use of two or more interferometers in series is possible but synchronising the scans of two interferometers with a spacing controlled to about 50 Å is not an easy task !

$$C_n = C_1^n \quad (7)$$

$$F_n = F_1 \left(2^{1/n} - 1\right)^{-1/2} \quad (8)$$

Indeed, the practical results are not as good as expected from (7) and (8). Nevertheless contrast approching 10^{10} can be obtained. Fig.(9) illustrates the improvement when using only a double pass instead of a single pass F.P. Thanks to the multipass interferometer BRILLOUIN scattering spectra have been recorded in opaque media.

The vernier tandem interferometer

As said before, the synchronous scanning of two interferometer is not an easy problem. However when two scanned F.P are placed in series the contrast and the finesse of the resulting filter are not only improved but also a significant increase in the free spectral range is obtained as shown in fig.(10). This latter advantage alleviates the problem of overlapping orders. Indeed, two cavities of different optical spacing e_1 and e_2 are placed in series, adjusted to transmit simultaneously on one order and then scanned so that at any time the change in optical cavity length Δe obeys the condition :

$$\frac{e_1}{e_2} = \frac{\Delta e_1}{\Delta e_2}$$

Sandercock introduced a very elegant design for a scanned tandem multipass F.P interferometer. His design makes use of a mechanical scan where one mirror of each interferometer is mounted on one and the same plate which in turn is scanned by one piezostack. The mirror pairs are set at an angle θ which determines the mirror spacing ratio of the two interferometers |16|.

The performance of the spectrometer are summarized in table II.

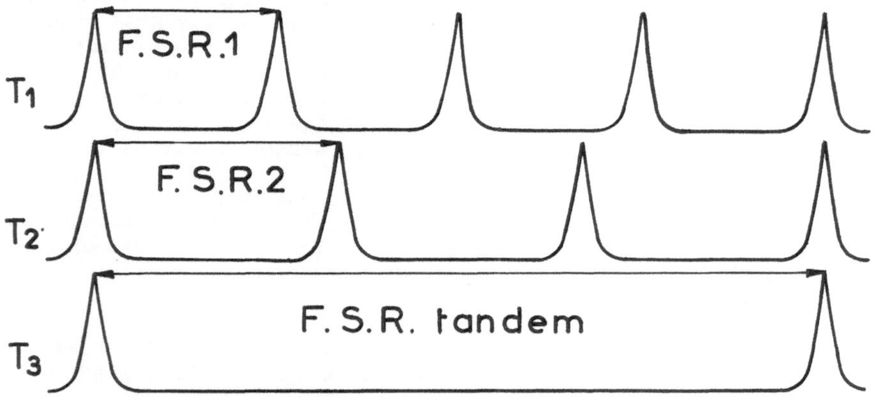

Figure 10 - Transmission functions (T_1, T_2, T_3) of P.F n° 1, P.F n° 2, and tandem. F.S.R = free spectral range.

	Calculated	Observed
Transmission through tandem	0.30	> 0.10
Contrast	4.10^{25}	$> 10^{17}$
Scanning range without loss of transmission	16 GHz to 1500 GHz	(30 MHz resolution) (2.4 GHz resolution)

Table II from |16|

BRILLOUIN scattering at high pressure in a diamond anvil cell

The knowledge of elastic constants at high pressure is of primary interest for the understanding of atomic scale forces ; such informations are also needed for interpreting data on earth's interior and for modelizing giant planets...

BRILLOUIN scattering is an ideal probe for measuring sound velocity in the small volume available in a diamond anvil cell. In a typical diamond anvil cell, a single crystal is placed in the desired orientation inside a 0.3 mm, 0.4 mm diameter hole in a 200 µm thick (at the start) hardened stainless steel foil. The diamonds are about 4 mm in diameter and 3 mm in height. The tip is cut to a 700 µm diameter flat. Anvil rest directly on maraging steel seats lapped with a 0.3 µm diamond grit. An alcohol mixture of 1 : 4 ethanol to methanol is placed in the hole in order to obtain an hydrostatic pressure. When the pressure is applied rapidly, it is possible to overcome the kinetics of the transition to the hard glassy state of ethanol-methanol mixture ; homogeneous pressure conditions can be reached even above 300 kb|17|

An arrangement for making B.S experiments at 90° in diamond anvil high pressure cell in shown fig. (11). Using a F.P in a 5 pass configuration, a contrast of 10^{10} at 4880 Å and 10^{13} at 5145 Å is reported|18| The weakness of the signal leads to the use of the multiscanning mode of a multichannel analyser.

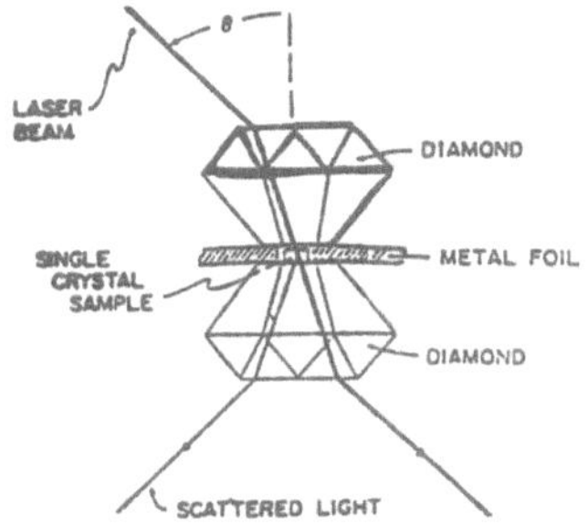

Figure 11 - BRILLOUIN scattering at 90° in a diamond anvil cell.

4. SOME SPECIFIC PROBLEMS ENLIGHTENED THANKS TO BRILLOUIN SCATTERING

Some examples of BRILLOUIN scattering can be used to illustrate the following statements.

- In inhomogeneous media nonlinear effects are often enhanced.
- The loss of translationnal invariance is "seen" by ultrasonic and hypersonic waves.
- In some specific situations BRILLOUIN scattering experiments are more efficient than ultrasonic experiments.

In all cases the examples will be borrowed to the physics of phase transition.

4.1 In inhomogeneous media nonlinear effects are often enhanced

The idea that inhomogeneous media are usually nonlinear is familiar to people dealing with parametric sonars. Parametric arrays used in underwater acoustics are based on the nonlinear properties of water, a material which is not known as a highly nonlinear ; in order to increase the nonlinearity of water it is usual to blow bubbles into the water. Water with bubbles is a soft inhomogeneous medium ; finite amplitude deformations occur when an elastic wave is propagated and the nonlinearity is enhanced.

When a crystal undergoes a structural phase transition at a critical temperature T_c it becomes soft for some mode of vibration. Fig. (12) shows the temperature dependence of C_{55} measured through BRILLOUIN scattering experiments near the critical temperature T_c for an ammonium hydrogen oxalate hemihydrate crystal |19| ; we remark the unusual low value of the shear velocity associated to C_{55}. ($V < 100\ m.s^{-1}$ at T_c).

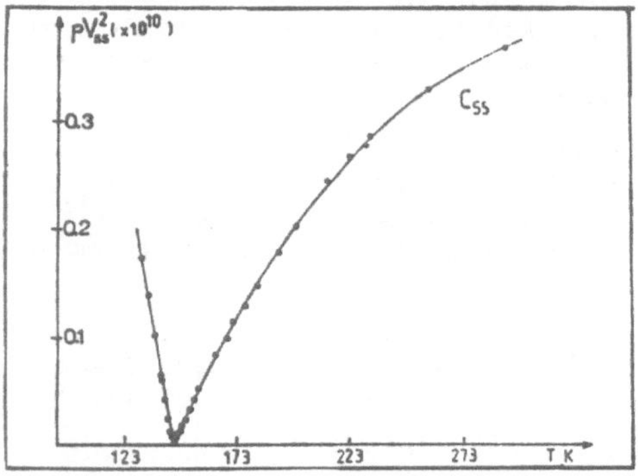

Figure 12 - Soft acoustic mode in an ammonium hydrogen oxalate hemihydrate crystal (A.H.O.H).

The distance of the BRILLOUIN lines in fig. (13) is only 0.081 cm^{-1}; the low intensity of the RAYLEIGH line is due to the exception al quality of the crystal.

Indeed crystals are generally splitted into domains below the critical temperature and large fluctuations often occur above the critical temperature ; in both cases the medium is inhomogeneous and soft.

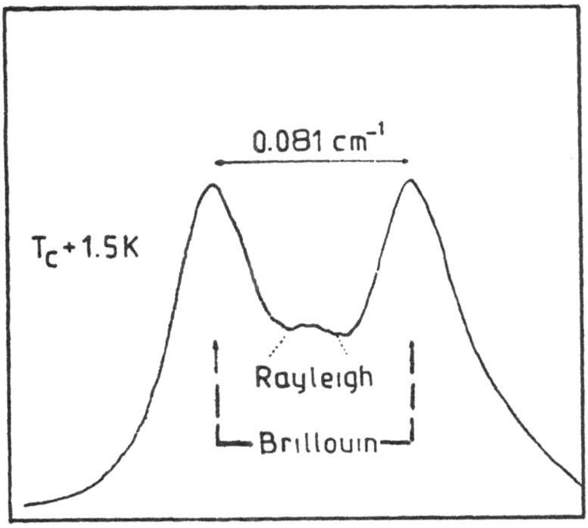

Figure 13 - BRILLOUIN and RAYLEIGH lines at T_c + 1,5 K in A.H.O.H.

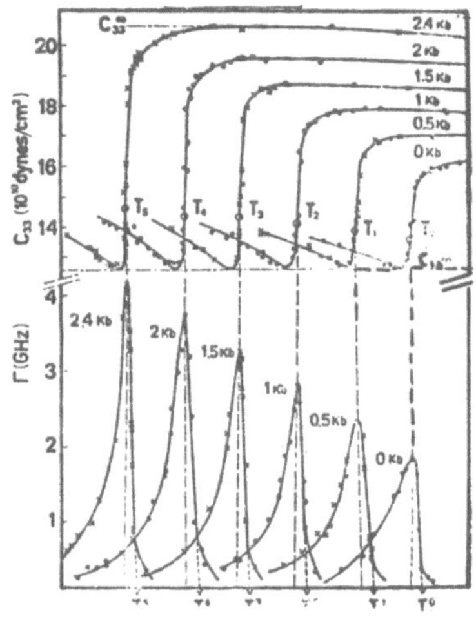

Figure 14 - Pressure and temperature dependences of C_{33} and Γ (width of the BRILLOUIN line) in thiourea.

The softening of an elastic constant when the critical temperature is approached is usual,even when the acoustic mode associated to that elastic constant is not responsible for the instability of the crystal. In fig (14) the elastic constant C_{33} of a thiourea crystal have been studied through BRILLOUIN scattering experiments as a function of temperature and pressure (the pressure dependence of the elastic constant is a nonlinear effect). This crystal undergoes a ferroelectric transition at $T_c = 169$ K at ordinary pressure P_o . At room temperature and for $P = P_o$

$$\frac{d(\log C_{33})}{dP} \sim 7.\ 10^{-5}.\ Pa^{-1}$$

Under pressure the critical temperature decreases and the pressure dependence of C_{33} becomes very large in the vicinity of T_c. This nonlinear effect can be 10 times larger than the previous effect. Such a behaviour has been observed in many cases. See for example |20|.

4.2 The loss of translational invariance is "seen" by ultrasonic and/or hypersonic waves

As explained in 2.1.3, incommensurate crystals allow us to study as a function of temperature the change from a commensurate phase (invariant under the translations of the lattice) to an incommensurate phase (which has lost any kind of translation al invariance). Many crystals exhibit such a behaviour - see for instance |21 , 22 , 23|. Ultrasonic waves (MHz frequency range) and/or hypersonic waves (GHz frequency range) are very sensitive to this changes in the organisation of the medium.

For example, in thiourea the structure is incommensurate (modulated) between 169 K and 202 K at ordinary pressure ; the wave vector $\underset{\sim}{q}$ incommensurate with the parameter b^* of the reciprocal lattice changes from $b^*/7$ at 202 K to $b^*/9$ at 169 K |24|. Fig. (14) shows B.S results obtained for this crystal ; the elastic constant C_{33} and the width of the corresponding BRILLOUIN line Γ have been plotted against temperature at different pressure in the vicinity of the upper commensurate - incommensurate phase transition. We can observe the drastic changes of C_{33} and Γ at T_c.

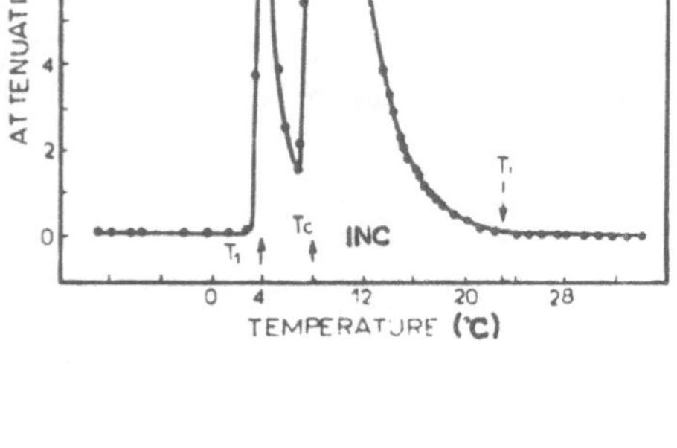

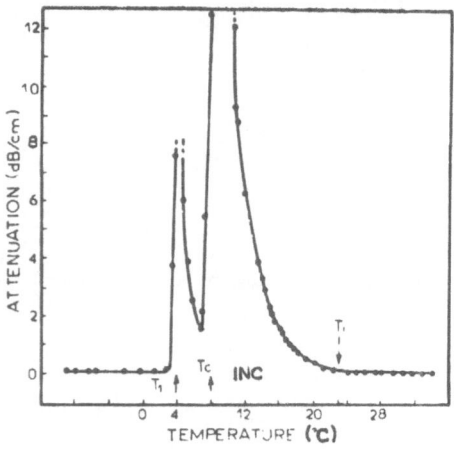

Figure 15 - Temperature dependence of the ultrasonic attenuation in the incommensurate crystal T.M.A.T.C. - Zn.

More complicated is the case of tetramethylammonium-tetrachlorozincate $|N(CH_3)_4|_2$ $ZnCl_4$ (T.M.A.T.C - Zn). T.M.A.T.C - Zn in its normal phase above 23° C belongs to the space group Pnam. The normal-incommensurate transition occurs at $T_i \simeq 23°$ C. Below T_i, a modulation develops along the pseudohexagonal $\underset{\sim}{a}$ direction. At the transition temperature $T_c \sim 8°$ C the crystal undergoes a lock-in transition to a commensurate ferroelectric structure ($Pna2_1$ symmetry). On further cooling, the crystal undergoes a first order transition at $T_1 \sim 4°$ C. These three transitions have been studied using ultrasonic measurements |25|. Fig. (15) shows the temperature dependence of the attenuation coefficient. It is remarkable that the attenuation coefficient, very high in the incommensurate phase (between T_i and Te) falls down to a low value at T_c and remains low in the commensurate phase below T_1.

4.3 <u>In some specific situations BRILLOUIN scattering experiments are more efficient than ultrasonic experiments.</u>

We have already mentionned that in the case of ferroelectric, ferroelastic, ferromagnetic materials, crystals are generally splitted into domains below the critical temperature T_c. In such inhomogeneous materials, the propagation of ultrasonic waves is impeded by scattering processes, by reflection on the domain walls etc, and it becomes difficult to obtain elastic data through the conventionnal pulse-echo ultrasonic technique. On the contrary, B.S acting as a local probe can give significant informations concerning the velocity and the attenuation of hypersonic waves inside a single domain. For example, fig. (16) shows B.S results above and below T_c for a ferroelastic crystal $RbCaF_3$. |26|

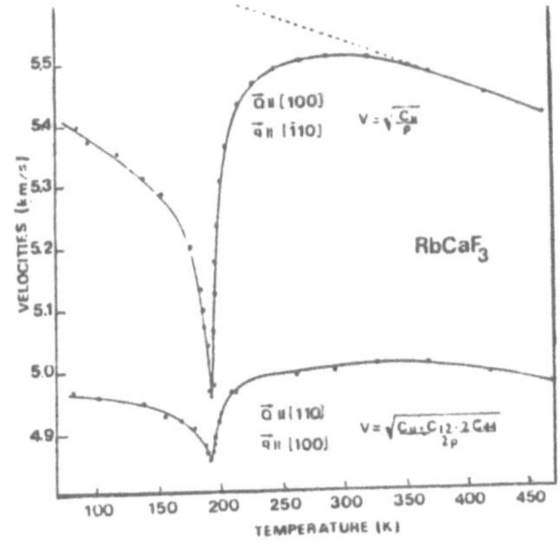

Figure 16 - Temperature dependence of longitudinal waves in the ferroelastic crystal $RbCaF_3$.

The interest of using B.S experiments is also self evident when one has to investigate the elastic properties of very small samples (e.g. thin crystal in a diamond anvil cell), or very fragile samples (e.g. layered crystals). These statements can be exemplified by gallium sulphide crystals.

Gallium sulphide is a hexagonal layer crystal. Each layer is made up of two planes of gallium atoms sandwiched between two planes of sulphur atoms. Such a structure is characterized by the contrast between the weak Van der Waals type interlayer bonds and the strong ionic-covalent bonds within the layers. As a consequence it is easy to propagate ultrasonic waves into a layer but difficult to propagate them in the direction perpendicular to the layers ; in particular it is very difficult to obtain under pressure ultrasonic pulse-echos for shear waves propagating perpendicularly to the layers (the corresponding elastic constant is C_{44}) ; only B.S measurements have been able to provide the pressuredependence of C_{44} |27|. Fig. (17) shows the BRILLOUIN lines corresponding to C_{44} at 0.461 GPa;an other illustration of the specific advantage of B.S is given in Fig. (18) which shows the BRILLOUIN lines corresponding to C_{33} in a diamond anvil cell at 9,6 GPa with a 30 microns thick crystal |28|.

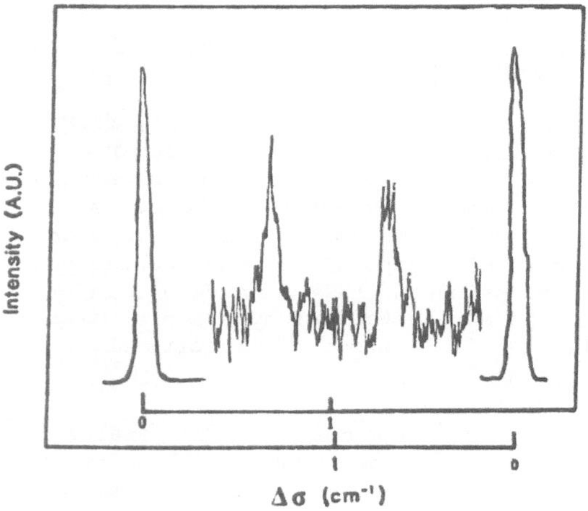

Figure 17 - BRILLOUIN lines corresponding to C_{44} at 0.461 GPa in the layered crystal GaS.

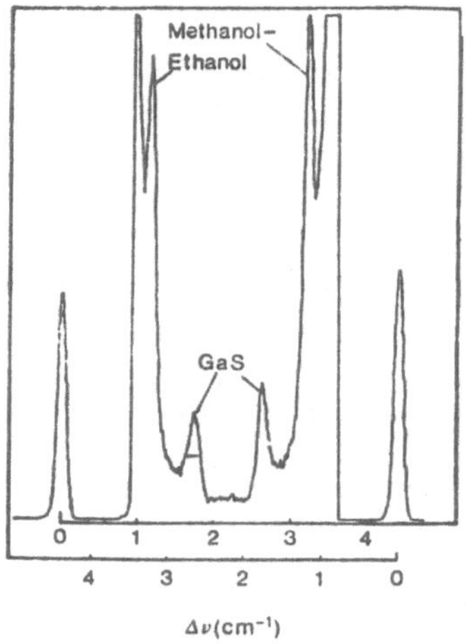

Figure 18 - BRILLOUIN lines corresponding to C_{33} in a diamond anvil cell at 9.6 GPa in GaS.

In conclusion, B.S is an alternative and a complementary approach of ultrasonic wavepropagation. This high frequency, local probe is well suited for investigating the elastic properties of many inhomogeneous media.

REFERENCES

| 1|. Michard F., Zarembowitch A., Vacher R., Boyer L., Proc. Int. Conf. Phonons, Rennes, France (1971).

| 2|. Gasse S., Zarembowitch A., Revue d'Acoustique. 16, 4, 263 (1971).

| 3|. Portigal D.L., Burstein E., Phys. Rev. 170, 673 (1968).

| 4|. Joffrin J. Levelut A., Solid State Comm. 8, 1573 (1970).

| 5|. Landau L., Lifchitz., Electrod. Cont. Media. Pergamon (1960).

|6|. Simondet F., Michard F., Torguet R., Optics Comm. 3, 16, 411 (1976)

|7|. Boyer L., Thèse Université de Montpellier (1973).

|8|. Vacher R., Thèse Université de Montpellier (1972).

|9|. Cummins H.Z., Proc. Intern. School "Enrico Ferni"., Course XL II Acad. Press. New York (1969).

|10|. Fleury P.A., Phys. Acoustics. Vol VI. Acad. Press. New York (1970).

|11|. Errandonea G., Thèse Université de Paris (1982).

|12|. Fabelinskii I.L., Usp. Fis. Nauk. 63, 355 (1957).

|13|. Sandercock J.R., R.C.A. Review. 36, 89 (1975).

|14|. Dufour C., Ann. de Physique. 6, 5 (1951).

|15|. Hariharan P., Sen D., J. Opt. Soc. Am. 51, 398 (1961).

|16|. Sandercock J.R., Lindsay S.M., Anderson M.W., Rev. Sc. Inst. 52, 10 (1981).

|17|. Besson J.M., Pinceaux J.P., Rev. Sc. Inst. 50, 5, 541 (1979).

|18|. Whitfield C.H., Brody E.M., Basset W.A., Rev. Sc. Inst. 47, 8, 942 (1976).

|19|. Benoit J.P., Berger J., Krauzman M., Godet J.L., J. Phys. tobepublished (1986).

|20|. Fischer M., Zarembowitch A., Breazeale M.A., I.E.E.E. Ultrasonics 999 (1980).

|21|. Pezelt J., Phase transitions. Vol 2, 155 (1981). Ed. Gordon and Breach.

|22|. Bruce A.D., Cowley R.A., Advances in physics 29, 1 (1981).

|23|. Dvorak V., Pezelt J., J. Phys. C 11, 4827 (1978).

|24|. Benoit J.P., Berger J., J. Phys. 44, 841 (1983).

|25|. Berger J., Benoit J.P., Garland C.W., Wallace P.W., J. Phys. (1986) inpress.

|26|. Berger J., Hauret G., Rousseau M., Solid State Comm., 25, 569 (1978).

|27|. Fischer M., Polian A., Chevy A., Chervin J.C., Solid State Comm., 56, 311 (1985).

|28|. Polian A., Besson J.M., Grimsditch M., Vogt H.,Phys. Rev. 25, 2767 (1982).

DISCUSSION

Comment: Socino
What is the accuracy in the determination of the phase velocity of acoustic phonons by means of Brillouin scattering experiments?

Reply: Zarembowitch
With accurate experiments, Brillouin scattering enables one to provide the absolute value of the phase velocity with an accuracy comparable to that obtained through ultrasonic measurements, i.e. 10^{-3}.
As a consequence, significant comparisons between the two techniques can be performed. For instance, in NaCl, the phase velocity is identical, (within a factor of 10^{-3}) at 10MHz (ultrasound) and 30 GHz (Brillouin scattering).

Comment: Sliwinski
Referring to measurements close to critical temperature, have you had difficulties with critical opalescence near the critical point? How close to that point were you able to perform measurements?

Reply: Zarembowitch
The answer obviously depends upon the magnitude of the fluctuations. In some crystals, large fluctuations develop in a wide range of temperatures. On the contrary, in some cases it happens that fluctuations are insignificant even close to the critical temperature (as in the case of oxalate ammonium hexahydrate where measurements can be performed in a temperature range $\Delta T/T < 10^{-2}$).

Comment: Bonnet
It is well known that measurements of the elastic constants performed with classical ultrasonic methods do not agree with those performed statically in the range of the MHz's. What differences are there observed between results in the MHz's region and those obtained through Brillouin scattering techniques?

Reply: Zarembowitch
In some compounds, no differences are observed (for instance: alkali halides). On the contrary, when the crystals show a relaxation mechanism with a relaxation time τ_c such that $\omega\tau_c \approx 1$ in the frequency range between ultrasounds (few MHz's) and hypersounds (few GHz's), the difference is large (for instance: NH_4Cl, thiourea, etc.).

Comment: Schmidt
Is it possible to obtain structural information from the intensity of the Brillouin lines?

Reply: Zarembowitch
Yes, one can get information about the individual components of a composite material, if one uses highly resolved Brillouin interferometry techniques.

CHARACTERISATION THROUGH THERMAL WAVES: APPLICATIONS OF OPTOACOUSTIC AND PHOTOTHERMAL METHODS

G. BUSSE

1. INTRODUCTION

From experience -especially at this summer school- we are used to various kinds of waves: to water waves that we enjoy here during some afternoon breaks, to acoustic waves transferring information from the speakers to our ears, and to light waves interacting with optical structures on the viewgraphs in such a way that we can see the equations and other results.

The waves dealt with in this contribution are quite different in nature. The physical quantity that behaves wave-like is not an elastic deformation or an electromagnetic field vector, but it is temperature or its deviation from the average value. These waves are known as thermal waves. They have already been described by Fourier in 1826 /1/. However, there is a considerable renewed interest in them, both for spectroscopy and non-destructive evaluation (NDE).

The purpose of this paper is to give an idea of the experiments and their applications to material characterisation.

A good illustration for the many possible experiments is the optoacoustic effect detected by Bell in 1881 /2/ (Fig. 1):

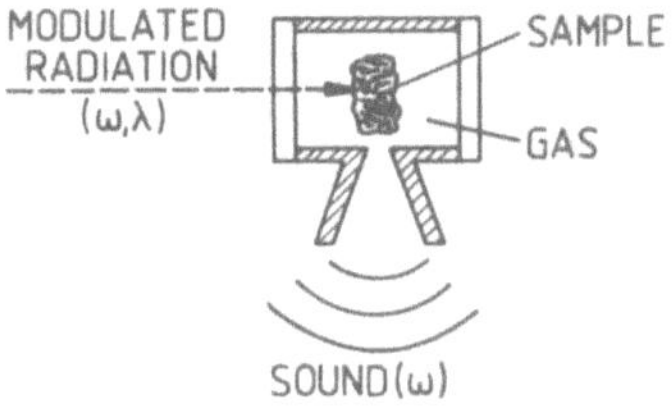

FIGURE 1.
Optoacoustic effect: intensity modulated light generates sound at the molation frequency ω /2/.

Light of wavelength λ shines on an absorbing sample (SPL). Due to the energy input the sample temperature increases. Switching off the light beam causes a corresponding decrease. If the light beam intensity is modulated at frequency ω, a temperature modulation ΔT results at this frequency. As the temperature of the gas adjacent to the sample changes, too, a pressure modulation is achieved which is heard as sound at the modulation frequency ω. Light eventually produces sound. This is the "optoacoustic" effect. To avoid confusion with the "acousto-optic" effect the term "photoacoustic" has become popular. Obviously the sound (to be detected with sensitive microphones and lock-in amplifiers) depends on optical input power, optical absorption at the wavelength λ, the kind of transitions between energy levels in the sample (e.g. fluorescence effects reduce the generation of heat), the modulation frequency ω, the thermal properties of the sample, and its geometry. Further it depends on the acoustic properties of the "optoacoustic cell" which is an airtight enclosure (containing the sample and the microphone) provided with a transparent window for the light.

The influence of these parameters indicates that one can perform many interesting experiments. By variation of the optical wavelength λ one obtains information on optical properties. Dependence on modulation frequency may display life-time effects or acoustic properties of the cell. Thermal properties of the sample are involved as well. If the sample is not homogeneous, the temperature modulation ΔT depends on the coordinates x and y where the optical beam hits the sample. Investigation of $\Delta T(\lambda)$ is "opto-(or photo-) acoustic spectroscopy", while $\Delta T(x,y)$ is the base for scanned imaging of thermal structures. These applications will be dealt with below.

Though acoustic detection of the temperature modulation ΔT is easiest to visualise, it is obvious that there should be other methods. They have been named " photothermal detection " to indicate that thermal effects of periodical optical illumination are analysed.

This paper is not intended to be a complete review. If one is interested should look into one of the review books /3,4/ or the proceedings of conferences /5,6/ dealing with recent developments of a rapidly growing field.

There is no just one obvious reason why a field like thermal wave applications suddenly becomes attractive after the degree of activity has been very low for a long time. Among the possible reasons could be that

- powerful sources have become generally available (lasers),
- sensitive detection techniques and electronics have been developed,
- the understanding of the potential (e.g. for NDE and sample characterisation in general) has been improved,
- the development of new materials with high specific strength requires a careful analysis of failure.

2.OPTO-(OR PHOTO-) ACOUSTIC SPECTROSCOPY

Special features of this technique can be derived from the comparison with conventional optical transmission spectroscopy.

Let us assume a light source of variable wavelength λ (e.g. dye-laser or monochromator) which after modulation (MOD) illuminates the sample (SPL) in an optoacoustic cell provided with a microphone (MIC).

It is assumed that the light intensity transmitted through the sample is measured with an optical detector (Fig. 2).

This detector needs to be matched to the source spectrum. It finds the arriving photons. If there is no absorption, one obtains the source spectrum. But there is no direct information of how the photons were lost in the sample. If the sample is not provided with flat and parallel surfaces, or if the scattered light has to be detected, one needs an integrating sphere. Fluorescent or highly absorbing samples need additional consideration. Highly transparent samples are difficult for optical spectroscopy since the change of signal that they cause may be smaller than the noise.

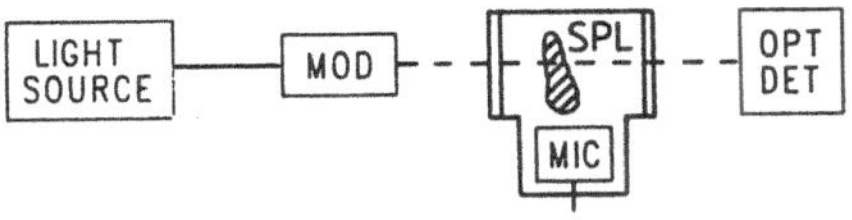

FIGURE 2.
Optical and optoacoustic spectroscopy.

Optoacoustic detection, however, is sensitive to what happened to the photons: only those give a signal that were absorbed and then produced heat (possibly after some delay). If there is no absorption, there is no signal, and zero is a well defined and stable reference. If the surfaces of the sample are not optically flat the microphone signal is not much affected. The care for sample preparation can be reduced. Even a scattering sample gives information. Such an example is optoacoustic spectroscopy of semiconductor powders (Fig. 3) /7/.

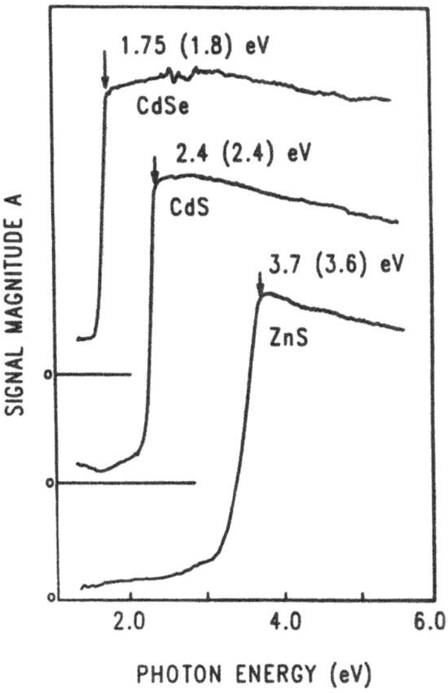

FIGURE 3.
Optoacoustic spectra of semiconductor powders. Numbers indicate bandgap energies /7/.

Absorption occurs only if the photon energy is larger than the bandgap. The arrows at the edges indicate the bandgap energies. Published data are given in brackets. Obviously optoacoustic detection provides the correct information on spectral features with no sample preparation at all.
A comparison of spectra obtained on nitrile plastic with four different morphologies is presented in Fig. 4 /8/. In many ap-

plications the information on peak position etc. is sufficient for substance identification (e.g. food, drugs).

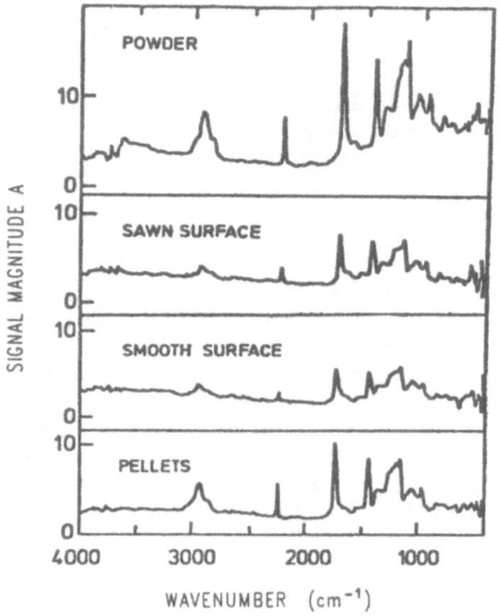

FIGURE 4.
Optoacoustic spectra of nitrile plastic for four different morphologies /8/.

2.1 Fluorescent samples

Non-radiative transitions from an excited level to a ground state level produce a temperature increase and hence an optoacoustic signal. If such a transition has to compete with a radiative transition, the signal decreases. Applications have been determination of quantum yields in dyes /9/ and of fluorescence quenching in Ho_2O_3 doped with Co and F as impurities /7/. In Fig. 5 the relative height of some maxima (positions indicated in the middle by bars provided with dots) is increased upon addition of these impurities thereby revealing that there had been radiative transitions before.

2.2 Lifetime effects

The radiation energy stored in an absorber may not come out as heat immediately. If the excited state is metastable it may take some time until the system returns to its ground state. As the optoacoustic signal is a response to a modulated input, it is described in the Gaussian plane by a complex signal vector characterised either by magnitude A and phase φ or by in-phase and out of phase (quadrature) components. Both sets are obtained with a lock-in amplifier used for analysis of such an ac-response. A time lag induced by life-time effects of metastable states rotates the complex signal vector. Therefore it is of interest to monitor signal phase and not only its magnitude while the optical wavelength is scanned.

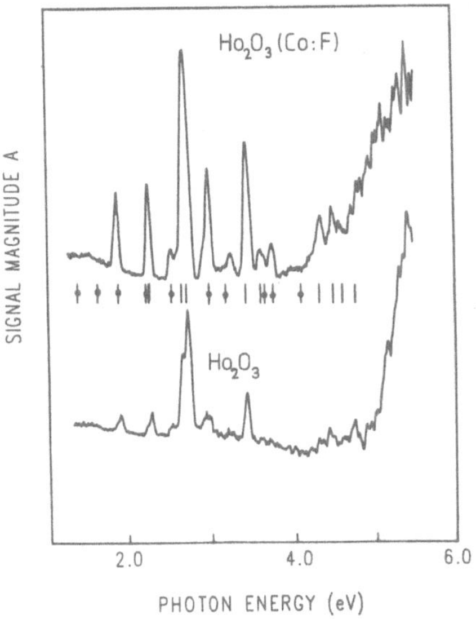

FIGURE 5. Influence of doping on radiative transitions /7/.

Fig. 6 is an example where two transitions in Biacetyl at 20 Torrs and 400 Hz modulation frequency are correlated with different phases and therefore with two different lifetimes /10/.

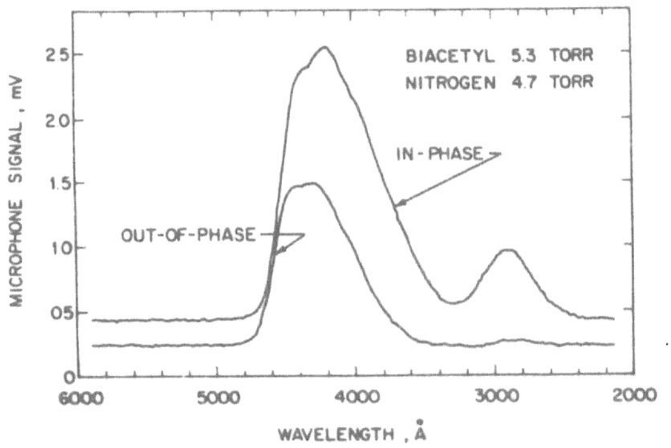

FIGURE 6. Lifetime effect revealed by signal phase /10/.

2.3 Weak absorption

An absorption line appears as a decreased signal in optical transmission spectroscopy. In optoacoustic spectroscopy, however, the line is not measured against the background of transmitted radiation but against zero. An absorption line observed with optoacoustic detection therefore appears like an emission line in conventional optical arrangements. The advantage is that non-relevant noisy information is suppressed. Thereby the signal to noise ratio can be improved.

Very weak absorptions have been investigated in gases /11/, liquids /12/, and solids /13/.

The noise equivalent absorption coefficients can be about 10^{-7} cm^{-1}. In a corresponding optical experiment, one would still find about one third of the original optical power behind a sample of a hundred kilometers thickness.

Optoacoustic spectroscopy at CO_2 -laser wavelengths has been used to identify explosives /14/ and to detect molecular lines that are suited for optical pumping and far-infrared laser emission /15/. Other applications are pump laser frequency stabilisation /16/ and power monitors by using the optoacoustic effect of optical components /17/.

2.4 Highly absorbing samples

Opto- (or photo-) acoustic spectroscopy is so simple and successful since the sample itself acts as a detector, therefore the sample characteristics determine the detector sensitivity. This loss spectroscopy has the advantage that signal background is always zero. But at the same time it has the disadvantage that quantitative information is difficult to obtain /8, 18, 19/. This is a drawback as compared to optical methods. However, in many situations the qualitative information obtained very easily with optoacoustics is quite sufficient, while optical detection gives the same results only with much more work.
Such an example is the inspection of highly absorbing samples. To get a spectrum of coal with optical means, one should prepare a very thin slab of homogeneous thickness. With optoacoustics, however, one does not actually need a thin sample: it can be simulated by using a high enough modulation frequency.

To understand this concept it might be helpful to look at two photoacoustic spectra of an apple /20/ (Fig. 7).

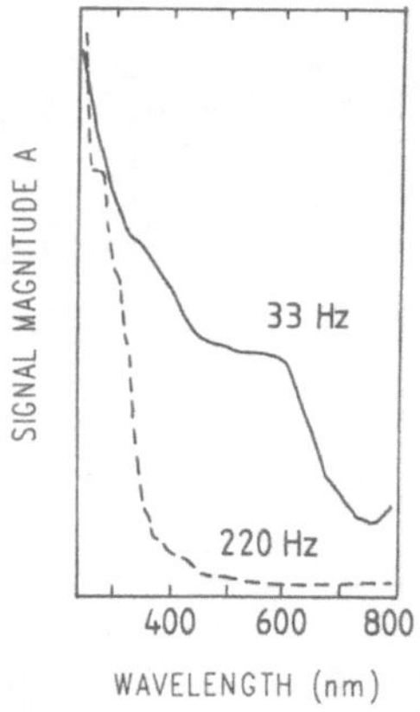

FIGURE 7.
Frequency dependent optoacoustic spectra of an apple /20/.

The spectrum at 220 Hz is not what one expects: there is no indication of the colour in which the apple appears to our eyes. The signal in the ultraviolet is due to the outer wax layer covering the apple. At the lower modulation frequency (33 Hz) a shoulder appears at the expected optical absorption of the apple peel. The modulation frequency evidently determines the thickness of the layer that contributes to the photoacoustic signal. The explanation is that the frequency dependent thermal diffusion length is involved /21/.

Depth resolved spectroscopy is certainly an application where opto-(or photo-) acoustics is superior to optical methods. The reason is that the thermal wave aspect /1, 13, 21, 23/ provides additional information. Therefore the next section will deal with basics of thermal waves.

3. THERMAL WAVES

3.1 Remarks on thermal wave physics

The essential features of thermal waves can be discussed on a one dimensional model where z = 0 describes the surface of a semi-infinite sample of thermal conductivity k, specific heat c, and mass density ρ . If there are no heat sources inside the sample - a good assumption if the sample is optically opaque like a metal with an optical penetration depth of less than 100 nm - temperature T as a function of coordinate z (extending into the solid) and time t is for $z > 0$:

$$\frac{\partial^2 T(z,t)}{\partial z} - \frac{\rho c}{k} \frac{\partial T(z,t)}{\partial t} = 0 \qquad (1)$$

If there is an harmonic heat generation at z = 0 due to absorption of optical radiation, heat flux j is modulated between $2 j_0$ and the minimum 0 with the average j_0 (see Fig.8). The surface temperature is then

$$T(z=0,t) = T(0) + \Delta T \sin \omega t \qquad (2)$$

This modulation decays exponentially with depth z, while there is a phase lag with respect to the surface temperature /23/.

$$T(z,t) = T(z) + \Delta T e^{-z/\mu} \sin(\omega t - z/\mu) \qquad (3)$$

with

$$\mu = (2k/\omega \rho c)^{1/2} \qquad (4)$$

Though heat propagation is a diffusion process, we have the result that periodically excited diffusion is described as a highly attenuated wave. The "thermal diffusion length μ " (equ. 4) is the distance where phase shift φ increases by 1 rad ($\approx$ 57 degrees) and where in a plane wave the modulation depth decreases to $1/e \approx 37$ % of its original value.

The thermal wave is the second term in equ. 3. The local average heat transport in a thermal wave is obviously zero, therefore the non-modulated local average temperature T(z) provides the background for conservation of total average heat flux. The depth of flux modulation changes along z according to Fig. 8.

The decay of modulation depth is a consequence of the diffusion mechanism. As always $j \geqq 0$ for optical generation of thermal waves total heat flow does not change sign.

Though velocity of energy transfer is not defined in a thermal wave, one can calculate phase velocity and group velocity since the wave number is $1/\mu$. As μ is around 1 mm for many metals at 10 Hz modulation frequency, one finds that a 10 Hz thermal wave in aluminium has a phase velocity of about 0.1 m/s which is very slow compared to optics and acoustics. From the parabolic dispersion curve follows that higher frequencies propagate faster and that group velocity is twice the phase velocity.

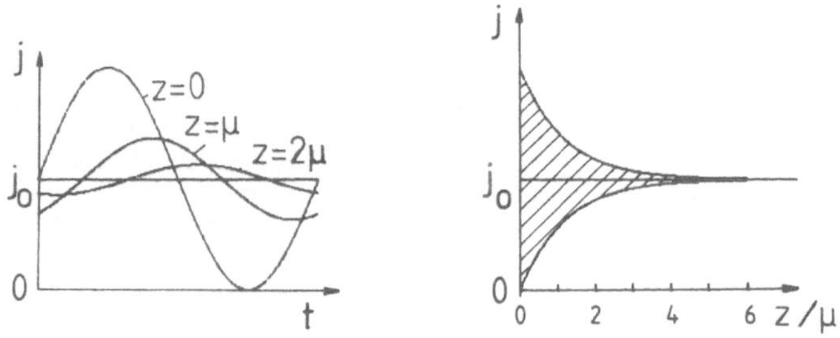

FIGURE 8. Heat flux modulation for various times. Modulation depth (shaded area) depends on normalised sample depth.

The phase velocity has been measured recently in a transmission arrangement with a moving aluminium sample /24/. Similar to time of flight measurements, the apparent thermal wave shift can be correlated to time (Fig. 9).

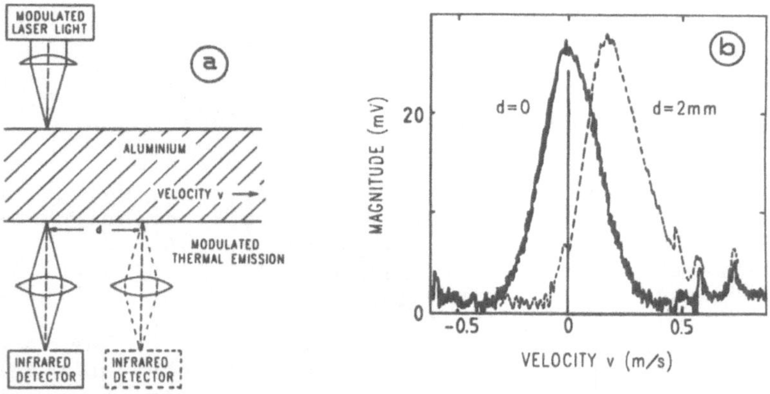

FIGURE 9. a) Experimental arrangement with spatially resolved generation and detection of thermal waves at 2o Hz.
b) Detector signal magnitude depending on sample velocity v for the two detector positions /24/.

If the detector offset d is equal to sample thickness the signal maximum is displaced. The maximum occurs when the time the thermal wave needs to move through the sample is equal to the time the sample needs to move to the offset detector position (Fig. 9, dashed). Therefore in this case the maximum is found when the sample velocity equals phase velocity. The observed value agrees well with the expected one.
A higher frequency with a correspondingly higher phase velocity would shift the maximum to higher sample velocities. This arrangement would therefore allow to perform a Fourier analysis based on thermal wave dispersion. The weaker maxima in Fig. 9 b, however, are due to a synchronisation effect (standing thermal wave patterns) since a rotating sample was used.

Thermal waves can be diffracted and reflected like other waves /25-27/. Reflection from the rear surface may affect the original thermal wave at the front surface of a sample. This interpretation /26/ is consistent with calculated /28/ and measured /29/ optoacoustic thickness dependent signals (Fig. 10).

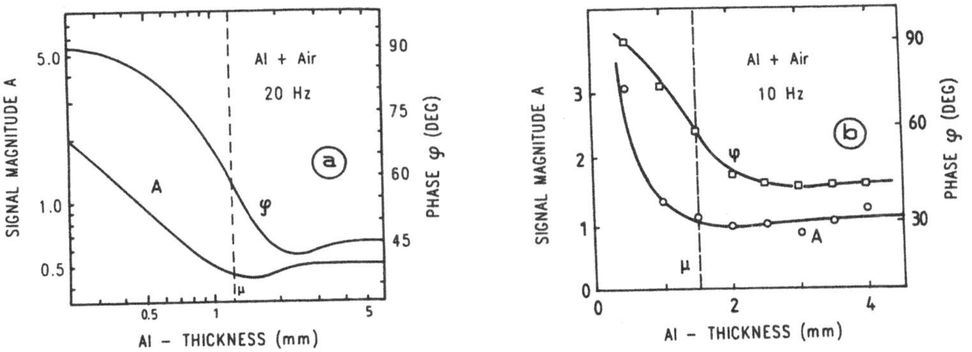

FIGURE 10. a) Expected thickness dependence for an opaque sample according to the Rosencwaig-Gersho theory /21,28/.
b) Experimental results for a 10 Hz modulation /29/.

The thickness dependences of signal magnitude and phase show minima which are interference effects. However, if sample thickness exceeds thermal diffusion length (dashed lines) the signal becomes constant. The interpretation is, that due to the attenuation the front surface signal has its origin in a sample layer whose thickness is comparable to the thermal diffusion length /21/.
With piezoelectric thermal wave detection the depth range is larger by a factor of two /30/.

3.2 Thermal wave generation and detection

After the thermal wave background of the optoacoustic effect had been understood it became evident that any kind of periodical energy deposition is suited to produce a thermal wave. Optical generation only has the advantage of being non-contacting and easy to modulate. However, particle beams instead of photons can be used as well /31/.
Acoustics is only one way to detect a temperature modulation. The disadvantage is that one needs a small optoacoustic cell or, alternatively, a piece of piezoceramic material to couple it to the sample /32,33/. Remote methods for analysis of temperature modulation have been investigated since 1979: the "mirage-effect" based on the probe beam deflection caused by the gradient of refractive index in the thermal wave /34, 35/, modulated infrared emission (a method already used in Fig. 9 a) /36/, and interferometric or reflection measurement of modulated thermal expansion /37, 38/. These different thermal wave detectors do not provide the same information since different processes of averaging are involved. The piezoelectric transducer signal depends on the volume integral of the thermal wave, while the microphone signal depends on the surface integral. This is the reason for the difference in depth range demonstrated in Fig. 11 where a hole is drilled into aluminium at such an angle that its distance to the surface increases gradually between successive scans. Comparison with the optical reference (OPT) shows in which depth the hole can still be detected at 20 Hz with the two optoacoustic detection methods (MIC and PZT) /39/.

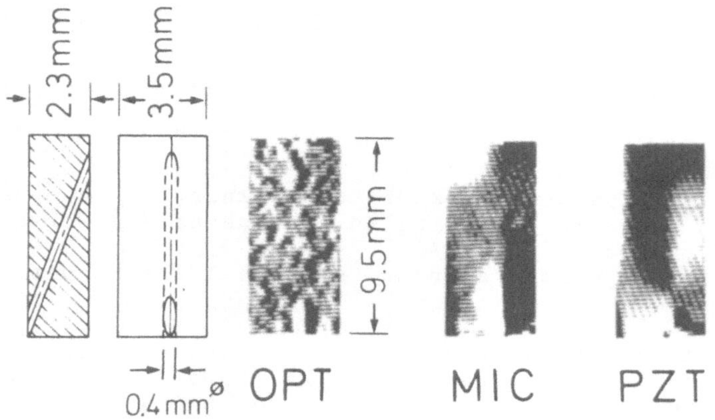

FIGURE 11. Comparison of piezoceramic and microphone detection of a subsurface hole /39/. Gray scale obtained by differentiation (ac filtering) of the phase angle signal.
Modulation frequency: 20 Hz

Because of the integration processes both kinds of optoacoustic detection are not spatially resolved. With optical beam deflection methods a line integral (along the probe beam or, in a good approximation, along the sample surface) is involved. True spatial thermal wave inspection is possible with phase sensitive infrared radiometry, with optical interferometry or with local optical reflection modulation.

In order to obtain information on local thermal properties one does not need spatially confined generation and detection of thermal waves. However, if both are localised one can perform stereoscopic measurements on subsurface structures /40/.

4. APPLICATIONS TO NON-DESTRUCTIVE TESTING

The properties of thermal waves - their phase lag proportional to the distance travelled, their reflection and interference ability - are very useful to monitor material properties in a remote and non-destructive way.

Both thermal wave transmission and reflection have been used. The transmission arrangement (Fig.12 a) has the advantage of a simple interpretation /41/, but it is not applicable to thick or hollow samples. Therefore one depends sometimes on one-sided methods (Fig. 12 b).

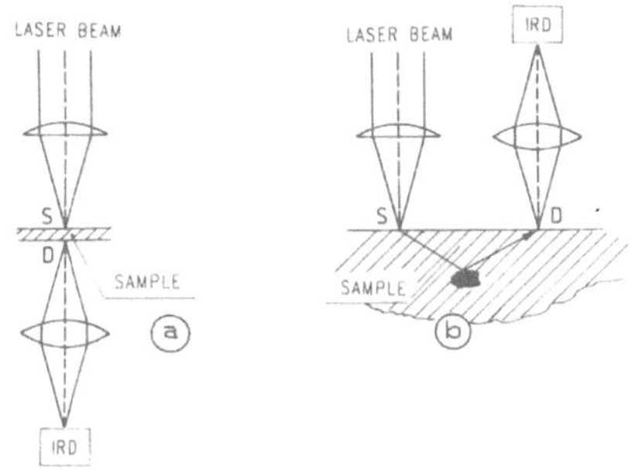

FIGURE 12. Thermal wave transmission (a) and reflection arrangements (b) /42/. Modulation of thermal emission is used for detection /38/.

Though the detector signal is dominated by the wave moving directly from the optical focus S to the detector spot D, there is a superposed contribution of the wave that is reflected from S to D by internal boundaries. The resulting signal is determined by geometry. The dependence on the depth of the defect is qualitatively described by Fig. 10.

4.1 Non-scanned monitoring of thermal properties

The observed thermal wave phase angle φ is related to normalised distance z/μ travelled, both in transmission (equ. 3) and in reflection (Fig. 9b). Therefore z can be determined if φ is known or vice versa. The accuracy of this kind of thickness measurements depends on normalised thickness and on the optical power level. In a transmission arrangement, a phase angle accuracy of ± 0.5 degrees results in a thickness accuracy of $\pm 5\,\mu m$

for metals at 2o Hz and $\pm 0.5\,\mu m$ for polymers at the same modulation frequency.
The method is of interest for remote thickness measurements of metal plates and of coatings of them. The mentioned accuracy is based on the assumption that μ is known well enough from calibration measurements.

Often one uses the information the opposite way to monitor in a known geometry the thermal diffusion length μ or changes of it. The idea is to relate the measured data to known information in such a way that one has a method for non-destructive and remote material inspection. Applications are determination of fibre content and orientation /42/ and the development of criteria for the quality of adhesive joints.

As another application, the time dependence of thermal wave transmission allows for monitoring of curing and aging or parameters that this process may depend on, e.g. chemistry, temperature history, and surface treatment /42, 43/.

As an example, Fig. 13 shows signal phase in thermal wave transmission of two 0.5 mm thickness aluminium plates with epoxy resin in between, applied at time 0.

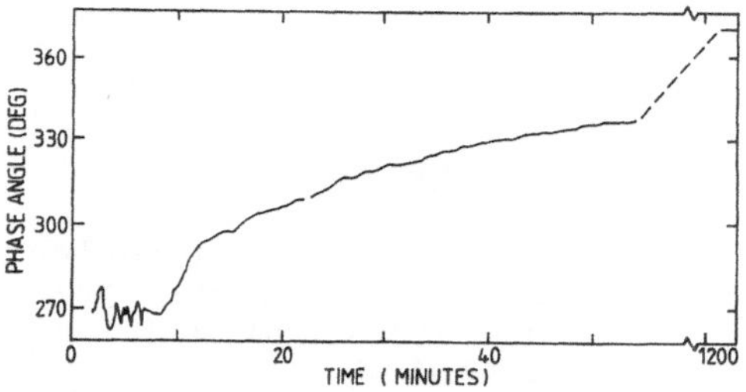

FIGURE 13.
Curing process of epoxy resin monitored with remote photothermal transmission /43/.

Also one can observe the effect of water diffusion into paint or along boundaries.

4.2 Scanned thermal wave inspection: imaging.

The general concept is that thermal wave propagation experiments are performed on a small spot ("pixel") whose size depends on the extension of the optical focus, on the detecting system, and on thermal diffusion length /44/. If detection is based on an integrating process (e.g. microphone- or piezoelectric detection), localisation depends only on the optical spot. Then essentially the relative motion is of interest, therefore one can move either the sample or the optical beam in a raster-like fashion in such a way that one obtains an image consisting of many pixels. If the scan field size is no longer small compared to the detector extension, spatial variations of sensitivity may affect the signal and the image /45/.

With localised detection, however, one must keep the relative orientation of source and detector constant during the scan. In that case it is easier to move the sample with respect to a stationary source and detector arrangement.

Thermal wave imaging is well suited for the search of thermal inhomogenities correlated with faults, e.g. thickness variations of coating, bonding defects, delaminations, and flaws in metal.

Many experiments have been performed on models of faults: delaminations /43, 46-48/, holes in metal / 28, 41, 46, 49, 50/, and regions of previous deformation /51, 52/. The reason why such model experiments are performed is that one knows the nature of the flaw: therefore the expected results can be compared with the obtained data. This way theoretical models or calibration procedures can be tested.

One result from these experiments is that depth range in front surface imaging can be varied via the modulation frequency /21, 45/. Thereby threedimensional information is obtained ("depth profiling"). Also it has been found that depth range is larger by a factor of 2 if signal phase φ is used for imaging, as expected from Fig. 10 /28, 49/.

Many realistic samples have optical surface structure in addition to their unknown thermal structure. An increased surface absorption of an opaque sample causes an increase of signal magnitude A. However, signal phase φ (as a normalised time shift) is not affected as long as the thermal contribution of the optical structure can be neglected. As an example, Fig. 14 shows an aluminium sample provided with black lines on top

FIGURE 14. Sample structure (left), magnitude (middle) and phase angle image (right)/53/.

to simulate optical structure and subsurface holes (dashed) as thermal structure. The photothermal transmission image of this sample shows that the optical structure is completely ignored in the phase angle image /54/.

A corresponding result for thermal wave microscopy with piezoelectric detection is shown in Fig. 15. While the magnitude image A is dominated by the optical properties of the metallised rectangular areas on the integrated circuit, the phase angle image φ is different in that it reveals only thermal structures even where they are hidden under metal /54/.

The ability of detecting hidden flaws is of potential interest for the inspection of welded seams. An example is presented in Fig. 16 both with thermal wave transmission and reflection (see Fig. 11), each for magnitude A and phase angle φ.

The defect region is found in all images, however, with transmission and signal phase one obtains best resolution /55/. The drawback of the transmission arrangement is the loss of depth information: one finds only the projection of structures (like in x-ray imaging), where the direction of projection is given by the orientation of the detector spot with respect to the optical focus. Using two different detector positions one obtains a pair of stereoscopic images which provide threedimensional information to a depth that can be larger than with front-surface detection /40,41/.

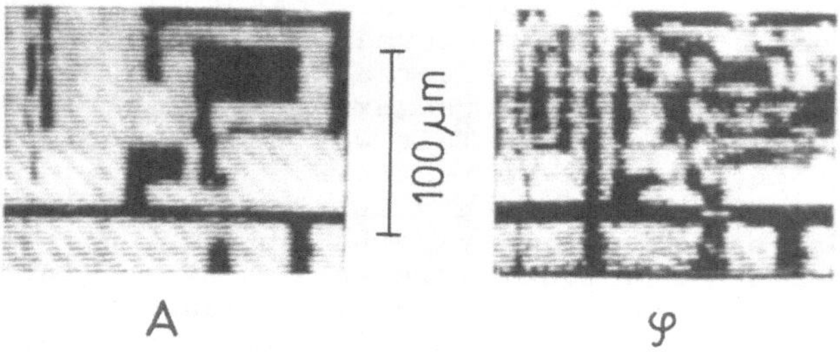

FIGURE 15. Thermal wave microscopy of an integrated circuit at 200 KHz /53,54/.

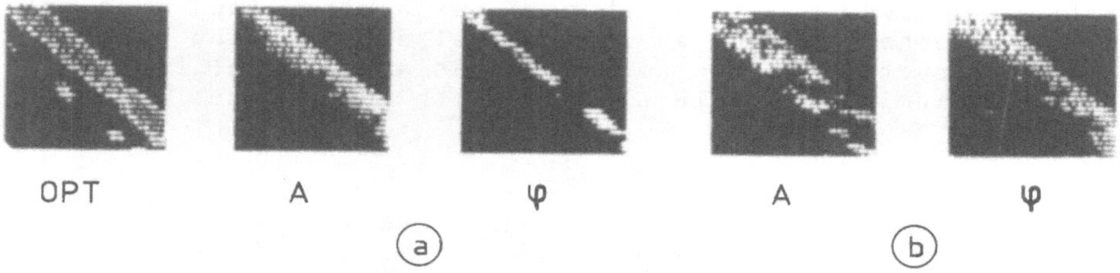

FIGURE 16. Comparison of seam inspection with thermal wave transmission (a) and reflection (b) (see Fig.12). Optical image for reference at left.

Besides application to metals and semiconductors another field of interest is polymers. The thermal diffusion length of these materials is by more than an order of magnitude smaller than for metals (at the same frequences), thereby allowing only for inspection of thin foils or near-surface regions. This is the situation of coatings. Looking for thickness variations is possibly not of that much relevance as monitoring the bonding and curing conditions.

A material with a strong thermal inhomogeneity is carbon fibre reinforced plastics (CFRP). Due to its high specific strength this material is important for air- and spacecraft industries. The quality depends on the fibre/matrix bonding. Not much thermal wave work has been done on the material until now /56,57/. An experimental result indicating the effect of debonding due to excessive local resistive heating is shown in Figure 17. Experiments of this kind will certainly be continued in the future.

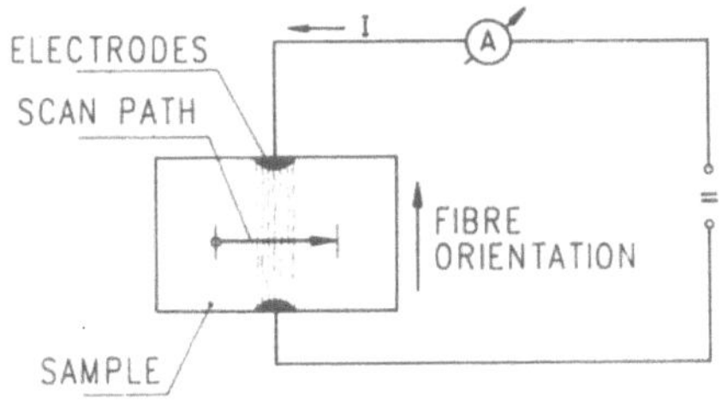

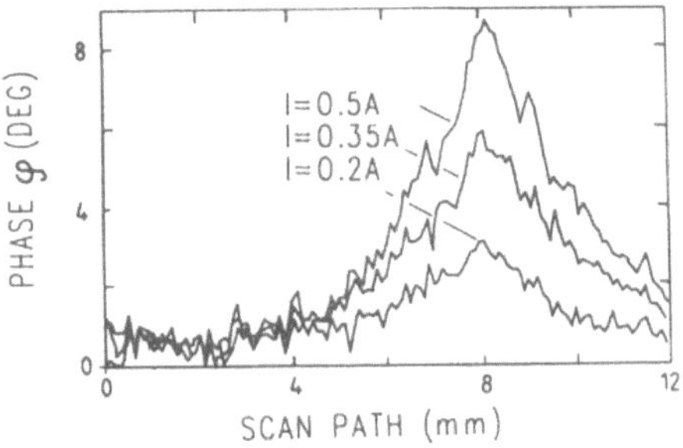

FIGURE 17. Thermal wave transmission of uniaxial CFRP with local resistive heating /57/.

5. CONCLUSION

Though thermal waves are known since more than a century, most of their applications to inspection and non-destructive testing have been investigated within about the last ten years.

Opto- or photoacoustic spectroscopy may still be applicable when optical spectroscopy is too difficult. In these situations one obtains in a short time at least qualitative results which may be sufficient for substance characterisation. Potential future applications are monitoring of food, drugs, and pollution.

On the other hand, if optical spectroscopy can be performed easily and if quantitative information is required, optical methods will certainly be preferred.

Non-spectral investigations like monitoring of sample properties that depend on coordinates, time, or other parameters will certainly find applications in non-destructive testing. As for scanned imaging, there are still problems that need to be solved. Compromises have to be made concerning optical power density in the focus, resolution, and time needed to obtain a result with an acceptable signal to noise ratio.

The information that is to be obtained concerns primarily thermal structures. However, optical and acoustical properties may be involved as well. The experimental conditions must be carefully analysed to understand what the result means.

From the high attenuation of thermal waves it is obvious that essentially near-surface regions are within their range. Therefore we may assume that coatings, seams and integrated circuits are fields of applications. Thermal waves are not suited to investigate the interior of thick samples.

We are still in a process where the potential of thermal waves has to be exploited. The aim will not be to replace existing techniques but to obtain supplementary information.

REFERENCES

1. Fourier J.,Memoires de l'Acadėmie des Sciences 4, 185 (1824).
2. Bell AG, Phil.Mag. 11, 510 (1881).
3. Pao Y-H (ed.): Optoacoustic Spectroscopy and Detection. New York: Academic Press, 1977.
4. Rosencwaig A: Photoacoustics and Photoacoustic Spectroscopie. New York: John Wiley & Sons, 1980.
5. Appl.Opt. 21, Nr.1 (1982).
6. J.de Physique 44, Suppl.Nr.10 C 6 (1983).
7. Rosencwaig A, Physics Today 9, 23 (1975).
8. Vidrine DW, Appl. Spectrosc. 34, 314 (1980).
9. Lahmann W and Ludewig HJ, Chem.Phys.Lett. 45, 177 (1976)
10. Kaya K, Harshbarger WR, and Robin MB, J.Chem.Phys. 60, 4231 (1974).
11. Kreuzer LB and Patel CKN, Science 173, 45 (1971).
12. Patel CKN and Tam AC, Rev.Mod.Phys. 53, 517 (1981).
13. Parker JG, Appl.Opt. 12, 2974 (1973).
14. Claspy PC, in Ref. 3.
15. Busse G and Thurmaier R, Appl.Phys.Lett. 31, 194 (1977).
16. Busse G, Basel E, and Pfaller A, Appl.Phys. 12,387 (1977).
17. Busse G and Perkowitz S, Int.J.Infr.and Submillimetre Waves 1, 139 (1980).
18. Teng YC and Royce BSH, J.Opt.Soc.,Am. 70, 557 (1980).
19. Rockley MG, Davis DM, and Richardson HH, Appl.Spectrosc. 35, 185 (1981).
20. Rosencwaig A, in Ref. 3.
21. Rosencwaig A and Gersho A, J.Appl.Phys. 47, 64 (1976).
22. Rockley MG and Devlin JP, Appl.Spectrosc.34, 407 (1980).
23. Carslaw HS and Jaeger JC: Conduction of Heat in Solids. Oxford: Clarendon Press. 1959.
24. Busse G, Paper Tu C 9, 4th Int.Top.Meeting on Photoacoustics, Thermal and Related Sciences, Montreal/Canada, 1985.
25. Burt JA, Proc. IEEE Ultrasonics Symp. 815 (1981).
26. Bennett CA and Patty RR, Appl.Opt. 21, 49 (1982).
27. Burt JA, in Ref. 6
28. Busse G, Appl.Phys.Lett. 35, 759 (1979).
29. Lehto A,Jaarinen J,Tiusanen T,Jokinen M, and Luukkala M, Electr.Lett. 17, 364 (1981).
30. Opsal J and Rosencwaig A, J.Appl.Phys. 53,4240 (1982).
31. Rosencwaig A, in "Scanned Image Microscopy" (Ash EA ed). London: Academic Press, 1980.
32. White RM, J.Appl.Phys. 34, 3559 (1963).
33. Hordvik A and Schlossberg H, Appl.Opt. 16, 101 (1977).
34. Boccara AC, Fournier D, and Badoz J, Appl.Phys.Lett.36, 130 (1980).
35. Jackson WV, Amer NM, Boccara AC, and Fournier D, Appl.Opt. 20, 1333 (1981).
36. Nordal P-E and Kanstad SO, Physica Scripta 20,659 (1979).
37. Ameri S, Ash EA, Neuman V, and Petts CR,

Electronic Lett. 17, 337 (1981).
38. Rosencwaig A, Opsal J, and Willenborg DL, Appl.Phys. Lett. 43, 166 (1983).
39. Busse G and Rosencwaig A, J.Photoacoustics 1,365 (1983).
40. Busse G, in Ref. 6.
41. Busse G, Infrared Phys. 20, 419 (1980).
42. Eyerer P and Busse G, Z.f.Werkstofftechn. 15, 140 (1984).
43. Busse G and Eyerer P, Appl.Phys.Lett. 43, 355 (1983).
44. Busse G, IEEE Trans.Sonics and Ultrasonics SU-32, 355 (1985).
45. Busse G and Rosencwaig A, Appl.Phys.Lett. 36, 815 (1980).
46. Busse G and Ograbek A, J.Appl.Phys.51, 3576 (1980).
47. Almond DP, Patel PM, and Reiter H, in Ref.6.
48. Nordal P-E and Kanstad SO, see Ref.31.
49. Thomas RL, Pouch JJ, Wong YH, Favro LD, Kuo PK, and Rosencwaig A, J.Appl.Phys. 51, 1152 (1980).
50. Luukkala M, see Ref. 31.
51. Luukkala M and Askerov SG, Electronic Lett.16, 84 (1980).
52. Busse G, Rief B, and Eyerer P, Paper WB 11, see Ref. 24.
53. Busse G, see Ref. 31.
54. Rosencwaig A and Busse G, Appl.Phys.Lett. 36, 725 (1980).
55. Busse G, in Ref. 5.
56. Inglehart LJ, Lepoutre F, and Charbonnier F, J. Appl.Phys. 59, 234 (1986).
57. Eyerer P, Busse G, and Rief B, paper WB 10, see Ref.24.

DISCUSSION

Comment: Alippi
You said that, in a certain sense, optoacoustics is a solution looking for a problem. Can't fundamental physics spectroscopy be considered such a problem?

Reply: Busse
This is an important suggestion and in fact there is a fine example: Patel and Tam four years ago measured very weak absorption caused by overtones and compared them with theoretical models.

CRYOGENIC ACOUSTIC MICROSCOPY

J.D.N. CHEEKE

INTRODUCTION

The field of microscopy as a whole can be said to be in the third and most productive phase of its long history. The first phase began with the work of Van Leehenhoek (1) on the optical microscope, on which refinements and improvements are continuing up to the present time. The second phase was the development of the electron microscope (2) which is again becoming a very mature field. Both of these microscopes have given us extraordinary insights into the world of the extra small and it is a certainty to say that they will continue to enjoy widespread use in the future. However it is really in the last ten years or so that the most exciting and innovative work has been done. It rapidly dawned on the scientific community that with suitable "optics" and mechanical or electronic scanning any incident radiation in the appropriate wavelength specimen could be used as the basis of a microscope. During this period we have seen the development of infrared microscopy (3), scanning Auger microscopy (4), Raman microscopy (5), scanning laser microscopy (6), ion microscopy (7), X-Ray microscopy (8), tunnelling microscopy (9) and the list seems to be growing every year. All of these microscopes have their special characteristics, their advantages and disadvantages, and they should be viewed as a complementary collection from which one should choose selectively for a given problem at hand. The scanning acoustic microscope (SAM) (10) is part of this collection and it would appear to be one of the most promising members for the study of the physical properties of materials.

This course of three lectures is focussed on one particular aspect of the SAM, that of low temperature microscopy. The first lecture deals with basic principles of the SAM; lens characteristics, choice of the liquid, imaging properties of the SAM, reflectance function, materials signatures or V(z) and ideas on the origin of contrast. These ideas are essential to understand the SAM's operation at all temperatures. The second lecture treats the question of nonlinear effects in fluids. This is particularly important in the SAM where a high concentration of acoustic power converges at the focus. In fact we feel that the SAM is a highly favoured situation for studying all nonlinear effects in acoustics. These effects are pronounced at low temperatures which is the subject of the third lecture. Throughout the discussion the outstanding contributions made by the Stanford group of C.F. Quate will become apparent.

1. SCANNING ACOUSTIC MICROSCOPY - GENERAL ASPECTS

The SAM was developed at Stanford University under the direction of Prof C.F. Quate. After several false starts (11) an exceedingly simple and effective acoustic lens was developed by Lemons and Quate (12). It consists of a spherical cavity ground in one face of a low attenuation dielectric crystal such as quartz or sapphire (Fig. 1). A piezoelectric transducer is

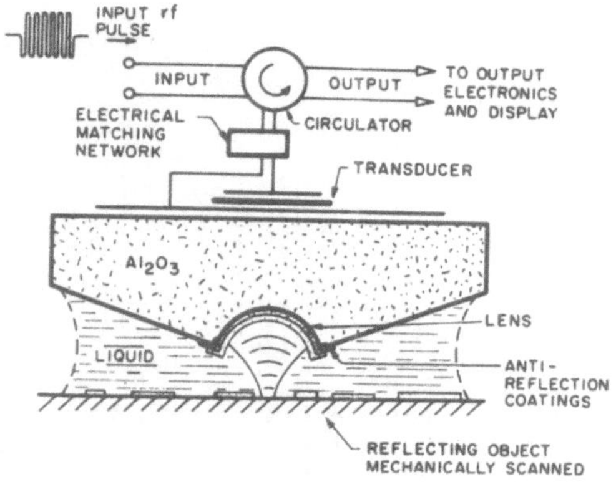

FIGURE 1. Typical configuration of the SAM lens in reflection (51).

placed opposite the cavity as shown in the figure. This transducer is usually PZT or $LiNbO_3$ for frequencies up to about 150 MHz and almost universally a thin film of RF sputtered ZnO for the higher frequency range. The transducer is excited by a tone burst of the appropriate frequency, which excites an ultrasonic beam in the crystal. This beam is focussed onto the surface of the reflecting object by a liquid (usually water) which forms the acoustical medium between cavity and object. The ultrasonic pulse is partially reflected at the interface and the echo thus produced traverses the system in reverse order and is video detected by an RF receiver connected to the transducer. The height of the detected pulse is directly related to the acoustic reflectance function at the specimen liquid interface, which is in turn directly related to the local acoustic properties of the sample. By scanning the focal point over the desired field of view an acoustic image can be built up in a few seconds and displayed on a TV monitor. The interpretation of such images in terms of the reflectance function is a major consideration of this chapter.

Considering the complex design needed to produce near perfect optical lenses it is an amazing fact that the simple acoustic lens shown in Fig. 1 has been capable of producing perfect diffraction limited images to resolutions of the order of 200 A! As the imaging is always done on axis at a single frequency, spherical aberration (SA) is the only possible source of focal point spreading other than diffraction. It is known that SA α D/n^2 (13) where D = lens aperture, $n = C_S/C_L$ and C_S, C_L are the sound velocities in solid and liquid respectively. In the optical case $n \sim 1$ so that SA is intrinsically important. For the acoustic lens we choose $C_S >> C_L$ and in practise it is easy in so doing to eliminate SA. Further SA scales down with D so that the problem does not get worse for smaller (high frequency) lenses.

Thus for a simple acoustic lens the resolution is given by $d_{min} = (D/F)\lambda$ (10) where F is the focal length and λ the wavelength in the liquid. For reference λ is about 15 microns for f = 100 MHz in water. The initial motivation to build an acoustic microscope was the realisation by Quate that submicron resolution would be possible for microwave frequencies. Everyone who has worked in RF and microwave ultrasonics knows that all of the difficulties encountered become increasingly severe as one goes up in frequency. The Stanford group has given proof of remarkable technical ingenuity in overcoming all of these problems for the SAM. Apart from transducer and electrode fabrication, the main problem is that the intrinsic attenuation in the liquid increases at least as the square of the frequency, and relaxation effects make the situation even worse. To optimise the resolution one wants a liquid with minimum wavelength and minimum absorption at a given frequency. To this end a figure of merit M has been defined (10) where

$$M = \frac{C_w}{C} \frac{[\alpha_w (f_w)/f_w^2]^{\frac{1}{2}}}{[\alpha(f)/f^2]^{\frac{1}{2}}} \qquad (1.1)$$

with C_w , C = velocity of sound in water, liquid
α_w , α = attenuation in water, liquid
f_w , f = highest attainable frequencies in water, liquid with the condition $\alpha(f) = \alpha_w(f_w)$.

Table 1 gives representative values for some liquids. The trend towards high figures of merit for the cryogenic liquids is clearly discernable.

TABLE 1. Acoustic Properties of various liquids (after 10).

LIQUID	T °C	α/f^2 x 10^{-17} $cm^{-1}sec^2$	C x 10^5 cm/sec	Z x 10^5 gm $cm^{-2}sec^{-1}$	M
Water	30	19.1	1.509	1.509	1.00
Methanol	30	30.2	1.088	0.866	1.10
Ethanol	30	48.5	1.127	0.890	0.84
Acetone	30	54.0	1.158	0.916	0.77
Carbon tetrachloride	30	538	0.930	1.482	0.30
Mercury	23.8	5.8	1.449	19.69	1.89
Gallium	30	1.58	2.87	17.48	1.82
Carbon disulfide	25	10(3GHz)	1.310	1.65	1.8(3GHz)
Nitrogen	77K	10.6	0.962	0.777	2.10
Helium	4.2K	260	0.183	0.027	2.23
Helium	1.95K	70	0.215		3.6

The most suitable liquid having been chosen, maximum resolution can then be attained by decreasing the lens radius, which can be done to the present technological limit of about 15 microns (15) (16). Lens grinding in itself is quite an art (13). For lenses down to about 200 microns in diameter precision ball bearings (or the equivalent) and diamond paste can be used, which a phonograph stylus is suitable for making the smaller lenses. There are also major problems with high speed switches to obtain sufficiently narrow pulses. These are needed in order to time gate the focus echo and distinguish it from spurious echoes in the lens. Finally there are all the usual problems with associated microwave circuitry and transducer optimisation (10). These techniques have been perfected and this has lead to the production of commercially available models (17).

The SAM has a number of advantages for imaging and material characterisation. Even with water at 300 K very high resolutions can be attained. The present record belongs to Hadimioglu and Quate (15) who obtained a resolution of about 2000 A at about 4.5 GHz using a nonlinear enhancement technique which will be discussed in the next chapter. From Fig. 2 we see that

FIGURE 2. a) SAM image in reflection of aluminum line in integrated circuit (15).

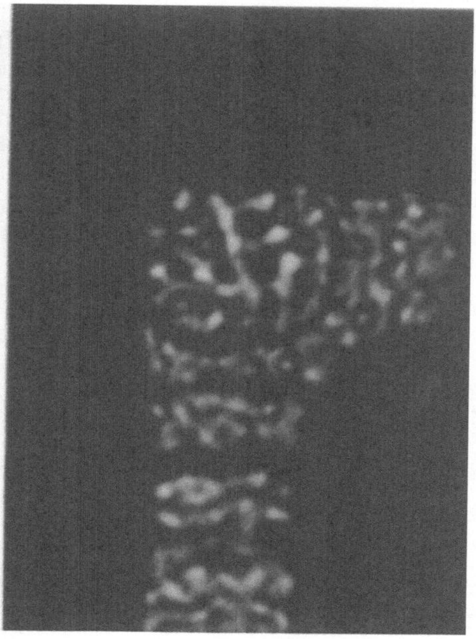

FIGURE 2. b) Optical image of the object in (a) (15).

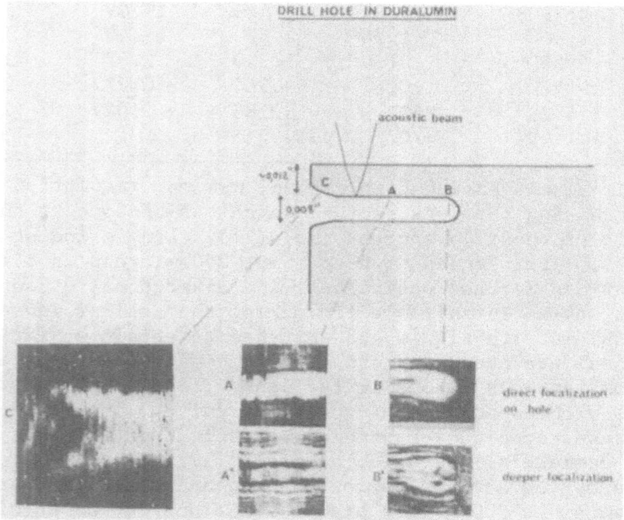

FIGURE 3. Example of SAM image of a drill hole in duraluminium at 150 MHz.

the resolution is significantly better than that obtained with an optical microscope. No serious effort has been reported to try to improve on this result, so it is too early to tell whether one can hope to obtain significantly higher resolution at 300 K.

A second feature of the SAM lies in its subsurface imaging capability. When the microscope is focussed inside the sample, ultrasonic waves penetrate and can be scattered off any existing microstructure. An acoustic image of a drill hole in an aluminium block is shown in Fig. 3 as an example. There is of course a trade off between resolution and subsurface imaging depth due to the increase in attenuation with frequency. At the moment the most promising frequency range for this type of materials analysis would appear to be 10 - 150 MHz, where penetration up to a few mm is easily attained. Typical problems involve the integrity of soldered, welded or glued joints (18), thermal contact of electronic devices to heat sinks (19) etc. Even when focussed on the surface the SAM is sensitive to subsurface detail due to penetration of Rayleigh waves as it will be discussed shortly.

Another interesting feature of the SAM is that the intrinsic contrast is quite large so that special staining and polishing techniques which are often necessary in optical microscopy are not needed here. The very fact that the SAM is looking directly at the acoustic properties of solids may be one of this chief attributes in the long term. This is in comparison to optical microscopy where minute changes in refractive index are detected. Examples abound in biology, for example the imaging of red blood cells (10) where the acoustic image is very contrasty compared to the optical one.

These considerations lead us to the central question of the interpretation of acoustic images; generally we see more detail than in optical images. Why is this so? As the reflection takes place at a liquid-solid interface we must examine this process in a little detail. The subject has been intensively studied and standard treatments are given in (20) - (22) amongst other excellent texts. The process is most simply described by use of the impedance concept as a complete analogy can be made with electromagnetic systems (20). We define the acoustic impedance for a mode in a given medium as $Z = \rho c/\cos\Theta$ where Θ is the angle of incidence to the interface normal. For normal incidence and $Z_L << Z_S$ we have for the energy flux reflection coefficient $R_E = [(Z_1 - Z_2)/(Z_1 + Z_2)]^2$ and transmission coefficient $T = 1 - R_E \simeq 4\, Z_L/Z_S$. This result gives already an important contribution to the insertion loss in the microscope where there is a large acoustic impedance mismatch.

Mathematically, incident, reflected and transmitted waves are defined in each medium and the equations of motion are solved by application of the boundary conditions of continuity of stress and normal displacement at the interface (20). Solving for the amplitudes we can define an amplitude reflection coefficient $R(\Theta) = re^{j\phi}$ for a wave incident from the liquid. Typical results for a water - aluminium frontier (29) are shown in Fig. 4. For $\Theta < \Theta_{CR}$ there is partial transmission into the solid but most of the energy is reflected back into a critical cone. The angles $\Theta \sim 15^o$ and $\Theta \sim 30^o$. correspond to setting up longitudinal and transverse modes along the surface. At $\Theta = \Theta_{CR} = \sin^{-1}(C/C_R)$ a Rayleigh wave of velocity C_R is excited on the surface. This is a resonant phenomenon and so we see a peak in the amplitude of the absorbed energy and a rapid phase change. For $\Theta > \Theta_{CR}$ there is total reflection of all of the incident energy.

It has been known for some time that the reflected sound field in the liquid for a bounded beam is rather more complicated (24) - (26). The excited Rayleigh wave propogates along the surface and continually reradiates into the liquid, giving rise to a so called "leaky" wave. The interference between this wave and the specularly reflected component gives rise to the Schoch displacement (24) - (26), analagous to the Goos-Haenchen effect in optics (27). The position of Θ_{CR} and the dip in r at Θ_{CR} have been used to study material properties (28) (29), in particular the velocity and attenuation of Rayleigh waves.

Leaky Rayleigh wave effects are of profound importance for the V(z) or acoustic materials signatures in acoustic microscopy. Rather than scan in the plane, we keep the lens and object at a fixed (x,y) position and translate the object towards the lens in the z direction. One then observes a series of fringes in the transducer video output as a function of z (Fig.5) (23) (30). Historically the effect was hard to pin down and work was pioneered experimentally by Weglein and coworkers (30) and theoretically by Atalar (31) (32). The simplest explanation is provided by a ray optic approach (33) which can be followed from Fig. 6. It has been shown that the interference is between the central specularly reflected part of the beam and the outer cone incident near the Rayleigh angle. The latter sets up leaky waves on the surface which are then reemitted and captured by the lens when it is in the defocussed condition. The phase difference between the axial reference beam and the set of leaky waves which appear to originate from the focus is a function of z, leading to a set of interference fringes with spacing between the minima given by

$$\Delta z_n = \frac{\lambda_R}{\sin \Theta_{CR}} \cdot \frac{1 + \cos \Theta_{CR}}{2} \qquad (1.2)$$

where λ_R is the Rayleigh wavelength using $\lambda_R C_R = f$ we can rewrite this in a more user friendly form:

$$C_R = C/\left[1 - (1 - C_L/2f \Delta z_n)^2\right]^{\frac{1}{2}} \qquad (1.3)$$

This phenomenon has been exploited a great deal for quantitative studies with the SAM. A good review is given in (30), with applications to surface wave velocity measurement, thin film thickness, layered systems, lamb waves, surface profiling, etc.

An interesting innovation is the line focus beam introduced by Chubachi and coworkers (34). A cylindrical lens is used so that directional information can be obtained from V(z). Anisotropy of the sound velocity for single crystal surfaces gives good agreement with theory. The method is proving useful for materials characterisation (34) and has even been used to study phase transitions in physical acoustics (35).

The original theoretical treatment of V(z) was given by Atalar (31) using an angular spectrum approach. This two dimensional Fourier analysis treatment is based directly on standard Fourier optics treatments (36) where each passage of the incident wave through a lens surface is represented by a multiplicative pupil function P(r) which takes into account the geometrical form of the lens as well as aberrations and the transmittance. The object acts on the wave by a multiplicative reflectance function R. Treating successive surfaces and using a paraxial approximation Atalar obtains:

$$V(z) = 2\pi \int_0^\infty r\, P_1(r)\, P_2(r)\, R(r/F) \exp\left(\frac{-jk_o Zr^2}{F^2}\right) dr \qquad (1.4)$$

where F is the focal length and $r = (x^2 + y^2)^{\frac{1}{2}}$.

This treatment has been simplified and extended to the non paraxial limit (37). Equation (1.4) gives much information about the intrinsic contrast mechanisms in the SAM. There is an implicit z dependence so the contrast is sensitive to surface topography. Any material changes will alter the directly reflected beam via R(r). In addition a material change will alter Θ_{CR} and we have seen that the phase of the leaky wave is highly dependent on the value of Θ_{CR}. The relative importance of these contributions will

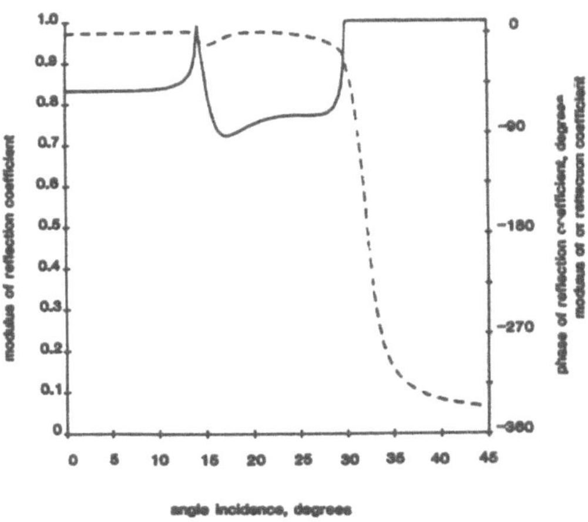

FIGURE 4. a) Reflectance function for water aluminium interface with no dissipative loss (29).

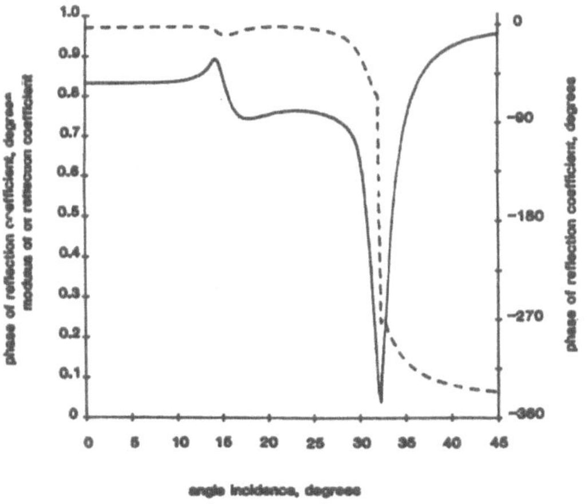

FIGURE 4. b) Same result as for Fig. 4a with an attenuation of 1.8 dB per wavelength (29).

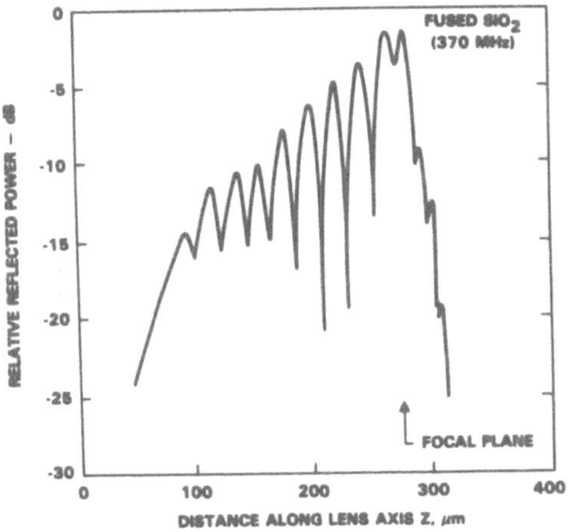

FIGURE 5. Example of acoustic materials signatures V(z) for a quartz water interface (30).

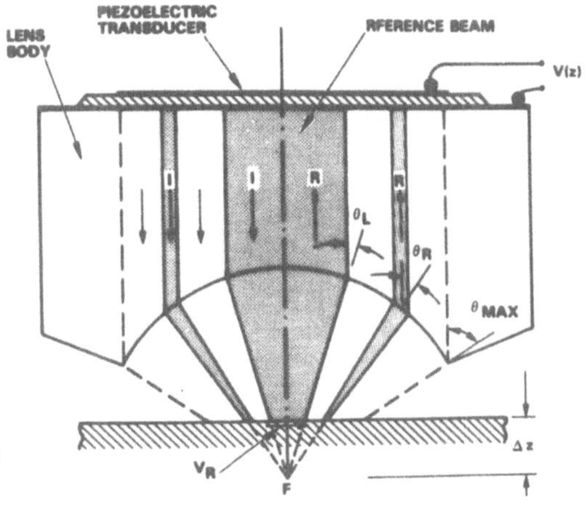

FIGURE 6. Ray analysis of V(z) curves (30).

depend critically on the position of the object plane and the relative acoustic impedances of the object and liquid. Thus while we understand the contrast mechanism in the SAM in principle it is by no means a trivial task and each individual case must be regarded in detail.

The Stanford group has more recently developed a new approach which seems to be much more powerful than the simple Δz_n measurement. Hildebrand et al (38) have used the Atalar approach to show that V(z) and R(Θ) are a Fourier transform pair. By measuring the amplitude and phase of V(z) they were able to invert V(z) to obtain the full reflectance function R(Θ) in amplitude and phase. As the lens pupil functions are in general unknown they calibrated the microscope by measuring V(z) for a material (lead) whose reflectance function is of uniform amplitude and phase for all angles excited by the acoustic lens.

Liang et al (39) have extended the theoretical approach by formulating a more rigorous treatment. They point out that both angular spectrum and ray optic approaches were limited by a paraxial approximation and that the ray optic approach is in any case very unrigorous (only part of the acoustic beam is used, diffraction effects are not included and it is invalid near the focus). They adopt a hybrid approach, valid at all angles and arrive at the general result

$$V(u) = \mathcal{F}\left(P^2(t)\, R(t)\right) \qquad (1.5)$$

where $u = z/\lambda$, $t = 2 \cos \Theta$, and $\mathcal{F}$ is the Fourier transform. A large number of experimental results are given on different materials and excellent agreement is obtained between experimental and theoretical values of R(Θ). In principle this is a most satisfying approach as all local acoustic material parameters, including the attenuation, can be obtained. The drawbacks include artefacts due to spatial truncation of V(z) which gives spurious attenuation values and the non triviality of the phase measurement. The phase measurement has also been applied to surface profiling and thin film measurement (40) and there seems little doubt that this approach will exploit the full potential richness of the V(z) phenomenon. It also explicitly shows the acoustic lens to be a very sensitive local probe capable of providing the full panoply of ultrasonic material information on a microscopic scale. Finally it should be noted that V(z) is a very general phenomenon and has been observed optically using a coherent source (41).

2. NONLINEAR ACOUSTIC EFFECTS IN FLUIDS

Most ultrasonic studies are done in the linear region where infinitesimal amplitudes are assumed in the theoretical calculations. This is obviously a poor approximation in the SAM, where a great deal of acoustic power converges on the focal point. A priori one thus expects that nonlinear effects will be very important in the SAM and this is indeed found to be the case experimentally. The present chapter gives a survey of the experimental and theoretical situation for both plane and focussed waves, which will serve as background material for the interpretation of the cryogenic images discussed in chapter 3.

The nonlinear behaviour of plane waves has been presented in some detail by Beyer and Letcher (42). For a fluid we can always write the adiabatic relation $p = p_o (\rho/\rho_o)^\gamma$ where p = pressure, ρ =density, $\gamma = C_p/C_V$ and subscript zero refers to equilibrium conditions. We expand $p = p(\rho,s)$ in a Taylors series up to the quadratic term for the isentropic case to obtain:

$$p - p_o = A \left(\frac{\rho - \rho_o}{\rho_o}\right) + \frac{B}{2} \left(\frac{\rho - \rho_o}{\rho_o}\right)^2 \qquad (2.1)$$

where $A = \rho_o C_o^2$ and $\frac{B}{A} = 2\rho_o C_o \left(\frac{\delta C}{\delta p}\right)_{s,\rho = \rho_o}$ (2.2)

with C = sound velocity and S = entropy.

By simple thermodynamic transformations

$$\frac{B}{A} = 2\rho_o C_o \left(\frac{\delta C}{\delta p}\right)_{T,\rho_o} + \frac{2\alpha T C_o}{C_p}\left(\frac{\delta C}{\delta T}\right)_{p,\rho_o} \quad (2.3)$$

where $\alpha = (1/V)\ (\delta V/\delta T)\ p$.

B/A plays the role of a nonlinear parameter for the fluid which is usually expressed as $\beta \equiv 1 + B/2A$.. Typical values are given in Table 2.

TABLE 2. Values of B/A (reference 42)

Substance	T(°C)	B/A	Substance	T(°C)	B/A
Distilled water	20	5.0	Methyl acetate	30	9.7
Seawater (3.5%)	20	5.25	Cyclohexane	30	10.1
Methanol	20	9.6	Nitrobenzene	30	9.9
Ethanol	20	10.5	Mercury	30	7.8
n-Propanol	20	10.7	Sodium	110	2.7
n-butanol	20	10.7	Potassium	75	2.1
Acetone	20	9.2	Tin	240	4.4
Benzene	30	9.0	Indium	160	4.6
Chlorobenzene	30	9.3	Bismuth	318	7.1

If we consider propagation with no dissipation the equation of motion for the local particle velocity u is

$$\frac{\delta^2 u}{\delta t^2} = \frac{C_o^2}{(1 + \delta u/\delta x)^{B/A + 2}} \cdot \frac{\delta^2 u}{\delta x^2} \quad (2.4)$$

with the implicit solution:

$$u(x,t) = u_o \sin\left[\omega t - \frac{\omega x}{C_o}\left(1 + \frac{B}{2A}\frac{u}{C_o}\right)^{-\frac{2A}{B} - 1}\right] \quad (2.5)$$

corresponding to wave velocity:

$$v = C_o\left(1 + \frac{B}{2A}\frac{u}{C_o}\right)^{\frac{2A}{B} + 1} \quad (2.6)$$

Equation (2.6) shows explicitly that large values of u (crests) are propogated more rapidly than small values (troughs) so that the wave form is deformed from a sinusoid to a sawtooth as it travels through the medium. When the slope of the front becomes infinitely negative we say that a shock wave has been created. This discontinuity occurs at a distance $\ell = 1/\beta M k$ from the source where $M = u_o/C_o$ is the acoustic Mach number and $k = \omega/C_o$. The situation for this shock build up is shown in Fig. 7. For water at room temperature and atmospheric pressure $\ell \sim 75$ m at 20 kHz and 7.5 mm at 200 MHz.

For low Mach numbers an explicit solution to (2.4) has been given by Fabini (44) as

$$\frac{u}{u_o} = 2\sum_{n=1}^{\infty} \frac{J_n(nx/\ell)}{nx/\ell} \sin n(\omega t - kx) \quad (2.7)$$

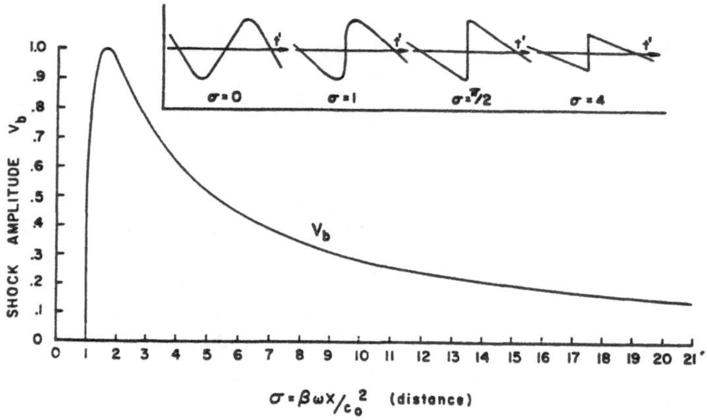

FIGURE 7. Shock front amplitude as a function of distance from the source. The wave front deformation is shown in the inset (43).

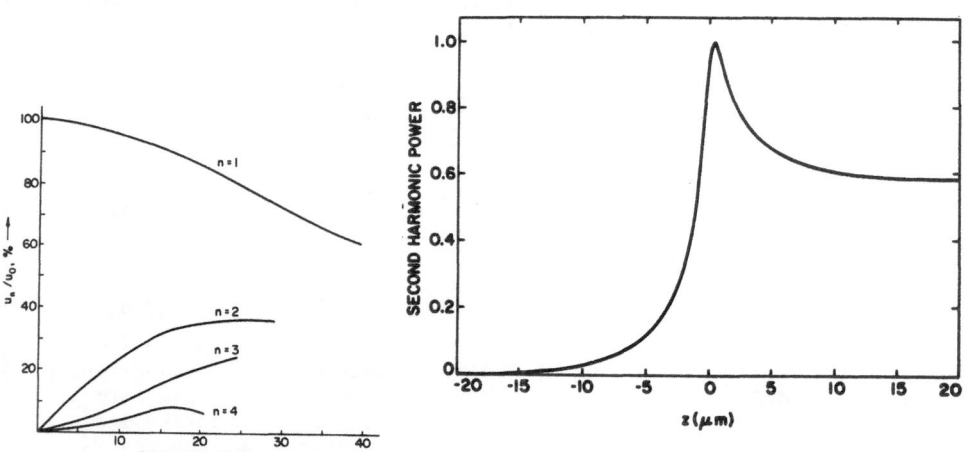

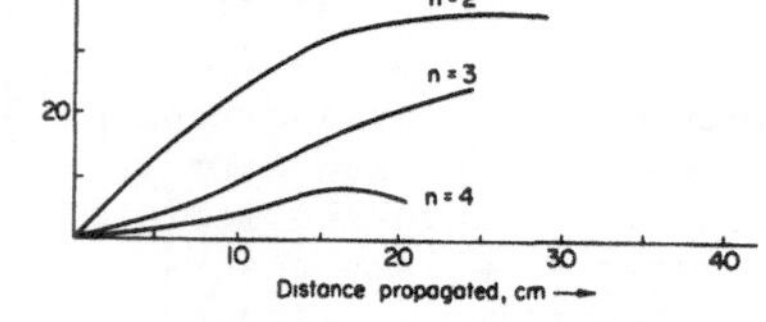

FIGURE 8. Harmonic growth and decay for 2.5 MHz plane sound waves in water (45).

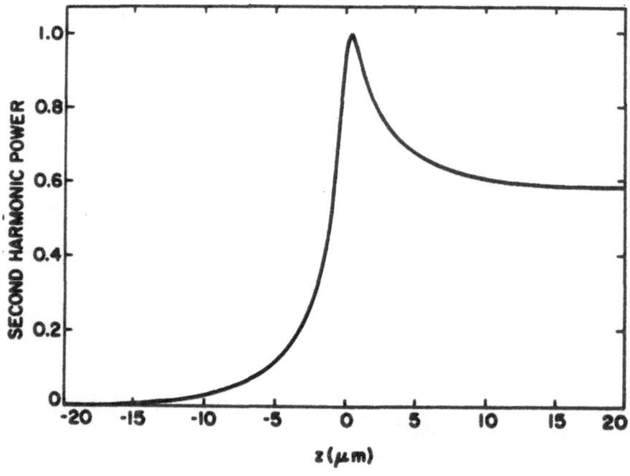

FIGURE 9. Growth of second harmonic power near focal point($z = 0$) for acoustic lens (48).

valid in the build up region. The depletion of the fundamental and the build up of the harmonics with distance is clearly seen in this result. If we now include dissipation we can expect that the harmonic build up and the appearance of the shock front will be delayed since the attenuation varies as u^2. Detailed calculations have been carried out for harmonics up to n = 6 (45) and some calculations for water at 2.5 MHz are shown in Fig. 8. Finally for completeness, we note that the steady state solutions far from the source have been obtained by Fay (46) and the transition between the Fabini and Fay regimes has been discussed in detail by Blackstock (43).

For applications to the SAM we need to consider the case where nonlinearity, dissipation and diffraction can be handled in the same formalism. This has been done in (47) where the wave equation is:

$$\left(\frac{\delta}{\delta t^2} - C_o^2 \nabla^2\right)\rho = \frac{D}{C_o^2}\frac{\delta \rho}{\delta t^3} + \frac{B}{\rho_o}\frac{\delta^2}{\delta t^2}(\rho - \rho_o)^2 + O\left[M^2 (\rho - \rho_o)\right] \tag{2.8}$$

where D is the sound diffusivity. This equation has been analysed by Rugar (48) for the case of no dissipation ($D \to 0$). Fourier analysing the density

$$\rho = R_e \left[\rho_o + \rho_1 \exp(-j\omega_1 t) + \rho_2 \exp(-j\omega_2 t)\right] \tag{2.9}$$

we obtain an infinite system of equations for the harmonics. Neglecting depletion of the fundamental and generation of harmonics higher than the second one obtains:

$$(\nabla^2 + k_1^2)\,\rho_1 = 0 \tag{2.10}$$

$$(\nabla^2 + k_2^2)\,\rho_2 = \gamma\,\rho_1^2 \tag{2.11}$$

where $k_1 = \omega_1/C$, $k_2 = \omega_2/C$ and $\gamma = 2B\,k_1^2/\rho_o$.

The first equation is for the non depleted fundamental while the second is for the second harmonic with a source term. Weak nonlinerarity is thus the essential assumption here. Equation (2.11) can be solved by a Greens function technique as shown by Rugar (48).

For application to the SAM we assume a focussed radiation field in the paraxial approximation and for simplicity a Gaussian beam. This can be written as:

$$\rho_1(x,y,z) = \frac{\rho_{10}\exp(jk_1 z)}{1 + j\,\xi}\exp\left(-\frac{x^2 + y^2}{\omega_o^2(1 + j\,\xi)}\right) \tag{2.12}$$

where $\xi = 2z/b$ is a dimensionless distance and $b = \omega_o^2 k_1$ is the confocal length. This is a paraxial solution to (2.10) and the width is Gaussian for all z with a minimum at the focal plane (z = 0) where the amplitude is ρ_{10}. The factor $(1 + j\,\xi)$ introduces the well known phase shift for propogation through the focal plane. From this solution we obtain the power in the beam as:

$$P_1(0) = \frac{C^3}{2\rho_o} / \rho_{10}/^2 \left(\frac{\pi\omega_o^2}{2}\right) \tag{2.13}$$

The solution for the second harmonic is more complicated and is based directly on the equivalent nonlinear optic problem (49). The solution is:

$$\rho_2(x,y,z) = \rho_{20}(z) \frac{\exp(jk_2 z)}{(1+j\,\xi)} \exp\frac{(-2\,(x^2+y^2))}{\omega_o^2\,(1+j\,\xi)} \tag{2.14}$$

where

$$\rho_{20}(z) = \rho_{10}^2 \frac{bk_1}{4\rho_o} \beta\left[\frac{1}{2}\log\left(\frac{1+\xi^2}{1+\xi_o^2}\right) + j\tan^{-1}\xi - \tan^{-1}\xi_o\right] \tag{2.15}$$

with $\xi_o = 2\,z_o/b$.

This result shows that the second harmonic is also a Gaussian focussed beam with a width at the focus of $\omega_o/\sqrt{2}$. This corresponds explicitly to the anticipated result that the focal width (hence resolution) would be finer than for the fundamental. From the solution we obtain directly the power in the second harmonic

$$P_2(z) = \frac{\omega_o^2 k_1^4 \beta^2}{8\pi\, C^3 \rho_o} P_1(0)^2 \left[\frac{1}{4}\log\left(\frac{1+\xi^2}{1+\xi_o^2}\right)^2 (\tan^{-1}\xi - \tan^{-1}\xi_o)^2\right] \tag{2.16}$$

A plot of this second harmonic beam power is shown in Fig. 9 for Rugar's work at 2 GHz in liquid Nitrogen. It is seen that the nonlinear effects are extremely concentrated in the focal region. The abrupt drop after the focal point is due to the phase shift while the further decrease for positive z is due to down conversion to the fundamental.

These results can now be applied directly to the acoustic lens. Within the above approximations the power lost to the second harmonic is given by (50):

$$P_2 = \frac{16\,\pi^3}{\rho_o\, C^5} \beta^2 F^2 f^2 \cdot P_1(0)^2 \tag{2.17}$$

where F is the f number of the lens. Arbitrarily defining onset of nonlinearirty P_o to be given by $P_2/P_1(0) = 0.1$ we obtain:

$$P_o = \frac{0.1}{F^2 f^2} L \tag{2.18}$$

where L is a nonlinearity parameter given by

$$L = \rho_o C^5/16\,\pi^3 \beta^2 \tag{2.19}$$

L is proportional to the power needed to drive a given liquid nonlinear. Representative values and other parameters are given for several liquids in Table 3. We note that the decreasing sound velocity of the cryogenic liquids is the main factor responsable for the appearance of nonlinear effects even at very low power levels.

A remarkably simple physical picture of the nonlinear effects in the acoustic lens has been given by Rugar (48) as shown in Fig. 10. Nonlinear generation near the focus leads to sawtooth generation followed by phase inversion on passing through it (or on reflection). The pressure peaks now lag and the troughs lead so that the distorted waveform progressively unravels, corresponding to power being fed back from the harmonics to the fundamental. This is the basis for the nonlinear resolution enhancement discovered by Rugar which is presented in the next chapter.

TABLE 3. Properties of Cryogenic Liquids. After (50)

Liquid	Temperature	Sound Speed (m/s)	1 GHz Loss (dB/mm)	Density (kg/m^3)	β	L ($W - GHz^2$)
Water	25^oC	1495	191	998	3.5	1.2×10^{-3}
Water	60^oC	1550	95	983	3.85	1.2×10^{-3}
Nitrogen	77 K	850	139	810	4.3	3.9×10^{-5}
Argon	85 K	840	132	1370	—	—
Helium	1.90 K	227	610	146	—	—
Helium	0.1 K	238	0.04	145	3.84	1.51×10^{-8}

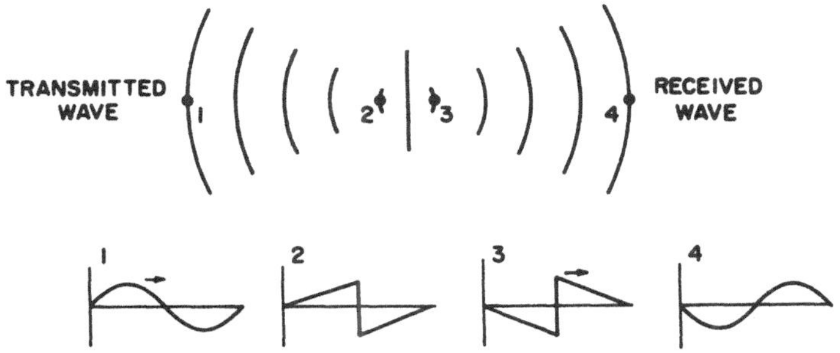

FIGURE 10. Deformation of sinusoidal signal in reflection acoustic lens. For simplicity the reflection path has been unfolded to the right (48).

There are as yet few experimental results to compare with the above calculations. The published work is exclusively due to the Stanford group, principally Rugar. The latter has quantified these effects by monitoring the power received as a function of lens input power. Results for water and liquid nitrogen at 2.6 GHz are shown in Fig. 11. The arrows indicate power levels at the focus for 1 db excess attenuation due to harmonic generation. As anticipated from the previous discussion the threshold value for liquid nitrogen is much smaller than that for water. Also the larger focal spot size for water means that the peak intensity is smaller for the same focal plane power. As we get into the nonlinear regime in nitrogen we notice that the excess attenuation increases linearly with the input power, i.e. the received signal no longer increases with input power but saturates. This effect is particularly important in liquid helium, as we see from Table 3 that the saturation effect occurs at extremely low power

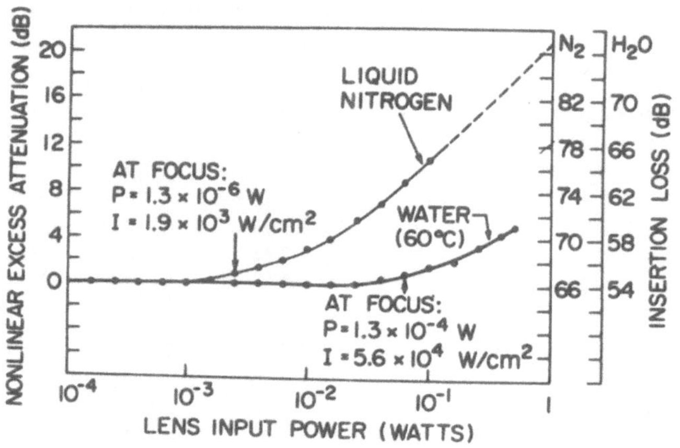

FIGURE 11. Insertion loss (or excess attenuation) as a function of incident power for SAM lens at 2 GHz in liquid nitrogen and water (48).

levels, which makes necessary the use of an extremely sensitive receiving system.

We have recently carried out extensive experiments on nonlinear effects in water, liquid nitrogen and liquid helium in the MHz frequency range. An example of nonlinear behaviour in water in an acoustic lens is shown in Fig.12. A straight line at 68° corresponds to linear behaviour, which is in fact never achieved at the power levels used in this case. The deviation of the points from linear behaviour is explicitly associated with generation of harmonics, and it is usually assumed that the second is dominant. We observed the second harmonic power as detected by a transducer plane at the focal plane (Fig. 13) and it is seen that a P^2 behaviour is observed at low powers in agreement with (2.17). Again a saturation effect is observed at higher powers which is associated with harmonic generation for $n > 3$.

Liquid nitrogen is a favorable liquid for observing higher harmonic generation as the attenuation is quite low as is the nonlinear threshold (Table 3). Higher harmonic generation up to $n = 10$ in liquid nitrogen is shown in Fig. 14 where it is seen that all of the harmonics appear to saturate with increasing abruptness at a given power level. The application of this technique to the study of shock front formation, cavitation and acoustic microscopy is presently under consideration.

Similar effects are presently under study in liquid helium although the results are rather preliminary. Fig. 15 shows the variations in received power at the fundamental in normal liquid helium ($T = 4.2$ K) and superfluid helium ($T = 1.9$ K). We believe that the decrease in power level in the normal liquid is due to cavitation effects. The behaviour in the superfluid is less clear as the relative decrease to saturation is much smaller and it may be due to quantum effects.

It is also possible to study the spatial variation of the harmonic power in the acoustic lens and this is shown in Fig. 16 for the case of water. The full line corresponds to the threshold result of Rugar (Fig. 9). The agreement is good and gives additional support to the already impressive experimental confirmation of Rugar's findings.

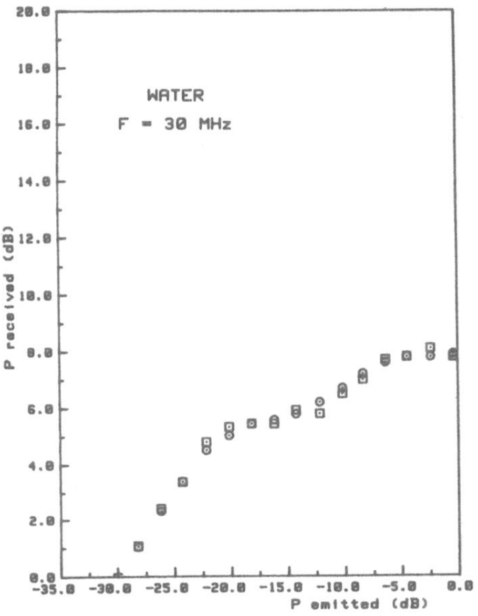

FIGURE 12. Received versus incident power for the fundamental for SAM lens in water.

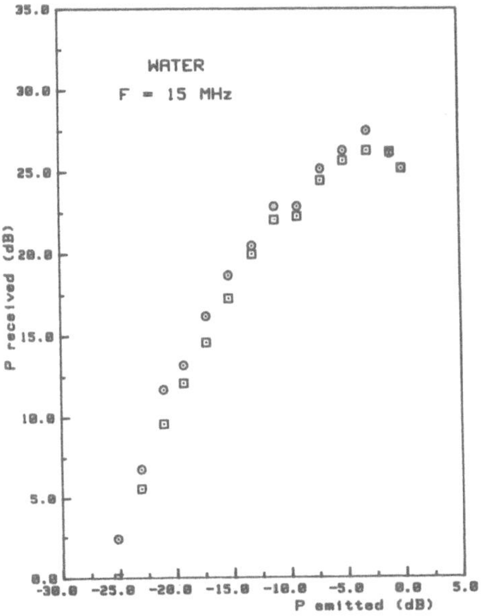

FIGURE 13. Received power in the second harmonic versus incident power for SAM lens in water.

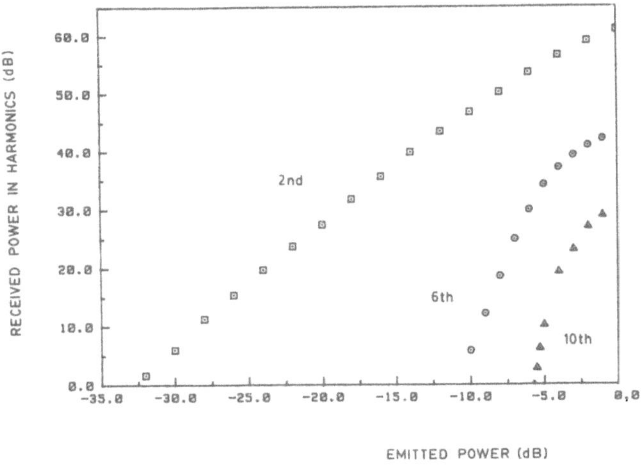

FIGURE 14. Received power for different harmonics for SAM lens in liquid nitrogen as a function of incident power.

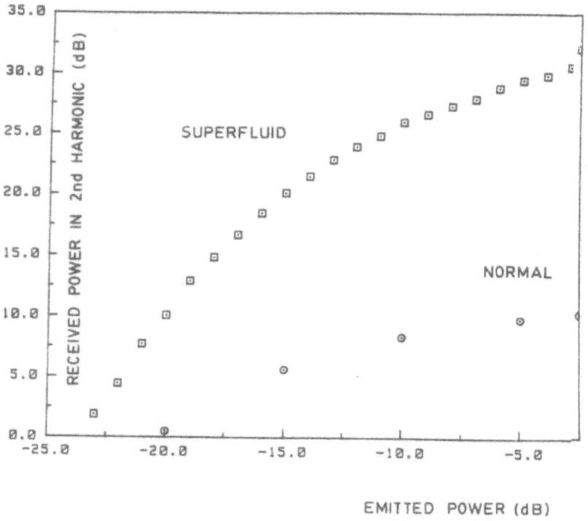

FIGURE 15. Received versus emitted power in the fundamental for the acoustic lens in liquid helium.

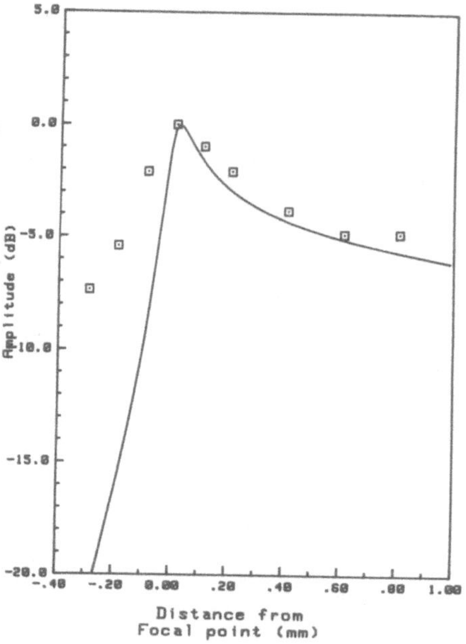

FIGURE 16. Second harmonic amplitude as a function of distance in the acoustic lens.

3. CRYOGENIC ACOUSTIC MICROSCOPY

The concept of a figure of merit M for fluids used as couplants in acoustic microscopy is based on the idea that an optimum fluid is one which has minimum attenuation and low velocity of sound for the highest attainable frequency. From Table 1 water is clearly a good choice but if we broaden our horizons then cryogenic liquids are clearly superior. Table 4 gives a summary for selected cryogenic liquids; in the classical regime the low attenuations are characteristic of those due to the low viscosity associated with small molecules (42). However it is the low velocity of sound and the vanishingly small attenuation characteristic of liquid helium in the quantum limit $T \to 0$ which is of central importance in what follows. There is of course a price to pay for these high figures of merit, principally the inconvenience associated with the use of cryogenic techniques and the extreme acoustic mismatch between crystalline solids and liquid helium, leading to an ultralow lens transmittance function.

Appreciation of the experimental difficulties and accomplishments in cryogenic acoustic microscopy depends critically on a kwnoledge of the attenuation mechanisms in liquid helium. Fig. 17 gives an overall view of the attenuation from the boiling point (4.2 K) down to very low temperatures ($T \ll 1$ K). Above $T \sim 2.2$ K we are in the hydrodynamic purely classical regime and $\alpha \sim \omega^2$. This behaviour can be extrapolated to give the minimum value at $T \sim 1.9$ K between the peak at the transition to superfluidity at $T = T_\lambda$ and the peak at $T \sim 1$ K (2). The peak at T_λ is in itself of interest as a perfect textbook example of a second order phase transition and has been studied in great detail (53). The broader peak near 1 K is of a more subtle origin, being due to a relaxation process between the equilibrium population of phonons and rotons (54). The peak occurs at a temperature where the viscous mean free path $\ell \sim \lambda$, the sound wavelength in liquid helium. As stated before for temperatures well above this peak ($\ell \ll \lambda$)

TABLE 4. Properties of selected cryogenic liquids compared to water (51)

Liquid	Temp (K)	$\alpha/f^2 \times 10^{17}$ (dBs^2/cm)	$c \times 10^{-5}$ (cm/s)	$Z \times 10^{-5}$ (g/cm^2s)	M
H_2O	25°C	191	1.5	1.5	1.0
	60°C	95	1.5	1.5	1.4
N_2	77	120	0.85	0.68	2.2
Ar	87	132	0.84	1.2	2.2
^4He	4.2	1966	0.183	0.027	2.5
	1.95	610	0.227	0.033	3.7
	0.4	15	0.238	0.035	23
	0	3.0	0.238	0.035	4000

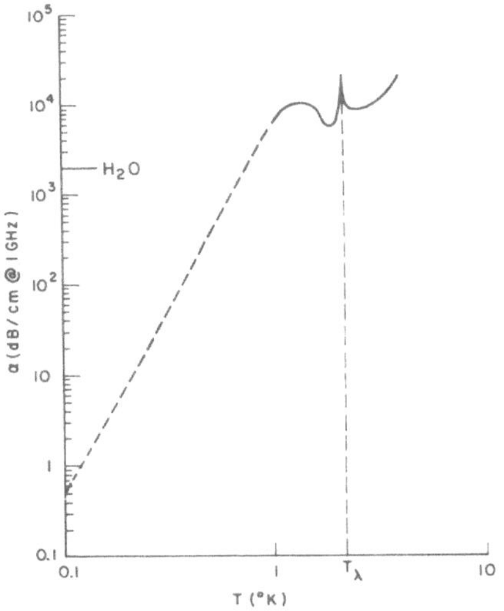

FIGURE 17. Attenuation of sound at 1 GHz in liquid helium as a function of temperature (51).

we are in the hydrodynamic regime where local thermodynamic arguments can be applied. At lower temperatures the phonon mean free path increases rapidly, primarily due to the exponential decrease in the number of rotons (essentially zero below T ~ 0.6 K) and we enter the collisionless regime where there is no thermodynamic equilibrium in the phonon gas during one acoustic wavelength.

The attenuation in the collisionless regime can be described by the scattering of an ultrasonic phonon by a thermal phonon, the so called three phonon process (55) leading to

$$\alpha_{3PP} = AfT^4 \qquad (3.1)$$

$$\text{where} \quad A = \frac{\pi^4 k_B^4}{15\hbar^3} \frac{(1+\mu)^2}{\rho C^6} \qquad (3.2)$$

with $\mu \equiv \rho/C\ (\delta C/\delta\rho)$ the Gruneisen constant.
This result has been verified semi quantitatively down to 0.1 K for frequencies up to 200 MHz (56) and more recently up to 1.25 GHz (51) as shown in Fig. 18. The functional form of (3.1) is explained by the scattering rate being proportional to the energy of each incident phonon and the number of thermal phonons Viz (hf) (k_Bt) (T^3). Putting in numerical values we obtain:

$$\alpha \simeq 5.10^{-6}\ f\ T^4\ \text{dB/cm} \qquad (3.3)$$

for f in Hz and T in Kelvin.

Heiserman (51) has given a useful discussion of the practical effects of attenuation on acoustic microscopy in the very low temperature region. Assuning a reasonable acceptable loss of 1000 dB/cm he plots the maximum usable frequency as a function of temperature as shown in Fig. 19. For low enough temperatures or high frequencies such that hf > $k_B T$ a new attenuation process, that of spontaneous decay of an ultrasonic phonon into subharmonic components becomes possible (55) (57) (58), leading to an attenuation

$$\alpha_{SD} = \frac{\hbar\,\pi^4}{30\,\rho C^6} (1+\mu)^2 f^5 \qquad (3.4)$$

$$\simeq 3.3 \times 10^{-50} f^5\ \text{dB/cm}.$$

which will become comparable to α_{3PP} for frequencies above 20 - 30 GHz at low enough temperatures.

The conditions for high frequency acoustic microscopy are thus very favorable in liquid helium at saturated vapor pressure, but dramatically more so under pressures of about 20 bar. Fig. 20 shows that the effect of pressure on the thermal phonon dispersion curve, which removes the anomalous dispersion and thus prevents attenuation by both spontaneous decay or three phonon processes (55). Experiments with thermal phonons at very high frequencies (60) have shown that the wave is totally stable against spontaneous decay above a critical frequency f_c(p) which is plotted in Fig. 21. So at sufficiently high frequencies and low enough temperatures the attenuation is totally eliminated even at saturated vapor pressure.

The particular characteristics of liquid helium require considerable changes in the techniques used for the SAM. The scanning system has to be modified to accommodate the cryogenic environment; the basic arrangement which has been used for all of the results published so far is shown in Fig. 22. For sufficiently small vibration amplitude the scanned area is approximately planar. For the fast axis the velocity signal from the pick up coil can be used in a servo loop to increase the stability of the system. A piezoelectric positioner in the z direction is also essential for high resolution work.

One of the main problems in helium microscopy is the small power transmission coefficient between solid and liquid, $T \sim 4\,Z\,/Z_s$ for approximately normal incidence and $Z_L \ll Z_S$. For a sapphire helium interface this leads to approximately 25 db of losses for each passage through the interface so the use of quarter wave matching layers is really essential. The subject has been developed in great detail by the Stanford group (62) and we will just present the basic principles here. For a single layer we need

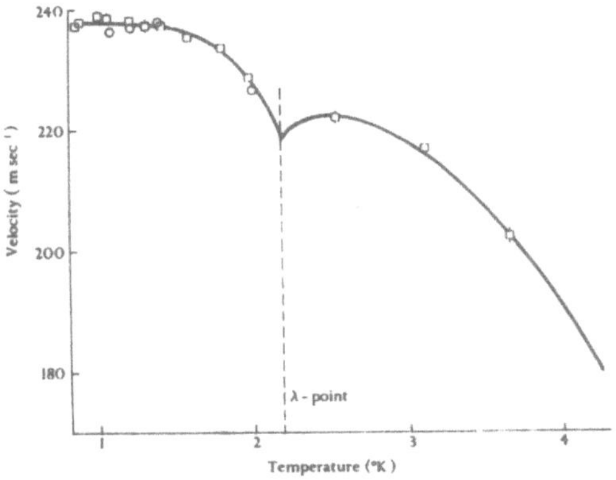

FIGURE 18. The velocity of sound in liquid ^4He under saturated vapor pressure (52).

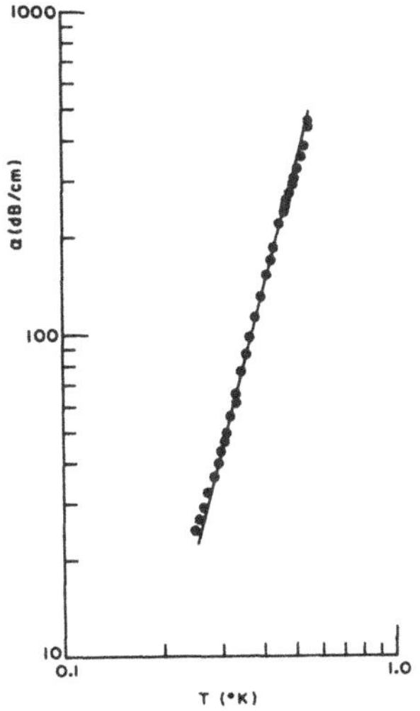

FIGURE 19. Attenuation of 1.25 GHz acoustic plane waves in ^4He. Solid line is T^4 (51).

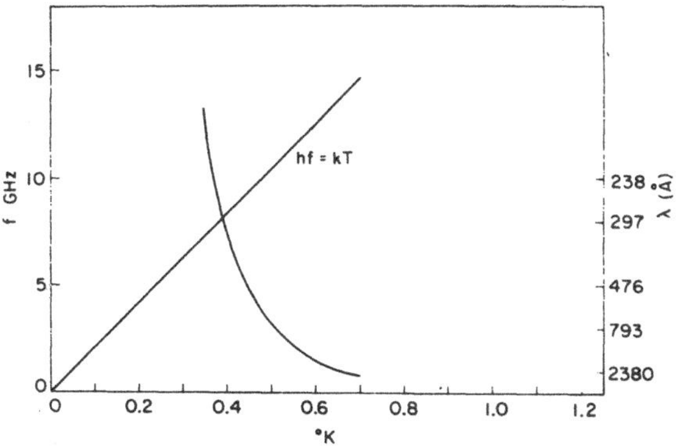

FIGURE 20. Frequency at which the attenuation constant α_{3PP} equals 1000 dB/cm as a function of temperature (51).

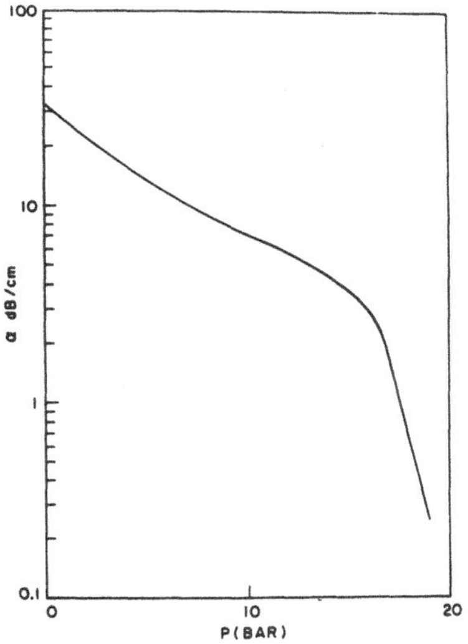

FIGURE 21. Attenuation of plane waves at 105 MHz in ^4He at T =0.45 K as a function of pressure (59).

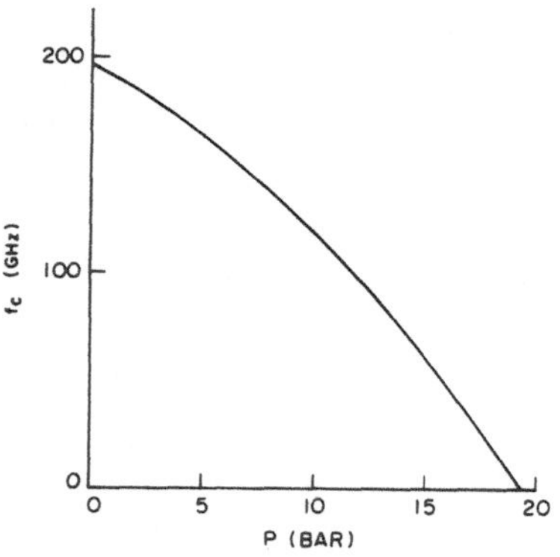

FIGURE 22. Critical frequency for the onset of normal dispersion and suppression of attenuation (51).

thickness $\lambda_c/4$ where λ_c is the sound wavelength in the layer and layer impedance $Z_c = Z_L Z_S$. For typical cristalline solids and liquid helium this gives $Z_c \sim 10^5$ g/cm^2 sec. Water and photoresist have impedances of the same order of magnitude but their application to small acoustic lenses has not proved successful. A solution adopted by Rugar is to use a second, very high Z quarter wave layer so as to raise Z_c to a manageable value. A Ti - C system deposited by the electron gun technique has been the most successful, with typical results shown in Fig. 23. An unexpected benefit of this technology is that quite by accident this combination gives a 100% tranmission into water. Thus with sophisticated deposition techniques the transmittance problem can be solved, at the cost (as usual) of band width.

A part from the usual technical complications associated with working in a cryogenic environment the remaining major problem is that associated with the saturation effect of the received power and at high enough frequencies, depletion of the fundamental by spontaneous decay. At the highest frequency reported (8 GHz) these two nonlinear processes restrict the maximum receivable power to -95 dBm (63). The improvements in sensitivity to the receiving system have been discussed in detail by Hadimioglu and coworkers (63) (64). The noise in the system is reduced by using a 20 db directional coupler at the input to reduce room temperature noise and putting the receiver preamplifier in the liquid helium. This consisted of three GaAS FETS at 4.2 K (65) giving a final noise power of - 102 dBm for a 60 MHz bandwidth. Further improvement in the signal to noise ratio (SNR) was obtained by using pulse compression techniques (16) (66) which enables using a small amplitude long duration pulse at the sender which is converted to a high amplitude short duration pulse in the receiver. The use of this technique requires larger lenses (125 micron diameter at 8 GHz) to enable time resolution, but the method is admirably suited to partially overcome the saturation effect in liquid helium. A net improvement of about 18 dB was obtained in the SNR.

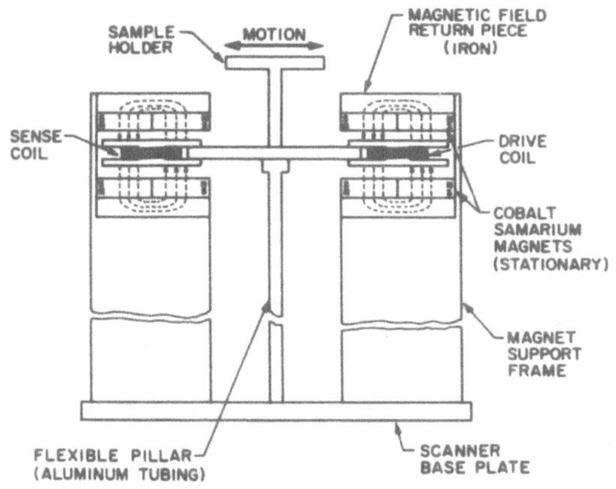

FIGURE 23. Schematic view of the mechanical scanning stage developed at Stanford for cryogenic microscopy (61).

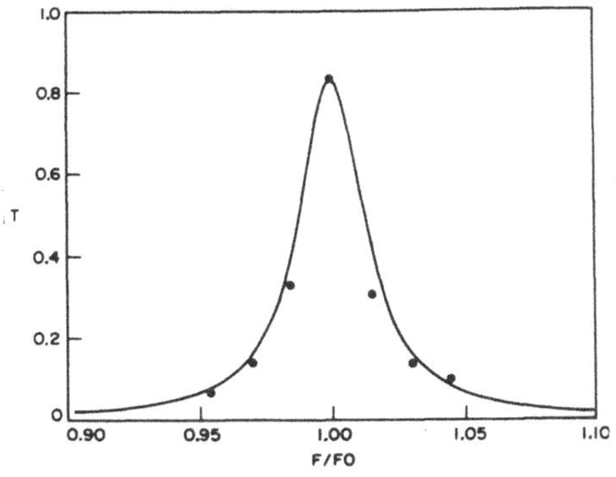

FIGURE 24. Transmission coefficient for a plane wave into liquid helium with a tungsten-carbon double quarter wave matching layer on sapphire (51).

A final improvement in receiver efficiency is made necessary as the relatively large electrodes at these frequencies have a very low radiation resistance, $R_0 \sim 0.1$ (64). This can easily be dominated by parasitic electrode and lead resistances. This can be circumvented by the use of superconducting cables and electrodes (63). The measured insertion loss of such a transducer is shown in Fig. 24.

In what would seem to be an unprecedented situation, all of the published results on the helium microscopy has come from Stanford. We will show some examples of this very beautiful work and mention some of the potential applications which have been explored by this group. Finally we will present some recent results obtained at Sherbrooke at much lower frequencies.

Rugar has given an impressive demonstration of resolution enhancement due to nonlinear effects (48). We saw in chapter two that strong nonlinear effects near the focal plane lead to an intense second harmonic signal with a spatial width a factor $\sqrt{2}$ finer than that associated with the fundamental. It is a remarkable fact that changes in intensity of the second harmonic modulate the fundamental intensity during the subsequent down conversion so that an improvement of resolution by $\sqrt{2}$ is obtained even when the fundamental is detected! A clear illustration is given in Figs. 25 et 26. The images in the linear (a) or near linear (b) regime are unable to detect the periodicity of the grating pattern. As the nonlinearity increases, (c) and (d), the pattern becomes clearly visible. It was confirmed in succeeding studies that the enhancement factor was at least 1.4.

Several impressive examples of helium imaging at 4.4 and 8 GHz have been given by this group (50, 63, 64, 66). A comparison of the acoustic and SEM image of an aluminium line in an integrated circuit is shown in Fig. 27 (50). Due to the high reflectivity the helium microscope is highly sensitive to topographical features and tilted surfaces, and we see that the two images give highly complementary information. The highest resolution image to date is shown in Fig. 28 (50). It is a 250 A resolution image of a Myxo bacteria taken at 8 GHz. Work is presently underway to extend this frequency range to 96 GHz (11).

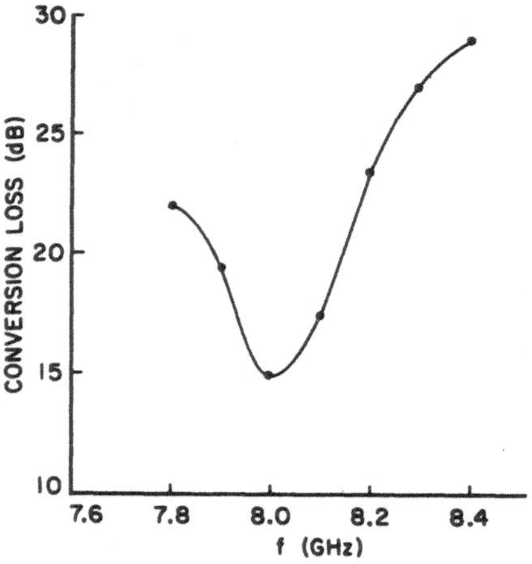

FIGURE 25. Conversion loss as a function of frequency for ZnO transducer with superconducting electrodes at 8 GHz (63).

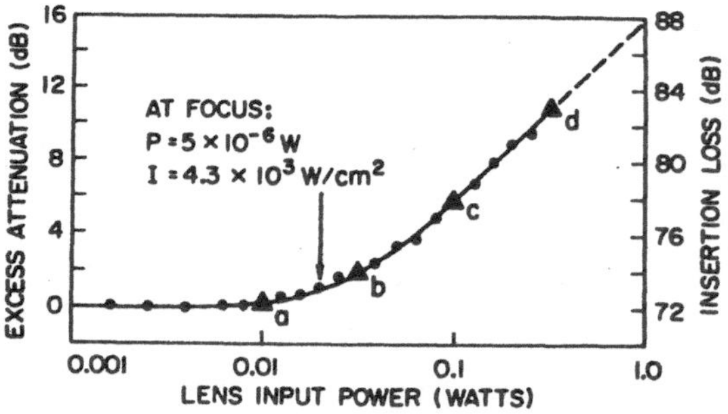

FIGURE 26. Insertion loss (excess attenuation) for a SAM lens in nitrogen at 2.0 GHz (48).

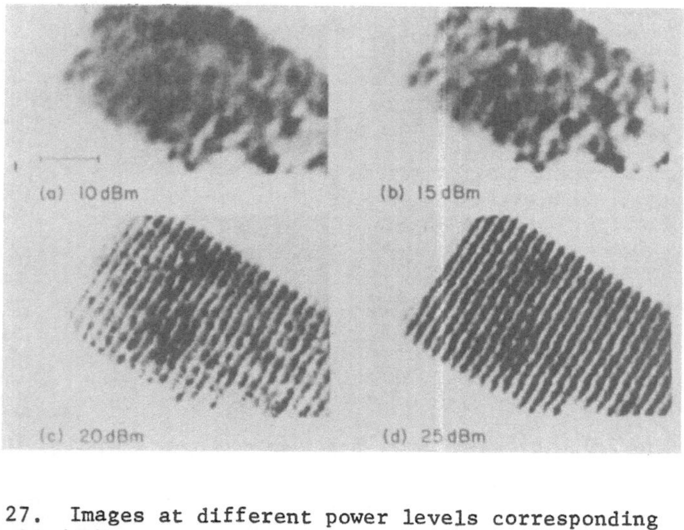

FIGURE 27. Images at different power levels corresponding to Fig. 26 (48).

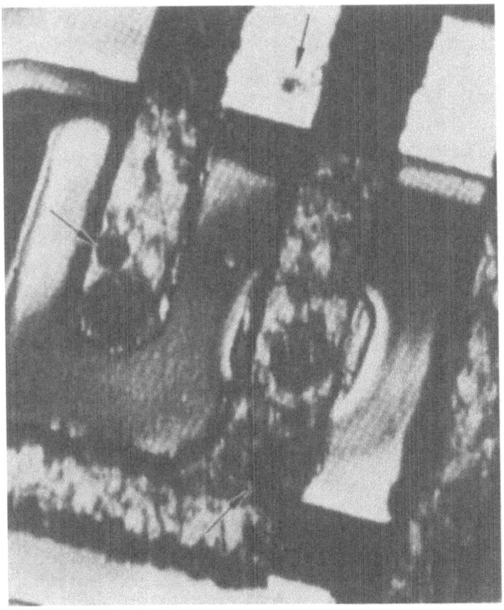

FIGURE 28. a) Cryogenic image of a 2 micron aluminium line at 4 GHz (66).

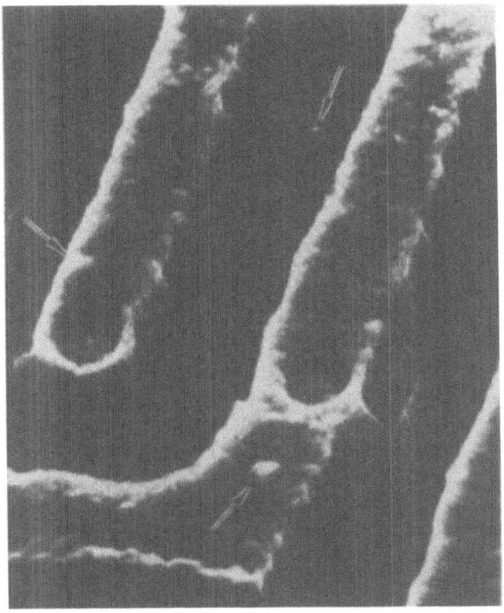

FIGURE 28. b) SEM image of the same region as Fig. 28.(a) (66).

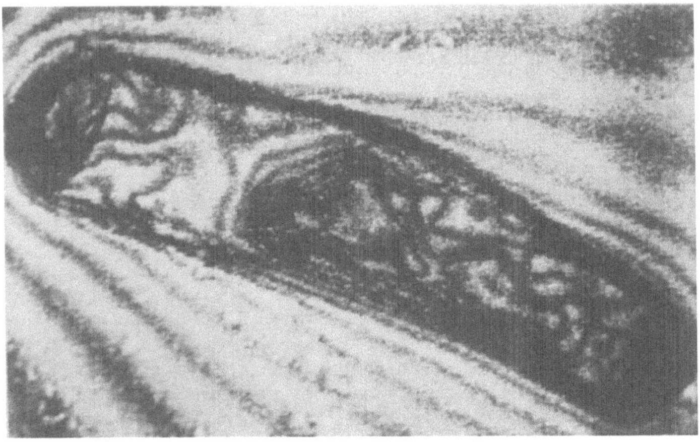

FIGURE 29. Cryogenic image of Myxo bacteria at 8 GHz with about 250 A resolution (50).

Since helium microscopy is a notoriously difficult experiment to do and the reflectance function is almost unity at normal incidence for nearly all solids, it is not surprising that the most common question asked is "What do you plan to use it for?" It would seem that the reply is obvious for the quest of the ultimate resolution, for it is surely worth building and characterising any microscope ultimately capable of a 1-3 nm resolution. But even at presently attainable frequencies there are useful applications. In the GHz range, Foster and Rugar (2-9) have shown that by imaging a heated film it may be possible to image subsurface structures and the spatial variation of the Kapiza thermal boundary resistance between solids and liquid helium (17). On a more fundamental level they have done a series of beautiful experiments on second harmonic detection at 2 GHz as a function of distance (18) to directly measure the anomalous dispersion in liquid helium (5). This experiment resolved the discrepancies between previous attempts to measure the dispersion parameters. Finally helium imaging would appear to be almost uniquely sensitive to surface topography (13). Due to the high acoustic mismatch at the interface the reflectance function is relatively insensitive to material properties. However any slightly tilted portion of the surface will reflect the beam out of the lens. Also with the short wavelengths involved interference effects should allow small height changes on the surface to be easily resolved. At 8 GHz for example, it is estimated that a 15 A step should be easily observable (13).

At Sherbrooke we have recently obtained results in helium microscopy in the MHz range above 1 K. Contrary to what one might at first think this is not trivial as it is not obvious how to construct the thick (5-10 micron) quarter wavelength matching layers and the attenuation in the liquid helium is quite high. Our first results in reflection for a Quartz lens at 30 MHz are shown in Fig. 29. Of more immediate interest is a new transmission configuration that we have developed (69) in which the sample is supported on the receiving transducer which is placed at the focal plane. The image (Fig. 30) corresponding to the reflection image is less clear but the resolution is essentially the same. The two approaches are essentially complementary as the reflection image is directly sensitive to the reflectance function and surface topography, while the transmission image is sensitive to the transmittance function and the attenuation in the specimen. The price to pay for this sensitivity to the specimen attenuation is of course the very small transmittance from the liquid helium to the sample. The technique should be particularly interesting for the study of the magnetic properties of superconductors and magnetic materials on a microscopic scale as well as numerous surface and interface effects at low temperatures.

The field of helium imaging is really only beginning. It is clear that the pioneering work on ultra high resolution helium imaging will be done at Stanford. Nevertheless groups lead by Wyatt (Exeter), Locatelli (Grenoble) and Ikushima (Tokyo) as well as our own at Sherbrooke are developing different approaches to the subject and these new results are already starting to be published. Apart from the instrumental development there is a lot of good low temperature physics to be done using the SAM as a new and highly sensitive local probe. Ultrasonics is known to be one of the most powerful experimental tools for studying superconductivity and liquid helium and the future will tell us whether or not the SAM will be able to open up new perspectives in these areas.

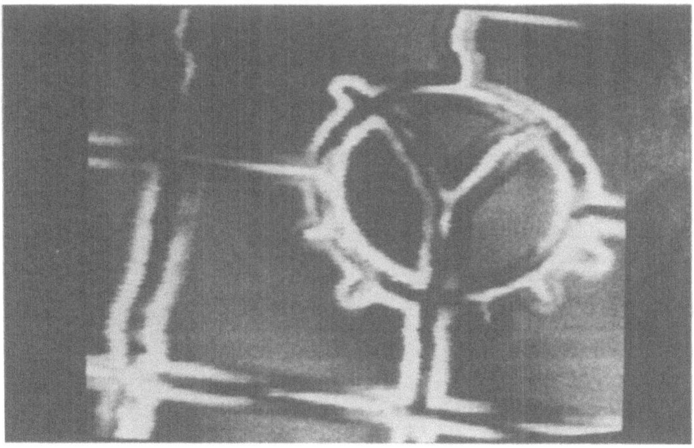

FIGURE 30. Reflection image at 1.9 K and 30 MHz of microscope grid (69).

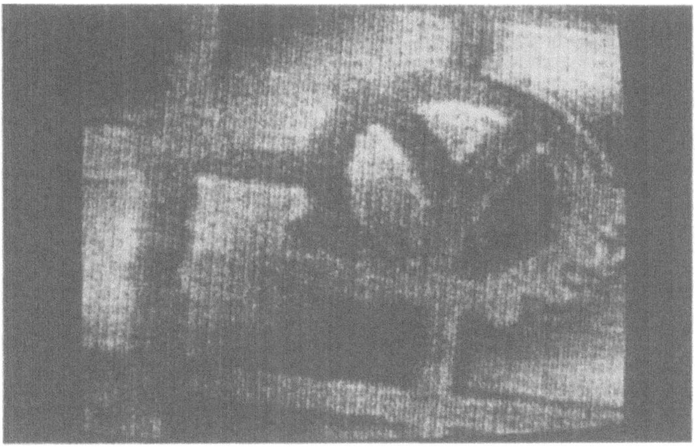

FIGURE 31. Transmission image corresponding to Fig. 30 (69).

ACKNOWLEDGEMENTS

The author wishes to thank past and present members of the acoustic microscopy group at Sherbrooke for their participation in this work, and in particular André Beauséjour and Luc Germain for the results presented here. This work was supported by the Natural Sciences and Engineering Research Council of Canada.

REFERENCES

1. van Zuylen J: Journal of Microscopy, 121 , 309-328, 1981.
2. Ruska E: The early development of electron lenses and Electron Microscopy, Heyden/Wisenschaftliche, 1980.
3. Cohen BG: Technical Bulletin no 1, Research Devices Inc. Berkeley Heights, New Jersey.
4. Stupian GW and Lering MS: Rev. Sci. Inst., 55, 92, 1984.
5. Dhamelincourt P: in "Microbeam Analysis 1982". Heinrich KHJ(ed): San Francisco Press, 1982, 261.
6. A commercial model of a scanning laser microscope is produced by Carl Zeiss, West Germany.
7. Robinson AL: Science 225, 1137, 1984.
8. Howells M, Kirz J, Sayre D and Schmahl G: Physics Today, 23, Aug 1985.
9. Binnig G, Rohrer H, Gerber CH and Weibel E: Phys. Rev. Lett. 50, 120, 1983.
10. Lemons RA and Quate CF: in Physical Acoustics, Mason WP and Thurston RN(ed): New York: Academic Press, 1979, XIV, 1.
11. Quate CF: IEEE Trans. SU-32, 132, 1985.
12. Lemons RA and Quate CF: Appl. Phys. Lett. 24, 163, 1974.
13. Lemons RA: PhD Thesis, Stanford University, 1975.
14. Attal J and Quate CF: Jour. Ac. Soc. Am. 59, 69, 1976.
15. Hadimioglu B and Quate CF: App. Phys. Lett. 43, 1006, 1983.
16. Nikoonahad M: Contemp. Phys. 25, 129, 1984.
17. Leitz, VG Semicon, Olympus, Hitachi and Honda Electronics.
18. Tsai CS, Wang SK and Lee CC: Appl. Phys. Lett. 31, 317, 1977.
19. Smith IR, Harvey RA and Fathers DJ: IEEE Trans. SU-32, 274, 1985.
20. Brekhovskikh LM: Waves in layered media, New York: Academic Press, 1960.
21. Ewing, Jardetsky and Press: Elastic Waves in Layered Media, McGraw Hill, New York, 1957.
22. Auld BA: Acoustic Fields and Waves in Solics, Vols 1 and 2, Wiley Interscience, 1973.
23. Atalar A, Quate CF and Wickramasinghe HK: App. Phys. Lett. 31, 791, 1977.
24. Diachok OI and Mayer WG: J. Acoust. Soc. Am. 47, 155, 1970.
25. Breazale MA, Adler L and Scott GW: J. App. Phys. 48, 530, 1977.
26. Neubauer WG: in Physical Acoustics, Mason WP and Thurston RN(ed): New York: Academic Press, 1973, X, 61.
27. Goos F and Hanchen H: Ann. Phys. (Leipzig) 1, 333, 1947.
28. Rollins FR: J. Acoust. Soc. Am. 44, 431, 1968.
29. Weaver JMR, Somekh MB, Briggs GAD, Peck SD and Ilett C, SU-32,302, 1985.
30. Weglein RD: IEEE Trans Sonics-ultrasonics SU-32, 225, 1985.
31. Atalar A: J. App. Phys. 49, 5130, 1978.
32. Atalar A: J. App. Phys. 50, 8237, 1979.
33. Bertoni HL: IEEE Trans Sonics-ultrasonics SU-31, 105, 1984.
34. Kushibiki J and Chubachi N: IEEE Trans Sonics-ultrasonics, SU-32, 189, 1985.
35. Fossum JO and Cheeke JDN: 8th Int. Conf. on Internal Friction and Ultrasonic Attenuation in Solids. J. de Phys (Colloques), Urbana, 1985.
36. Goodman JW: Introduction to Fourier Optics. New York: McGraw Hill, 1968.
37. Sheppard CJR and Wilson T: App. Phys. Lett. 38, 858, 1981.
38. Hildebrand JA, Liang K and Bennett SD: J. App. Phys. 54, 7016, 1983.

39. Liang KK, Kino GS and Khuri-Yakub BT: IEEE Trans Sonics-ultrasonics SU-32, 213, 1985.
40. Liang KK, Bennett SD, Khuri-Yakub BT and Kino GS: ibid, p. 266.
41. Cox IJ, Hamilton DK and Sheppard CJR: Appl. Phys. Lett. 41, 604, 1982.
42. Beyer RT and Letcher SV: Physical Ultrasonics, Academic Press, New York, 1969.
43. Blackstock DT: J. Ac. Soc. Am. 39, 1019, 1966.
44. Fabini E: Alta Frequenza 4, 539, 1935.
45. Ryan RP, Lutsch A and Beyer RT: J. Acoust. Soc. Am. 34, 31, 1962.
46. Fay R: J. Acoust. Soc. Am. 3, 222, 1931.
47. Tjotta J and Tjotta S: J. Acoust. Soc. Am. 69, 1644, 1981.
48. Rugar D: J. Appl. Phys. 56, 1338, 1984.
49. Kleinman DA, Ashkin A and Boyd GD: Phys. Rev. 145, 338, 1966.
50. Foster JS and Rugar D: IEEE Trans. SU-32, 139, 1985.
51. Heiserman JE: Physica 109B, 1978, 1982.
52. Wilks J: The properties of liquid and solid helium, Clarendon Press, Oxford, 1967.
53. Stanley HE: Introduction to phase transition and critical phenomena, Oxford University Press, New York, 1971.
54. Khalatnikov IM: Introduction to the theory of superfluidity, Benjamin WA: New York, 1965.
55. Maris HJ: Rev. Mod. Phys. 49, 341, 1977.
56. Abraham BM, Eckstein Y, Ketterson JB, Kuchnir M and Vignas J: Phys. Rev. 181, 347, 1969.
57. Jäckle J and Kehr KW: Phys. Rev. Lett. 24, 1101, 1970.
58. Cabot M and Putterman S: Phys. Lett. 83A, 91, 1981.
59. Roach PR, Ketterson JB and Kuchnir M: Phys. Rev. A5, 2205, 1972.
60. Dynes RC and Narayanamurti V: Phys. Rev. B12, 1720, 1975.
61. Heiserman J, Rugar D and Quate CF: J. Acoust. Soc. Am. 67, 1629, 1980.
62. Rugar D: PhD thesis Stanford, 1981.
63. Hadimioglu B and Foster JS: IEEE ultrasonics Symposium, Dallas, 1984.
64. Hadimioglu B and Foster JS: J. Appl. Phys. 56, 1976, 1984.
65. Weinreb S: IEEE Trans Microwave Theory Tech MTT-28, 1041, 1980.
66. Foster JS and Rugar D: Appl. Phys. Lett. 42, 869, 1983.
67. Anderson AC: in Nonequilibrium Superconductivity, Phonons and Kapitza Boundaries, NATO Advanced Study Institute, Gray K(ed): Plenum Press, 1981.
68. Rugar D and Foster JS: Phys. Rev. B30, 2595, 1984.
69. Germain L, Beauséjour A and Cheeke JDN: 1985 Ultrasonics Symposium, San Francisco, Oct 1985.

NOTE 2: Quantitative considerations of contrast in helium microscopy leading to the same general conclusions as those presented here have been given by S.Christie and A.F.G.Wyatt in Acoustical Imaging, E.A.Ash and C.R.Hill, Eds. New York, Plenum 1982.

DISCUSSION

Comment: Busse
From the V(z) curve I understand that there are two contrast mechanisms in acoustic microscopy; therefore the obtained image shows both topographic and elastic structures. Both contain information. Is it possible to eliminate the topographic influence by other ways than comparison with images taken with optical microscopy?

Reply: Cheeke
This is a good question and to my knowledge there is no imaging technique yet existing to do this. I think the profiling techniques developed by Kino and coworkers may provide a solution, although to what extent these work on inhomogeneous structures I don't know.

Comment: Socino
Have you also analyzed the V(z) behaviour of thin film overlays in the case of layer mode generation?

Reply: Cheeke
This question has been regarded by several workers and the agreement with the expected behaviour is good. The effect can then be used to determine thin film thickness.

COMPOSITE MATERIAL SYSTEMS FOR SURFACE WAVE INTEGRATED OPTIC BRAGG CELLS

J D Skinner

Abstract

Integrated optics has tremendous potential for utilisation of the enormous bandwidth made available by optical fibres. One of the principal problems in realising these devices is the lack of a suitable material as waveguide - none possesses all the desireable properties. One way of circumventing this problem is to use a combination of materials, whereby advantage is taken of the best properties of each. As an example, a surface wave Bragg Cell has been fabricated using a combination of an excellent but inactive waveguide, and a piezo-electric, but optically scattering zinc oxide overlay. The results are compared with a similar device in crystal quartz.

1.1 INTRODUCTION Optical fibre technology has now reached the point where it is possible to transmit information with bandwidths of several tens of gigahertz over distances of tens of kilometers without repetition. The attenuation of fibres is now as low as 0.01 dB/Km. However, with present integrated circuit technology, electronic handling of this huge bandwidth is clumsy, the optical signal having to be converted back into electrical energy at each signal processing/re-amplification stage. The present range of transducers (i.e. microphones, thermocouples) are unsuitable for interfacing with optical fibres, because an LED or laser is then needed to convert this energy into light, introducing another, unnecessary component. One way around this problem is to use all bulk optical sensors and signal processing, and there is considerable effort in making suitable optical transducers (ref. 1). Optical signal processing is not as straightforward, since many components (e.g. optical amplifiers, transistors) are still at a very early stage in research (ref. 2-3). Bulk optical systems are not likely to provide a long term answer, since optical components are heavy, intrinsically expensive and require vibration insensitive environments or good mechanical isolation. These problems could be overcome if optics were to move in the same direction as electronics moved in the 1960's, losing the third dimension, with all components being fabricated on a single substrate. In direct analogue to the name given to electronics of this sort, this field of research has been given the name 'Integrated Optics' (ref 4).

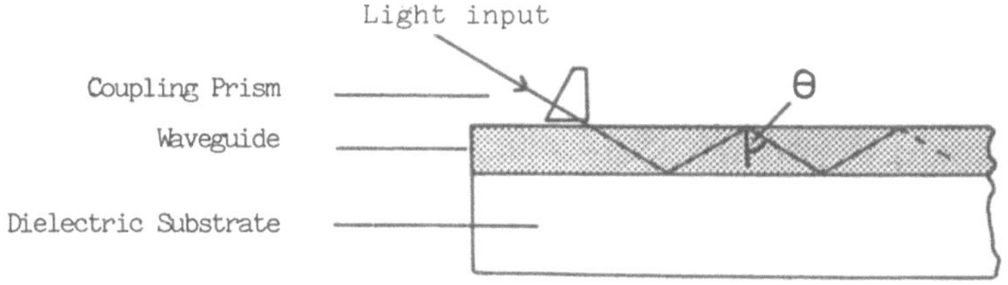

Figure 1 - cross section of integrated optic waveguide

2. OPTICAL WAVEGUIDES

Confinement of the optical energy to the surface of the substrate (a transparent dielectric) is achieved using total internal reflection (figure 1). A high index layer is formed on the surface of the substrate and light, once introduced into this layer, is totally reflected at each boundary. The high index layer may be made by several methods depending on the material. Diffusion of metallic ions (ion exchange) works well in sodalime glass, titanium metal indiffusion is equally effective in lithium niobate. Other methods are shown in table 1. In each case, the index change is confined to the top few microns of the surface.

METHOD	EXAMPLES
Diffusion	Ti in $LiNbO_3$ Ion exchange in glass
Deposition	Spinning plastics RF Sputtering, evaporation
Crystal growth	Molecular beam epitaxy Hydrothermal growth
Ion bombardment	Nuclear damage in crystal quartz Dopants in fused silica

TABLE 1 - Waveguide Fabrication Techniques

Light may be coupled into the waveguide through high index prisms which are pressed into contact with the surface of the waveguide (ref. 4). Evanescent field coupling takes place between prism and waveguide. In a simple model, light can be thought of as bouncing its way along the waveguide. Obviously, because of the small dimension of the guide, the light will interfere with itself since the beam width is of the same order as the depth of confinement. This interference is only constructive if:

$$khn_f \cos\theta + \phi_c + khn_f \cos\theta + \phi s = 2m\pi$$

upwards path + phase change at top surface + downwards path + phase change at substrate = $2m\pi$

where n_f = waveguide bulk index, h = thickness of guide

$$k = \frac{2k}{\lambda_0} \; ; \; \lambda_0 = \text{wavelength of light}$$

Consequently, light may only travel in one of a number of modes. Figure 2 shows the allowed propagation of a two mode waveguide, with electric field distribution created by each mode.

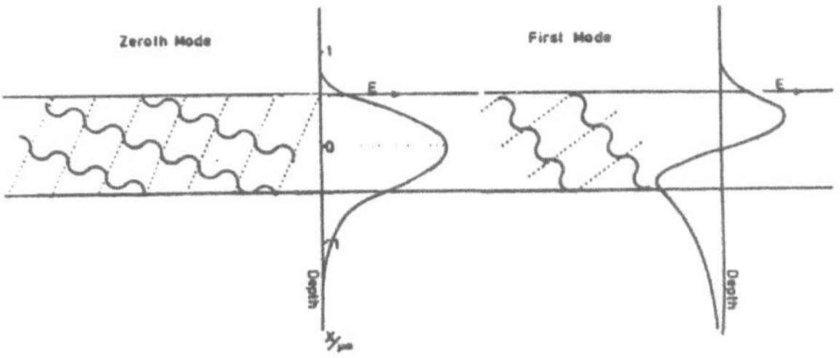

Figure 2 - electric field profiles in a two mode waveguide

In the equation above, it is assumed that the waveguide is of a constant index, This need not be so, a gradually varying index (e.g. a diffusion profile) will have the same effect: new light is gently deflected to follow a curved path at the base of the waveguide in the same way as a mirage is formed. Light may also be confined in the plane of the substrate surface by fabrication of high index channels, rather than over the entire surface. In the case of diffusion this may be done by photolithography masking all of the substrate from the diffusing material except for those tracks in which it is desired to guide light. Alternatively, a planar waveguide maybe etched (preferably by reactive ion etching) with the tracks photolithographically masked from the etchant. This provides an attractive simple way of attaching optical fibres to the waveguide, positioning blocks also being formed by the same etching procedure. It is possible to realise many components on an integrated optic waveguide. Passive lenses, mirrors, gratings and 3dB splitters have been demonstrated (ref.6), but in all cases, performance is much poorer than for state of the art bulk components. Active components such as the fabry perot interferometer (which may be the basis of a sensor, by altering the phase in one arm) optical switches, acousto-optic modulators using surface waves have also been demonstrated (ref. 7).

3. INTEGRATED/ACOUSTO-OPTIC SPECTRUM ANALYSER One complete signal processing device that has been proposed (ref. 7) is the integrated acousto-optic spectrum analyser, shown in figure 3. A solid state laser emits divergent waveguiding light, which is collimated by a lens. The laser is bonded onto the end of the waveguide, though in future this could be fabricated in GaAs substrate to give a totally integrated device.

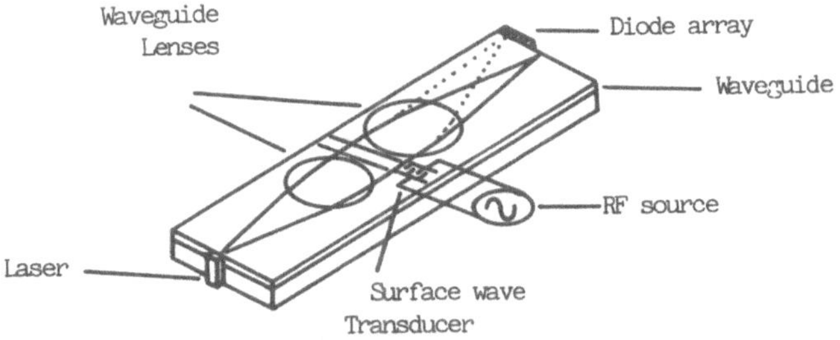

Figure 3 - the integrated acousto-optic spectrum analyser

The light is then incident upon a surface acoustic wave generated by an inter-digital transducer which is excited by an R.F. source. This assumes that there is some piezo-electric activity, either in the substrate, or locally, by depositing an overlay of some piezo-electric material, e.g. ZnO, in the region of the transducer. Pressure variations in the waveguide form a phase grating and light is diffracted at an angle θ' being dependent on the wavelength of the SAW (i.e. the frequency of the electrical R.F). The amplitude of the diffracted light is proportional to the RF power input for low power inputs. Acousto-optic interaction with the surface wave is an efficient one, since both the acoustic and optical energies are confined to the very surface of the structure. If more than one frequency is present in the R.F. source, then light is diffracted at a plurity of angles, the intensity of light of each angle being weighted by the magnitude of the corresponding frequency component in the R.F. signal. Finally, the angular distribution is converted to spatial distribution by a second waveguiding lens, and a linear diode array detects the light, each pixel of the array corresponding to a small range of frequencies. This extremely simple device thus acts as an R.F. spectrum analyser. The resolution of the device is at present limited by the lens performance and in plane scatter in the waveguide, which tends to blur the angular frequency spectrum. Present devices can resolve 100 frequencies over a bandwidth of (ref. 9).

4. WAVEGUIDE MATERIALS The properties that are essential and some that are desirable for a waveguide are given in table 2. Low optical scatter and loss are most important, and demand that the waveguide is totally homogeneous on a microscopic scale. This implies that the waveguide should be either totally amorphous or single crystal. The properties that make a film suitable for integrated optic application are quite different to those that are important for bulk applications (e.g. anti-reflective coatings). A slowly varying film thickness introduces distortions in the wavefront, and thus a distorted image. However, microscopic cracks, defects and pinholes perturb only a tiny proportion of the total incident light, and have little effect. An integrated optical film is not adversely affected by a slowly spatially varying substrate thickness, however, defects in the surface act as scattering centres, and any cracks penetrating deeper than the waveguide will completely destroy guiding action.

ESSENTIAL :	DESIREABLE :
- Low optical scatter	- Chemically stable
- Low loss	- Mechanically robust
- High Δn	- Long term stability
- Good uniformity	- No photorefractive effects
- Single mode	- Not birefringent
	- Low temperature coeffieients

TABLE 2 - Properties of waveguide materials

Single mode waveguides are essential for most components, since each mode propagates with a slightly different effective refractive index, and the characteristics of I-O components are a function of the R.I of the guiding light. No material possesses all the characteristics which are both essential and desirable for a waveguide. One solution to this problem is to use a combination of materials on the one integrated optical chip. A component on the chip that requires electro-optical activity can be made using an overlay of highly electro-optical material (e.g. liquid crystal), even though the bulk properties of this material make it wholly unsuitable as a waveguiding material in its own right. Very high index materials may be used as overlay materials to fabricate passive lenses.

In the following section, the fabrication of surface wave Bragg cell is described, using two types of quartz waveguides. The first is based on crystalline quartz which exhibits piezo-electric activity, but is unsuitable as a base for a waveguide, since the guides formed have high losses (ref. 10).

The second is based on amorphous quartz-fused silica. On this substrate it is possible to make low loss waveguides by ion implantation, but it is totally passive. In order to fabricate a Bragg cell it is necessary to use a piezo-electric layer over the surface wave transducer (figure 4). ZnO is a material ideally suited to this application, having large piezo-electric constants, and being readily deposited using a variety of techniques (ref. 11). Depsite having a high index, it is of little use as a waveguide, since the material grows in polycrystalline form, and attempts to guide light in ZnO films have shown highly scattery, attenuating propagation.

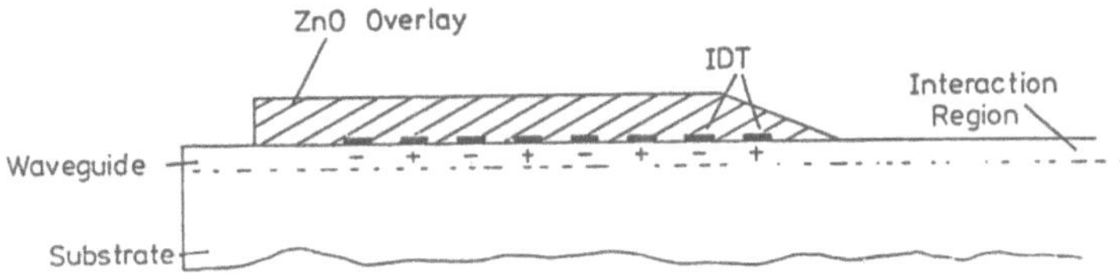

Figure 4 - ZnO on glass surface wave transducer

By using this composite, we may obtain the best of every world, good waveguide characteristics and high efficiency transduction of surface waves, a combination of properties possessed by no material alone.

5. SURFACE WAVE BRAGG CELLS ON α QUARTZ AND ZnO/SILICON There are two factors which determine the efficiency of acousto-optic interaction per watt of electrical input power. The overall efficiency is the product of the two. Dealing with each separately:

5.1 Efficiency of interaction per watt acoustic power The efficiency of a phase grating (figure 11) is given by

$$I/I_0 = \sin^2 \frac{\Delta\phi}{2}$$

where $\Delta\phi$ is the amplitude of phase change caused by the surface wave. This is related to the amplitude of the index change, Δn, caused by the surface wave.

$$\Delta\phi = \frac{2\pi}{\lambda_0} \frac{L}{\cos\theta'} \Delta n$$

λ_0 = wavelength of guided light, other symbols given in figure 5.

Figure 5 - detail of interaction

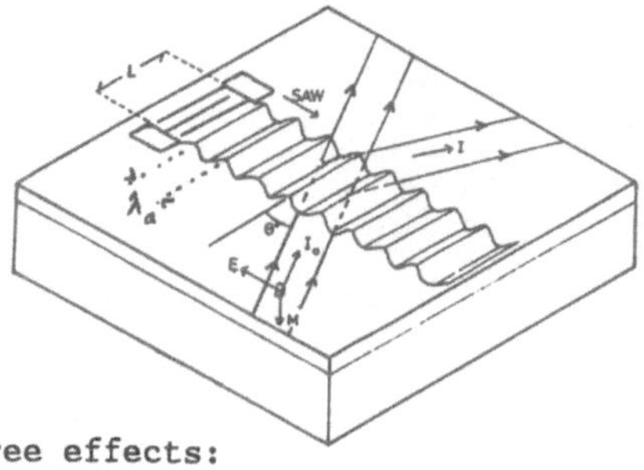

The index of change, Δn is a sum of three effects:

i. Acousto-optic : variation of index with pressure.

ii. Electro-optic : variation of index with electric field, which is carried along with the SAW in piezo-electric materials.

iii. Surface ripple : variation of guide thickness caused by surface ripple (small effect).

The relationship between acoustic power and Δn is a complex one, involving calculation per watt of acoustic power P_{ac} of the strain in the waveguide as a function of depth and multiplication by the relevant photo-electric and electro-optic sensors. A more convenient, if approximate method of evaluation of a substrate/ waveguide for acousto-optic interaction is to use the acousto-optic figure of merit derived from bulk Bragg cell work. In this case:

$$\Delta\phi \propto \frac{1}{\lambda_0}\sqrt{L.\ M_2.}\ P_{ac}$$ ref. 12

MATERIAL	M_2	K^2	ε_r	Acoustic Attenuation/dB/μm
Fused quartz	1	0	3.1	1.5
ZnO/fused quartz	N/A	0.015 (max)	8.58	N/A
Xcut α quartz	1.04	0.0022	4.5	0.45
Ycut $LiNbO_3$	4.6	0.05	50.2	0.19

TABLE 3 - Acousto-optic and electro-mechanical properties of materials

The figure of merit, M_2, for crystal and fused quarts (also for $LiNbO_3$ as a comparision) are shown in table 3. Figures for X cut crystal quartz and silica are very similar. Titanium indiffused $LiNbO_3$ has been the most widely used material because of its high electro-optic and piezo electric contents, but it also suffers from poor temperature stability, birefringence, mechanical fragility, poor stoichometry and a low waveguide index .

5.2 ELECTRO-MECHANICAL CONVERSION EFFICIENCY OF TRANSDUCER

The electro-mechanical conversion efficiency of the transducer can be found from the insertion loss of a transducer pair. The insertion loss is the total loss of power when an R.F. electrical signal is converted to SAW energy by the interdigital transducer, propagates across the acousto-optic interaction region, and then is reconverted to electrical power by a second transducer. It is composed of five parts:

1. Electrical mismatch
2. Tuning circuit parasytic resistance
3. Transducer finger parasytic resistance
4. Bidirectional loss (6dB)
5. Propagation loss (negligible)

The first three of these are associated with the transducer, and are dealt with in turn. The transducer may be tuned so that it is totally matched to the electrical supply at the centre frequency of operation. However, away from this frequency, the transducer is not matched. The degree of mismatch, and thus the bandwidth, is dependent on the electrical/piezo-electric properties of the transducer. Considering a simple equivalent circuit for the transducer (ref.13).

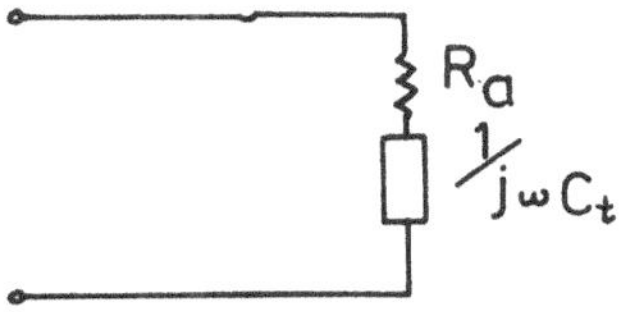

Ra is given by

$$\frac{4}{\pi^2} \cdot K^2 \cdot \frac{1}{\omega_0 C_t}$$

C_t is the capacitance of the transducer and is proportional to the dielectric constant of the substrate material, and the number of finger pairs N in the transducer.

K is the electromechanical coupling coefficient, a material constant, defined as the mechanical energy stored in the transducer divided by the total energy stored. For a non piezo-electric material it is zero.

For interdigital transducers or a piezo-electric substrate, K^2 is a slowly varying function of frequency, but for ZnO or silica the structure acts as an acoustic waveguide (ref. 14) and K^2 is strongly dependent h/λ_a, where h is the thickness of the ZnO layer.

For a large bandwidth device, the resistive part of this impedance should be as large as possible compared to the reactive part i.e. K^2 should be large, ε_r small. A second beneficial effect of large Ra is that the tuning circuit need be less complex, and this will reduce parasytic losses.

The mechanical bandwidth of the transducer is given by 2/N, so for a high bandwidth, a small number of finger pairs are used. Unfortunately, this has the effect of increasing the parasytic resistance of the transducer, since the fingers are electrically in parallel. This can be overcome by using one of the configurations in figure 6, where the finger periodicity varies along or across the transducer. The launch angle of the SAWs vary as a function of frequency in order to preserve the Bragg angle of interaction with the light.

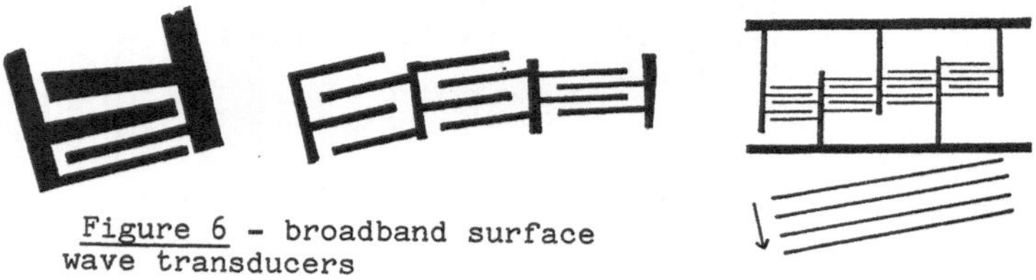

Figure 6 - broadband surface wave transducers

Values of K^2 and ε_r are given in table 3.

It can be seen that the crystal quartz has a much lower K^2 value than $LiNbO_3$ and thus will be suitable only for narrow bandwidth applications. A ZnO on silica configuration offers K^2 values approaching those of $LiNbO_3$.

6. ACOUSTO-OPTIC INTERACTION RESULTS For both quartz and ZnO/silica substrates, an interdigital transducer with 8 uniformly spaced finger pairs was used. They were photolithographically defined on the substrate in Al, and gold bonded to microstrip lines, on which tuning components were mounted.

The ZnO was deposited on the fused silica substrate by RF reactive magnetron sputtering in an oxygen atmosphere. The ZnO was sputtered over the entire substrate, and etched away except in the region of the IDTs.

Light was coupled to the substrate using prisms. The acousto-optic interaction was monitored using the experimental apparatus shown in figure 7. A charged coupled diode linear array is used for initial location of the interaction, a photodiode for accurate quantitative measurement of interaction efficiency.

Table 4 shows details of interaction efficiency bandwidth, dynamic range and sensitivity of the two devices. Both showed linear diffracted light intensity against power characteristic across their dynamic range.

The results for ZnO on quartz are similar to those on crystal quartz in all respects except that of the dynamic range it is lower. This is due to heating effects in the ZnO silica sandwich which cannot escape as easily as from an open structure. This causes a limitation in the maximum power input into the device, which limits the dynamic range of the device.

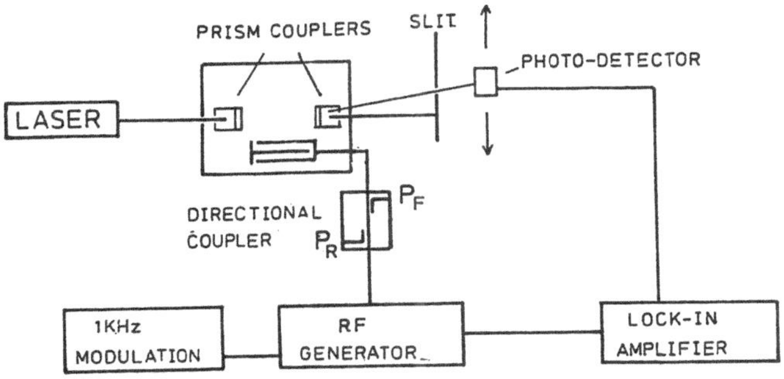

Figure 7 - measurement of interaction efficiency

Substrate	X quartz	Silica and ZnO
f_o	113MHz	57MHz
Insertion loss	13.5dB	13dB
Measured k^2	0.0017	0.0042
Sensitivity	2mW	2mW
Dynamic range	24.5dB	12dB
Diffraction efficiency/mW acoustic power	0.87%	0.66%
Bandwidth	16.5%	16%

TABLE 4 - Acousto optic Interaction Performance

7. CONCLUSIONS Is possible to enhance the piezo-electric properties of fused silica using a ZnO overlay. The performance of a surface wave Bragg cell made from this structure has a performance similar to that of a similar device made on piezo-electric crystal quartz, but has much better waveguiding properties.

A device built from a combination of materials may have a performance superior to that of any of the constituent materials.

References

1. D.E.N. Davis
Making measurements with light
Proc. IEE Vol. 129 pt. A No. 1 (1982).

2. E. Abraham, C. Secter, S.D. Smith
Optical Computers
Scientific American February 1983 p63.

3. P.K. Tien
Integrated Optics
Scientific American p.28-35 April 1974.

4. S. Miller
Integrated Optics - An Introduction
Bell Systems Technical Journal Vol. 48, 7 p.2059-69.

5. J.H. Harris
Beam coupling to films
J. Opt. Soc. Am. Vol. 60 No. 8 p.1007 (1970).

6. D.B. Anderson et al
Composition of optical waveguide lens technologies
IEEE J. Quantum Electronics QE13 No. 4 p.275-28 (1977).

7. J.S. Wilkinson
Integrated optical devices
Physics Technology (GB) Vol. 14 No. 4 p.190-3 (1983).

8. D.B. Anderson et al
Integrated approach to the Fourier Transformer
IEEE Journal of Quantum Electronics QE-13 No.4 p.268 (1979).

9. T.R. Joseph
Performance of the integrated optic spectrum analyser
1981 IEEE Ultrasonics Symposium P.721-26.

10. C. Pitt, J. Skinner, P. Townsend
Acousto-optic interaction on low broadband crystal quartz waveguides
Electronics Letters 13 January 1984 p. 14.

11. F.S. Hickernall
2-D processing for bulk and surface wave devices
1980 IEEE Ultrasonics Symposium p.785-94 (1980).

12. D.A. Pinnow
Electro-optical materials
Ed Pressley Laser Handbook CRC Press Inc. p.487 (1973).

13. W Smith et al
Analysis of interdigital surface wave transducers by use of an equivalent circuit model
IEEE MTT-17 No. 11 p.864 (1969).

14. G. Kino and R. Wagers
Theory of interdigital couplers as non-piezo-electric substrates
J. Appl. Physics Vo. 44 No. 4 p.1480 (1973).

DISCUSSION

Comment: Busse
I would like to simply add a comment to you paper.
You mentioned the thermal wave imaging capabilities of the laser probe. As this detector is sensitive to displacement and not to temperature itself, the structure that you find is actually a combination of a thermal wave image and an image of the local thermal expansion, where in addition to the thermal expansion coefficient the local mechanical constraint is involved. Therefore physical properties other than thermal (in thermal wave imaging: only thermal diffusion length) contribute, e.g. elastic, like in the acoustic microscopy. This aspect might be helpful in interpretation of structures found in laser probe scanning.

Comment: Cheeke
It would seem that the waveguide formed by the ion bombardment method would not be very sharply defined. Is this a problem in practice?

Reply: Skinner
The sharpness of the interface between two regions of different indexes is irrelevant. If the boundary is gradual, then light slowly bends (mirage effect), rather than being sharply reflected, but the angle of deviation is the same.

QUANTUM FIELD THEORY METHODS IN STUDIES OF ULTRASONIC WAVES PROPAGATION IN RANDOM MEDIA

E. SOCZKIEWICZ

1. INTRODUCTION

Several mathematical methods are used in the studies of ultrasonic wave propagation in random media [1-6]. The length of acoustic waves in comparison to the dimensions of inhomogeneities, as well as the intensity of fluctuations of medium refractive index for acoustic waves, determine the choice of the method actually used. When medium inhomogeneities are large in comparison with the wavelength, the method of optical geometry (or ray theory) can be used [1-3]. In the case of weakly inhomogeneous media the method of small perturbations [1-3] and the so called smooth perturbations method (Rytov's method) are employed in studies of acoustic wave propagation. If the wavelength is small in comparison with the correlation distance of medium refractive index fluctuations, so that the Fresnel approximation can be used, the method of parabolic equation is preferred [1,3,7]. Recently, progress in the theory of ultrasonic wave propagation in strongly inhomogeneous random media has been achieved by the application of a method elaborated in the quantum field theory [2,4,8].

Acoustical properties of random media are characterized not only by the mean value of the medium index of refraction for ultrasonic waves, but they depend also on the form of the autocorrelation function of refractive index fluctuations [1,3,9,10]. The mean and fluctuating parts of the square of the refractive index n^2 are usually separated as follows: $n^2(\vec{r}) = 1 + \epsilon(\vec{r})$, where $\epsilon(\vec{r})$ is a stochastic function of the radius vector r of a point (xyz). The autocorrelation function defined by:

$$\Psi_\epsilon(\vec{r}_1, \vec{r}_2) = \langle \epsilon(\vec{r}_1)\epsilon(\vec{r}_2) \rangle$$

is a measure of the size and shape of a typical inhomogeneity in a medium. It can have varius functional forms depending on the nature of the irregular medium. Some examples are: the Gaussian form:

$$\Psi_\epsilon(\vec{r}_1, \vec{r}_2) = \langle \epsilon^2 \rangle \, e^{-\alpha^2 (\vec{r}_1 - \vec{r}_2)^2} \tag{1}$$

and the exponential form:

$$\Psi_\epsilon(\vec{r}_1, \vec{r}_2) = \langle \epsilon^2 \rangle \, e^{-\alpha |\vec{r}_1 - \vec{r}_2|} \tag{2}$$

where α^{-1} denotes the correlation distance, i.e. the average distance over which the refractive index fluctuations remain correlated. One can prove that for the autocorrelation function given by (1), the refractive index changes in a continuous manner with the changes in coordinates, while for the exponential formula (2) it changes in a discontinuo fashion [1].

For inhomogeneities caused by turbulences in fluid media, von Karman proposed the correlation function in the form:

$$\Psi_\epsilon(\vec{r}_1, \vec{r}_2) = \frac{<\epsilon^2> \alpha^2 (\vec{r}_1 - \vec{r}_2)^a}{2^a(a-1)!} \, \mathrm{Kg}\,(\alpha |\vec{r}_1 - \vec{r}_2|) \tag{3}$$

where a is some number and $\mathrm{Kg}\,(\alpha |\vec{r}_1 - \vec{r}_2|)$ denotes the McDonald function [11]. Sometimes in studies of wave propagation in random media the so-called structure function:

$$D(\vec{r}_1, \vec{r}_2) = <[\epsilon(\vec{r}_1) - \epsilon(\vec{r}_2)]^2> \tag{4}$$

is preferable instead of the correlation function, since fluctuations of great space extension do not affect significantly the structure function in contrast to their influence on the correlation function, yet at the same time such kind of fluctuations are practically undistinguishable from small gradual changes of statistical characteristics of a medium, i.e., from the stochastic inhomogeneity of a medium. The following relations take place between structure and correlation functions:

$$\begin{aligned} D(\vec{r}_1, \vec{r}_2) &= 2[<\epsilon^2> - \Psi_\epsilon(\vec{r}_1, \vec{r}_2)] \\ D(\infty) &= 2 <\epsilon^2>. \end{aligned} \tag{5}$$

2. GREEN FUNCTION METHOD IN CALCULATION OF ACOUSTIC FIELD IN RANDOM MEDIA

The problems met in the theory of acoustic wave propagation in random media are similar to those in the quantum field theory, where the field equation is considered for a medium with external sources interacting with the field [4]. Application of the methods developed in the quantum field theory, i.e. Green functions and Feynman diagrams, to the study of wave propagation in random media make it possible to investigate wave propagation also in strongly inhomegeneous media [6] contrary to the methods mentioned in section 1 of this work. Feynman diagrams provide a clear interpretation of the obtained results in the case of multiple scattering of waves [12], [13]. The wave equation describing the propagation of acoustic waves in a random medium can be written in the following form:

$$\frac{1 + \epsilon(\vec{r})}{w_o^2} \frac{\partial^2 p}{\partial t^2} - \Delta p = 0, \tag{6}$$

where p denotes the propagating changes of pressure, w_o the velocity of acoustic waves in homogeneous medium and Δ is the operator of Laplace. The Helmholtz equation results from (6) for harmonic waves:

$$\Delta\Psi(\vec{r}) + k^2 \, (1 + \epsilon(\vec{r}))\, \Psi(\vec{r}) = 0, \tag{7}$$

where k is the so-called free-space wave number $k = {}^{\omega}/w_o$, denotes the circular frequency of acoustic waves. One can write the equation (7) in the integral form:

$$\Psi(\vec{r}) = \Psi_0(\vec{r}) - k^2 \int G_0(\vec{r}, \vec{r_1})\, \epsilon(\vec{r_1})\, \Psi(\vec{r_1})\, d^3\vec{r_1} \tag{8}$$

where $\Psi_0(\vec{r})$ satisfies the equation:

$$\Delta \Psi_0(\vec{r}) + k^2 \Psi_0(\vec{r}) = 0 \tag{9}$$

and $G_0(\vec{r}, \vec{r_1})$ is the free-space Green function:

$$G_0(\vec{r}, \vec{r_1}) = -\frac{e^{ik|\vec{r} - \vec{r_1}|}}{4\pi |\vec{r} - \vec{r_1}|} \tag{10}$$

obeying the equation:

$$\Delta G_0(\vec{r}, \vec{r_1}) + k^2 G_0(\vec{r}, \vec{r_1}) = \delta(\vec{r} - \vec{r_1}), \tag{11}$$

where $\delta(\vec{r} - \vec{r_1})$ denotes the Dirac distribution. Formula (10) has the obvious interpretation of an acoustic field at point $\vec{r}$ n a homogeneous medium, that is generated by a point source point $\vec{r_1}$.

The acoustic field in a random medium generated by a source at $\vec{r_0}$ obeys the equation:

$$\Delta G(\vec{r}, \vec{r_0}) + k^2 \left(1 + \epsilon(\vec{r})\right) G(\vec{r}, \vec{r_0}) = \delta(\vec{r} - \vec{r_0}) \tag{12}$$

or in the integral form:

$$G(\vec{r}, \vec{r_0}) = G_0(\vec{r}, \vec{r_0}) - k^2 \int G_0(\vec{r}, \vec{r_1})\, \epsilon(\vec{r_1})\, G(\vec{r_1}, \vec{r_0})\, d^3\vec{r_1} \tag{13}$$

where the integration is over the space including irregularities. Solving (13) by iteration, one obtains the following series

$$\begin{aligned} G(\vec{r}, \vec{r_0}) = {} & G_0(\vec{r}, \vec{r_0}) - k^2 \int G_0(\vec{r}, \vec{r_1})\, \epsilon(\vec{r_1})\, G_0(\vec{r_1}, \vec{r_0})\, d^3\vec{r_1} + \\ & + (-k^2)^2 \iint G_0(\vec{r}, \vec{r_1}) \epsilon(\vec{r_1}) G_0(\vec{r_1}, \vec{r_2}) \epsilon(\vec{r_2}) G_0(\vec{r_2}, \vec{r_0}) d^3\vec{r_1}\, d^3\vec{r_2} + \ldots \end{aligned} \tag{14}$$

3. FEYNMAN'S DIAGRAMMATIC REPRESENTATION OF THE MEAN ACOUSTIC FIELD IN A RANDOM MEDIUM

In the case of the Gaussian form of the inhomogeneities autocorrelation function, it results from (14) the following series for the mean acoustic field in a random medium:

$$\begin{aligned} \langle G(\vec{r}, \vec{r_o}) \rangle = {} & G_o(\vec{r}, \vec{r_o}) + k^4 \iint G_0(\vec{r}, \vec{r_1})\, G_0(\vec{r_1}, \vec{r_2}) G_0(\vec{r_2}, \vec{r_0}) \langle \epsilon(\vec{r_1}) \epsilon(\vec{r_2}) \rangle d^3\vec{r_1}\, d^3\vec{r_2} + \\ & + k^8 \iiint G_0(\vec{r}, \vec{r_1}) G_0(\vec{r_1}, \vec{r_2}) G_0(\vec{r_2}, \vec{r_3}) G_0(\vec{r_3}, \vec{r_4}) G_0(\vec{r_4}, \vec{r_0}) \Big[\langle \epsilon(\vec{r_1}) \epsilon(\vec{r_2}) \rangle \langle \epsilon(\vec{r_3}) \epsilon(\vec{r_4}) \rangle + \\ & + \langle \epsilon(\vec{r_1}) \epsilon(\vec{r_3}) \rangle \langle \epsilon(\vec{r_2}) \epsilon(\vec{r_4}) \rangle + \langle \epsilon(\vec{r_1}) \epsilon(\vec{r_4}) \rangle \langle \epsilon(\vec{r_2}) \epsilon(\vec{r_3}) \rangle \Big] d^3\vec{r_1}\, d^3\vec{r_2}\, d^3\vec{r_3}\, d^3\vec{r_4} \end{aligned} \tag{15}$$

Employing the Feynman diagrams, the structure of the above series is represented as follows:

(16)

where the heavy solid line denotes the mean Green function, the light solid lines correspond to free Green functions; the lines denote correlation functions and the dot denotes k^2. The particular terms of series (16) represent the wave scattering of various orders as, for instance, in the following examples:

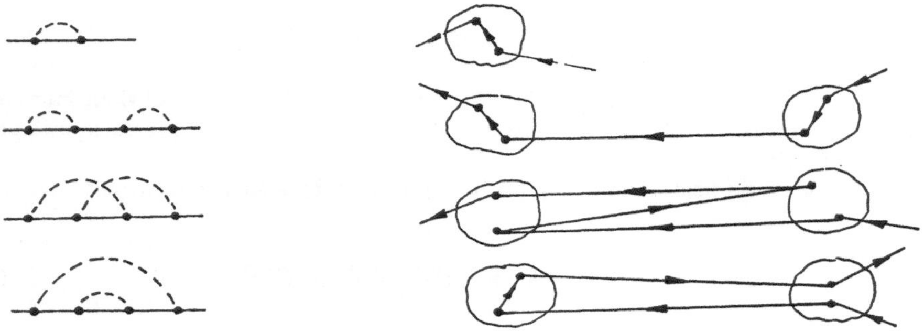

Collecting the strongly connected diagrams together in (16) (i.e.those ones that cannot be split into two distinct diagrams by division of one $G_0(\vec{r}_1, \vec{r}_2)$ line only) and introducing the so-called mass operator [8,14]:

$$Q(\vec{r}', \vec{r}'') = k^4\ G_0(\vec{r}', \vec{r}'')<\epsilon(\vec{r}')\epsilon(\vec{r}'')> + h^8 \iint G_0(\vec{r}', \vec{r}_1)G_0(\vec{r}_1, \vec{r}_2)G_0(\vec{r}_2, \vec{r}'')\cdot$$
$$\cdot<\epsilon(\vec{r}')\epsilon(\vec{r}_2)><\epsilon(\vec{r}_1)\epsilon(\vec{r}'')>d^3\vec{r}_1\, d^3\vec{r}_2 + k^8 \iint G_0(\vec{r}', \vec{r}_1)G_0(\vec{r}_1, \vec{r}_2)G_0(\vec{r}_2, \vec{r}'')\cdot \tag{17}$$
$$\cdot<\epsilon(\vec{r}')\epsilon(\vec{r}'')><\epsilon(\vec{r}_2)>d^3\vec{r}_1\, d^3\vec{r}_2 + \ldots$$

or in the diagrammatic form:

(18)

the mean Green function may be expressed in the form of the Dyson integral equation:

$$<G(\vec{r}, \vec{r}_0)> = G_0(\vec{r}, \vec{r}_0) + \iint G_0(\vec{r}, \vec{r}_2)Q(\vec{r}_1, \vec{r}_2)< G(\vec{r}_2, \vec{r}_0)>d^3\vec{r}_1\, d^3\vec{r}_2 \tag{19}$$

that has the following diagrammatic representation:

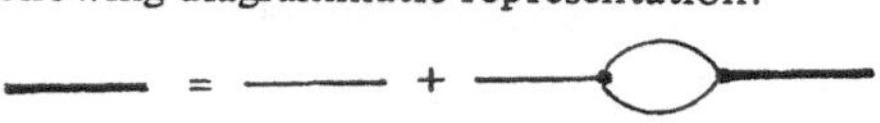

(20)

4. STOCHASTIC HOMOGENEOUS AND ISOTROPIC MEDIA

In the case of stochastic homogeneous media, the statistical characteristics: mean square fluctuations of the medium refractive index for acoustic waves and autocorrelation function of those fluctuations, do not change with the translation of the coordinate system. Green functions and the mass operator are functions of the differences of coordinates only and the Dyson equation

$$<G(\vec{r}-\vec{r}_0)> = G_0(\vec{r}-\vec{r}_0) + \iint G_0(\vec{r}-\vec{r}_1)Q(\vec{r}_1-\vec{r}_2)<G(\vec{r}_2-\vec{r}_0)> d^3\vec{r}_1 d^3\vec{r}_2 \tag{21}$$

can be solved effectively by means of the Fourier transformations:

$$\begin{aligned} <G(\vec{r}-\vec{r}_0)> &= \int <g(\vec{\varkappa})> e^{i\vec{\varkappa}(\vec{r}-\vec{r}_0)} d^3\vec{\varkappa}, \\ G_0(\vec{r}-\vec{r}_0) &= \int g_0(\vec{\varkappa}) e^{i\vec{\varkappa}(\vec{r}-\vec{r}_0)} d^3\vec{\varkappa}, \\ Q(\vec{r}_1-\vec{r}_2) &= \int g(\vec{\varkappa}) e^{i\vec{\varkappa}(\vec{r}_1-\vec{r}_2)} d^3\vec{\varkappa}. \end{aligned} \tag{22}$$

From (11), (21), (22), one obtains:

$$g_0(\vec{\varkappa}) = \frac{1}{(2\pi)^3(k^2-\varkappa^2+i0)} \tag{23}$$

$$g(\vec{\varkappa}) = \frac{1}{(2\pi)^3\,[k^2-\varkappa^2-(2\pi)^3 q(\vec{\varkappa})+i0]} \tag{24}$$

where i0 denotes some infinitesimal imaginary number. The inverse Fourier transformation gives:

$$<G(\vec{r}-\vec{r}_0)> = \frac{1}{(2\pi)^3}\int \frac{e^{i\vec{\varkappa}(\vec{r}-\vec{r}_0)}}{k^2-\varkappa^2-\int Q(r)e^{i\vec{\varkappa}\vec{r}}\,d^3\vec{r}+i0}\,d^3\varkappa. \tag{25}$$

If one confines oneself to isotropic random media, after introducing the spherical coordinate system and integrating over the angles, Eq. (25) takes the form:

$$<G(R)> = \frac{1}{4\pi^2 iR}\int_{-\infty}^{\infty} \frac{e^{i\varkappa R}}{k^2-\varkappa^2+\frac{4\pi}{\varkappa}\int_0^{\infty} Q(r)\sin(\varkappa r)r\,dr}\,d\varkappa, \tag{26}$$

where $R=|\vec{r}-\vec{r}_0|$. In order to calculate $<G(R)>$ and the so-called effective wave number, the form of the mass operator should be specified. To first approximation, as one can see from the series (17), the mass operator is:

$$Q(r) = k^4 G_0(r)\Psi_\epsilon(r), \tag{27}$$

where $\Psi_\epsilon(r) = <\epsilon(0)\epsilon(r)>$. The approximation (27) corresponds to the summation of the following series of Feynman diagrams from those of (16):

From (26) the following formula for the mean Green function results:

$$<G_1(R)> = \frac{1}{4\pi^2 iR}\int_{-\infty}^{\infty} \frac{e^{i\varkappa R}\varkappa}{k^2-\varkappa^2+\frac{k^4}{\varkappa}\int_0^{\infty}\Psi_\epsilon(r)e^{ikr}\sin(\varkappa r)dr}\, d\varkappa \approx -\frac{e^{i\varkappa_1 R}}{4\pi R} \tag{28}$$

where $\varkappa_1$ denotes one of the poles of the integrand function that has the smallest imaginary part. The poles of the integrand function in (28) satisfy the equation:

$$k^2-\varkappa^2+\frac{k^4}{\varkappa}\int_0^{\infty}\Psi_\epsilon(r)e^{ikr}\sin(\varkappa r)dr = 0. \tag{29}$$

From this it results to the first approximation [13]:

$$\varkappa_1 = k_{eff} = k_1+ik_2 = k\left[1+\frac{k}{4}\int_0^{\infty}\sin(2kr)\Psi_\epsilon(r)dr + \frac{ik}{2}\int_0^{\infty}\sin^2(kr)\Psi_\epsilon(r)dr\right], \tag{30}$$

where k_{eff} denotes the effective complex wave number of the mean acoustic field. Introducing the spectral function:

$$\Psi_\epsilon(r) = \frac{4\pi}{r}\int_0^{\infty}\sin(kr)\,\Phi_\epsilon(k)\,kdk, \tag{31}$$

Eq. (30), after integration, takes the form:

$$k_{eff} = k\left[1+\frac{\pi k}{4}\int_0^{\infty}\varkappa\ln\left(\frac{2k+\varkappa}{2k-\varkappa}\right)^2\Phi_\epsilon(\varkappa)\,d\varkappa + \frac{i\pi^2 k}{2}\int_0^{\infty}\Phi_\epsilon(\varkappa)\varkappa\, d\varkappa\right] \tag{32}$$

and the effective complex refractive index may be calculated from the expression:

$$\frac{k_{eff}}{k} = n_{eff} = n_1+n_2, \tag{33}$$

where:

$$n_1 = 1+\frac{\pi k}{4}\int_0^{\infty}\ln\left(\frac{2k+\varkappa}{2k-\varkappa}\right)^2\Phi_\epsilon(\varkappa)\varkappa\, d\varkappa, \tag{34}$$

$$n_2 = \frac{\pi^2 k}{2}\int_0^{2k}\Phi_\epsilon(\varkappa)\varkappa\, d\varkappa .$$

5. CALCULATION OF THE REFRACTIVE INDEX OF THE MEAN ACOUSTIC FIELD

It is easy to calculate the complex index of refraction from (34) in the case $k \gg \varkappa_1$, i.e. for the coarse grained random media. Using the approximation: $\frac{1}{4}\ln\left(\frac{2k+\varkappa}{2k-\varkappa}\right)^2 \approx \frac{\varkappa}{2k}$

one can write formulae (34) in the form [13]:

$$n_1 = 1 + \frac{\pi}{2}\int_0^\infty \varkappa^2 \Phi_\epsilon(\varkappa)\, d\varkappa ,$$
$$n_2 = \frac{\pi^2 k}{2}\int_0^\infty \varkappa \Phi_\epsilon(\varkappa)\, d\varkappa . \tag{35}$$

The spectral function of the autocorrelation function is given by the expression:

$$\Phi_\epsilon(\varkappa) = \frac{1}{2\pi^2 \varkappa}\int_0^\infty \sin(\varkappa r)\, \Psi_\epsilon(r)\, r\, dr. \tag{36}$$

which in case of Gaussian autocorrelation function of medium irregularities (1), becomes

$$\Phi_\epsilon(\varkappa) = \frac{<\epsilon^2>}{8\pi^{3/2}\alpha^3}\, e^{-\frac{\varkappa^2}{4\alpha^2}}, \tag{37}$$

while in case of the exponential form (2), it becomes:

$$\Phi_\epsilon(\varkappa) = \frac{<\epsilon^2>\alpha}{\pi(\alpha^2 + \varkappa^2)} . \tag{38}$$

The following expressions for the real part of the medium refractive index for acoustical waves result from (35):

$$n_1 = 1 + \frac{<\epsilon^2>}{8} \tag{39}$$

for the Gaussian form of the autocorrelation function and:

$$n_1 = 1 + \frac{\pi <\epsilon^2>}{8} \tag{40}$$

for the exponential one. For the imaginary part of refractive index we obtain:

$$n_2 = \frac{\sqrt{\pi}<\epsilon^2> k}{8\alpha} \tag{41}$$

in the case of spectral function given by (37) and

$$n_2 = \frac{\pi <\epsilon^2> k}{4\alpha} \tag{42}$$

for the spectral function given by Eq. (38). Between the complex part of the refractive index and the attenuation coefficient of the mean acoustic field the following relation holds: $\gamma = 2kn_2$. For the Gaussian form of $\Psi_\epsilon(r)$ there is:

$$\gamma_1 = \frac{\sqrt{\pi} <\epsilon^2> k^2}{4\alpha} \tag{43}$$

and:

$$\gamma_2 = \frac{\pi <\epsilon^2> k^2}{2\alpha} \tag{44}$$

for the exponential one.

In the work [8] the author has calculated attenuation coefficient of the mean acoustic field using the method of parabolic equation. The following expressions have been obtained for γ_1 and γ_2 respectively:

$$\gamma_1 = \frac{\sqrt{\pi} <\epsilon^2> k^2}{4\alpha} \tag{45}$$

$$\gamma_2 = \frac{<\epsilon^2> k^2}{2\alpha} \tag{46}$$

and, as one can see, the expressions obtained by both the methods in the case of the Gaussian autocorrelation function are close one to the other.In the case of fine grained inhomogeneities, i.e. for $\frac{k}{\alpha} \ll 1$, we have obtained the following expressions for attenuation coefficient of the mean acoustic field, from (35):

$$\gamma_1 = \frac{\sqrt{\pi} <\epsilon^2> k^4}{4\alpha^3}, \tag{47}$$

$$\gamma_2 = 2\pi <\epsilon^2> \frac{k^4}{\alpha^3} \tag{48}$$

for the Gaussian (1) and exponential (2) forms of the autocorrelation function of refractive index fluctuations, respectively.

6. EVALUATION OF THE EFFECT PRODUCED BY HIGHER TERMS IN Q(r) ON THE VALUE OF THE COMPLEX WAVE NUMBER OF THE MEAN ACOUSTIC FIELD

In the second approximation of the mass operator Rytov [13] has assumed the following expression:

$$Q(|r' - r''|) = k^4 <G_1(|r' - r''|)> \Psi_\epsilon(|r' - r''|) \tag{49}$$

that corresponds to the following series of the Feynman diagrams:

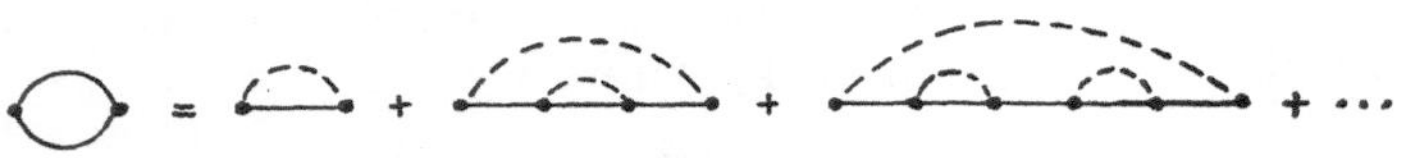

(50)

For formulae (26) and (49) it follows that the poles of the mean Green function in the second" approximation $<G_2(r)>$ should be estimated from the equation analogous to that of (29):

$$\varkappa_2^2 = k^2 + \frac{k^4}{\varkappa_2}\int_0^\infty \Psi_\epsilon(r) e^{ik_{eff}r} \sin(\varkappa_2 r) dr \tag{51}$$

or approximately:

$$\varkappa_2^2 = k^2 + \frac{k^4}{k_{eff}}\int_0^\infty \Psi_\epsilon(r) e^{ik_{eff}r} \sin(k_{eff}r) dr. \tag{52}$$

Putting $k_{eff} = k + \Delta k$, where $\Delta k \ll 1$, i.e. $\frac{1}{k_{eff}} = \frac{1}{k} - \frac{\Delta k}{k^2}$, from (29) and (52) it results:

$$\varkappa_2^2 - \varkappa_1^2 = \frac{k^3}{2i}\int_0^\infty \Psi_\epsilon(r)\left(e^{2ik_{eff}r} - e^{2ikr}\right) dr - \frac{k^2\Delta k}{2i}\int_0^\infty \Psi_\epsilon(r)\left(e^{2ik_{eff}r} - 1\right) dr. \tag{53}$$

From the above equation we have calculated the differences $\varkappa_2^2 - \varkappa_1^2$ for the two forms (1), (2) of the autocorrelation function $\Psi_\epsilon(r)$ of the medium inhomogeneities. In the case of the Gaussian form of $\Psi_\epsilon(r)$ the following equation has been obtained:

$$\varkappa_2^2 - \varkappa_1^2 = \frac{k^3}{2i}\left[\frac{<\epsilon^2>\sqrt{\pi}}{2\alpha}\left(e^{-\frac{k_{eff}^2}{\alpha^2}} - e^{-\frac{k^2}{\alpha^2}}\right) + \frac{i<\epsilon^2>}{\alpha^2}(k_{eff} - k)\right] + $$
$$- \frac{k^2\Delta k}{2i}\left[\frac{<\epsilon^2>\sqrt{\pi}}{2\alpha}\left(e^{-\frac{k_{eff}}{\alpha^2}} - 1\right) + \frac{i<\epsilon^2>k_{eff}}{\alpha^2}\right], \tag{54}$$

while for the exponential form (2) from (53) it results:

$$\varkappa_2^2 - \varkappa_1^2 = \Delta k\left[\frac{k^3<\epsilon^2>}{(\alpha - 2ik)(\alpha - 2ik_{eff})} - \frac{k^2 k_{eff}<\epsilon^2>}{\alpha(\alpha - 2ik)}\right]. \tag{55}$$

Confining oneself to the linear terms in Δk and assuming $\varkappa_1 + \varkappa_2 \approx 2k$ from (54) one has:

$$\left|\frac{\varkappa_2 - \varkappa_1}{\Delta k}\right| = k^3<\epsilon^2>\frac{\sqrt{\pi}}{8\alpha^3} \tag{56}$$

and from (55):

$$\left|\frac{\varkappa_2 - \varkappa_1}{\Delta k}\right| = \frac{k^3<\epsilon^2>}{\alpha(\alpha^2 + 4k^2)} \quad . \tag{57}$$

7. CONCLUSIONS

The following conclusions may be drawn from the above considerations.

a) The coefficient of attenuation of the mean acoustic field in random media depends on the form of the autocorrelation function $\Psi_\epsilon(r)$ of the medium inhomogeneities. For the Gaussian shape of $\Psi_\epsilon(r)$ this coefficient is smaller than for the exponential one given by the formula (2).

b) The effective velocity of the acoustic waves in random media, depending on the form of $\Psi_\epsilon(r)$, is greater in the case of tha Gaussian form of $\Psi_\epsilon(r)$ than for the exponential one given by (2).

c) The correction to the complex effective wave number of acoustic waves resulting from the second approximation of the mass operator Q(r) is smaller for the Gaussian form of $\Psi_\epsilon(r)$ than for the exponential one.

d) The values of the attenuation coefficient of the mean acoustic field, calculated by the approximation (27) and the Gaussian autocorrelation function, are very close to those obtained by the method of parabolic equation if $k/\alpha \gg 1$ and by means of the perturbation method of Chernov if $k/\alpha \ll 1$.

REFERENCES

1. Chernov L.A.: Wave Propagation in Random Media. Moscow: Nauka, 2975.
2. Sobczyk K.: Stochastic Waves. Warsaw: PWN, 1978.
3. Uscinski B.J.: The Elements of Wave Propagation in a Random Media. New York: McGraw-Hill, 1977.
4. Tatarsky W.I.: Wave Propagation in a Turbulent Atmosphere. Moscow: Nauka, 1967.
5. Ishimaru A.: Theory and Application of Wave Propagation and Scattering in Random Media. Proc' IEEE **65**, 1030 (1977).
6. Barananenkov Y.N., Kravtsov Y.A., Ryton S.M., Tatarsky W.I.: State of the Theory of Wave Propagation in a Random Medium. Usp. Fiz. Nauk **102**, 1 (1970).
7. Candel S.: Numerical Solution of Wave Scattering Problems in the Parabolic Approximation. J. Fluid Mech. **90**, 465 (1979).
8. Soczkiewicz E.: Propagation of Acoustic Waves in Random Media, the Parabolic Equation Method. Akustyka Molekularna i Kwantowa 5, 139. Warsaw: IPPT-PAN, 1984.
9. Soczkiewicz E.: Propagation of Ultrasonic Waves in Random Media. Proc. of the XXXth Open Seminar on Acoustics, p. 249. Gdansk: Komitet Akustyki PAN, 1983.
10. Soczkiewicz E.: Propagation of Acoustic Waves and Stochastic Properties of Random Media. Proc. of the XXVIIth Open Seminar on Acoustics, p. 87. Gliwice: Polish Acoustical Society, 1981.
11. Janke E., Emde F., Losch F.: Special Functions. Moscow: Nauka, 1977.
12. Soczkiewicz E.: Application of the Methods of Quantum Field Theory to Investigation of Acoustic Waves Propragation in Random Media. Akustyka Melokularna i Kwantowa 3, 25. Warsaw IPPT-PAN, 1982.
13. Rytov S.M., Kravtosov Y.A., Tatarsky W.I.: Introduction to Statistical Radio-Physics. Moscow: Nauka, 1978.
14. Frish U.: Wave Propagation in Random Media. Probabilistic Methods in Applied Mathematics, Bharucha-Reid AT ed. New York: Academic Press 1968.

DISCUSSION

Comment: Chivers
What are the assumptions of your analysis?

Reply: Soczkiewicz
About the random inhomogeneous medium, the following assumptions have been made:
1. The statistical characteristics of the medium (i.e. the mean square fluctuation of the medium refractive index and the autocorrelation function of the medium inhomogeneities) do not change with time and with translation of the coordinate system.
2. The medium is isotropic and infinite.
The source of the ultrasonic waves is a point one and the study was performed in scalar approximation.
The paper is related to two topics: first, to calculate the effective complex wave number of the mean acoustic field and afterwards to determine the effective attenuation coefficient and the mean refractive index for ultrasonic waves and, secondly, to determine the dependence of γ and n on the form of the medium inhomogeneities autocorrelation function and the approximation assumed for the so called mass operator.

CHARACTERISATION OF MICROSTRUCTURES USING ULTRASONICS

C M SAYERS

1. INTRODUCTION

The scattering of an ultrasonic wave in an elastically inhomogeneous medium results in a frequency dependent velocity and attenuation of the wave. The ultrasonic attenuation and dispersion are therefore sensitive to the microstructure of the sample. Since the microstructure of a material also has an important effect on material properties there is considerable interest in the development of ultrasonic techniques for the non-destructive determination of fracture toughness, hardness, impact strength, yield strength and tensile strength for example (1). Variations in the microstructure within a sample and from sample to sample can arise from composition fluctuations, inclusions, grain growth due to faulty heat treatments, incorrect fibre fraction in composites and high porosity in ceramics for example. In contrast to the elastic constants of the material, which can be obtained from ultrasonic velocity measurements at a single frequency, the determination of the above mentioned properties requires the measurement of the frequency dependence of the velocity or attenuation. A recent application is the prediction of yield strength in plain carbon steel using an ultrasonic determination of grain size (2). Klinman et al (2) base their discussion on the Hall-Petch relations, which relate the yield strength and impact transition temperature to the mean grain size, an important parameter determining the frequency dependence of the ultrasonic attenuation.

In general, the interpretation of frequency dependent velocity and attenuation data requires the use of multiple scattering theory. This is introduced in section 2 and the simplifications which occur at low volume fractions are discussed. Section 3 presents an application of the theory to transducer backings and to porous media. The sensitivity of the ultrasonic attenuation and velocity dispersion to the microstructure is emphasised in section 4 by comparing the scattering from the same volume fraction of MnS in iron distributed either as isolated spherical inclusions or as a thin layer around the iron grain boundaries.

2. SCATTERING THEORY

Consider a medium with n_0 scatterers per unit volume. The response of the jth scatterer may be obtained from the total field at j. This consists of the incident wave together with waves scattered from all other scatterers in the medium and this may be described on the average by a complex propagation constant $\beta = \omega/v' + i\alpha$ where ω is the angular frequency, v' the resultant wave velocity and α the attenuation.

If the volume fraction of scatterers is sufficiently low, multiple scattering can be ignored and it may be assumed that the distribution of scatterers is statistically independent, ie

$$p(\underline{r}_1, \underline{r}_2, \ldots, \underline{r}_N) = p(\underline{r}_1)\, p(\underline{r}_2) \ldots p(\underline{r}_N) \tag{1}$$

Here $p(\underline{r}_1, \underline{r}_2, \ldots, \underline{r}_N)$ is the probability of a configuration with scatterer 1 at $\underline{r}_1$, 2 at $\underline{r}_2$ etc and $p(\underline{r}_j)$ is the probability of j being at $\underline{r}_j$ independent of the positions of the other scatterers. For an incident longitudinal wave with wave number $k_1 = \omega/v_{11}$ in the medium surrounding the scatterers, the wave number β in the inhomogeneous medium is given by

$$\left(\frac{\beta}{k_1}\right)^2 = 1 + \frac{4\pi\, n_0\, f(0)}{k_1^2} \tag{2}$$

for spherical scatterers in the low volume fraction limit. In this expression spherical polar coordinates (r, θ, ϕ) have been used with origin at the centre of the sphere and θ and ϕ measuring rotations from and about the z axis which is taken to lie along the direction of propagation of the incident wave. The far field amplitude $f(\theta,\phi)$ relates the scattered longitudinal wave ψ_{sca} to the incident wave ψ_{inc} at j:

$$\psi_{sca}(r) = \psi_{inc}\, f(\theta, \phi) \exp(i\, k_1\, r) / r$$

For spherical scatterers $f(\theta, \phi)$ is independent of ϕ. $f(0)$ in equation (2) is the forward scattered amplitude.

Figure 1 shows a comparison between single scattering theory (3) and the experimental measurements of Kinra et al (4) for samples with 5% and 15% volume fraction of lead spheres of radius 660μm in an epoxy matrix. $f(\theta)$ was calculated using the theory of Ying and Truell (5). The open circles represent the zero frequency longitudinal velocity, which may be used experimentally for determining the volume fraction of lead present. At low frequencies ($k_1 a \ll 1$), equation (2) takes the form (6).

$$\left(\frac{\beta}{k_1}\right)^2 = 1 - \frac{4\pi\, a^3}{3}\, n_0\, (A + B\, k_1{}^2 a^2 - i\, C\, k_1^3 a^3) \tag{3}$$

Here a is the radius of the scatterers and A, B and C are functions of ρ_1, ρ_2, k_1, k_2, K_1 and K_2; ρ_i, k_i and K_i being the density, longitudinal wave number and transverse wave number in medium i with i = 1, 2. At low frequencies, equation (3) gives

$$\frac{v_1'}{v_{11}} = 1 + \frac{1}{2}\frac{\Delta V}{V}\, (A + B\, k_1^2 a^2) \tag{4}$$

$$\alpha = \frac{2\pi\, n_o}{3}\, k_1^4 a^6\, C \tag{5}$$

where $\Delta V/V = 4\,\pi\, a^3 n_0/3$ is the volume fraction of phase 2.

For solid inclusions in a solid matrix

$$A = 2 - \frac{\rho_2}{\rho_1} - \frac{3\,(K_1/k_1)^2}{4(1-p) + 3p\,(K_2/k_2)^2} + \frac{5(p-1)}{(p-1) + 3(p+3/2)\,(K_1/k_1)^2/2} \tag{6}$$

where $p = \mu_2/\mu_1$. For lead spheres in epoxy resin A = -7.20 (3).

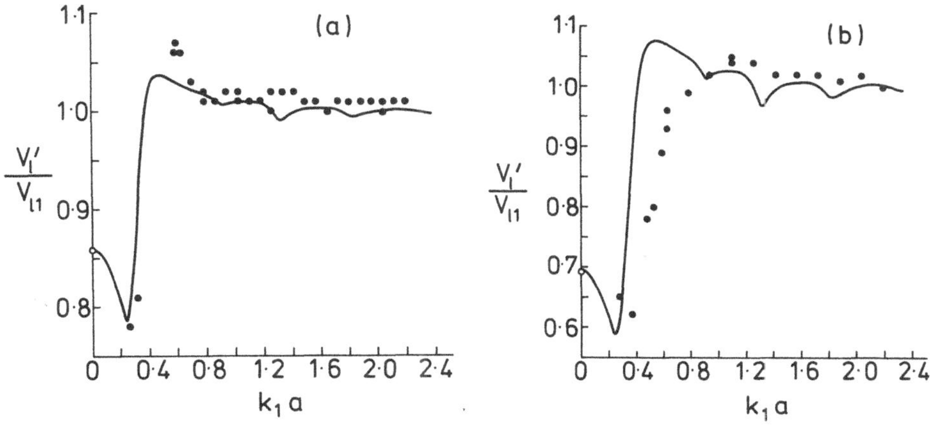

Figure 1. Normalised longitudinal velocity v_1'/v_{11} versus k_1a for (a) $\Delta V/V$ = 5%, (b) $\Delta V/V$ = 15% (3). v_1' is the longitudinal velocity in a sample with volume fraction $\Delta V/V$ of lead inclusions with radius a = 660 μm. $k_1 = \omega/v_{11}$, v_{11} being the longitudinal velocity in the epoxy matrix. The full circles give the experimental results of Kinra et al (4), the open circle the $\omega = 0$ result of equation (4) and the curve the results obtained from equation (2).

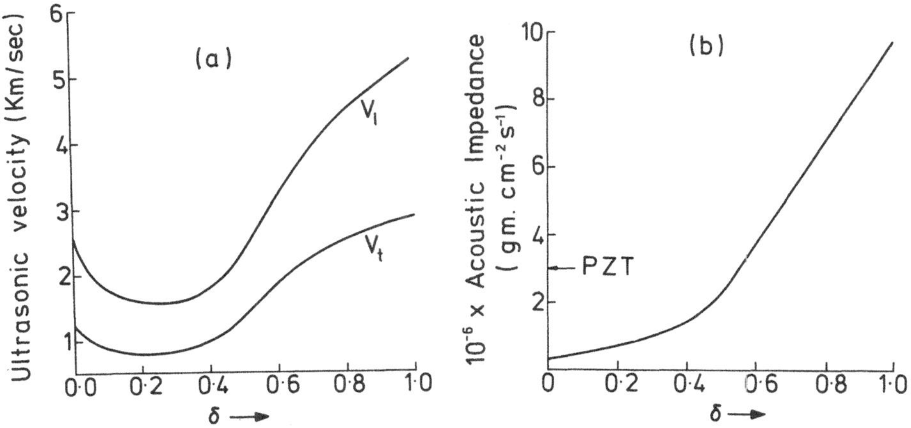

Figure 2. Variation of (a) ultrasonic longitudinal v_l and transverse v_t velocities; (b) longitudinal acoustic impedance Z_1 of a W/ Araldite composite as a function of the volume fraction of tungsten δ. The acoustic impedance of PZT is indicated.

It is seen from figure 1 that single scattering theory gives a good description of the velocity dispersion of an epoxy matrix with 5% volume fraction of lead spheres with radius 660μm. The main discrepancy, between theory and experiment is an underestimate of the 'overshoot' of v_1'/v_{11} after the rapid increase in v_1'/v_{11} which occurs as the frequency is increased beyond the Rayleigh or small frequency region. The position of this rapid rise in velocity may be used to determine the size of the particles in the material. A strong peak occurs in the ultrasonic attenuation (3) in this frequency range and this may be used for particle sizing. In view of the agreement seen in figure 1a it seems reasonable to interpret the disagreement between theory and experiment for $\Delta V/V = 15\%$ as arising from multiple scattering effects neglected in the single scattering approximation of equation (2). If this interpretation is correct, the effect of multiple scattering is to shift the rapid rise in v_1'/v_{11} from $k_1 a = 0.3$ to $k_1 a = 0.5$ and to suppress the magnitude of the 'overshoot' which occurs after this rise in velocity. This deviation from single scattering theory will be sensitive to the scatterer pair distribution function and information on this important function should therefore be obtainable from the velocity dispersion curve (7).

3. Application to transducer backings and porous materials

3.1 Transducer backings

Most ultrasonic transducers have, as their active element, a disc of a piezoelectric ceramic such as PZT with low ultrasonic absorption. To obtain a short ultrasonic pulse, a backing with a high ultrasonic attenuation and an acoustic impedance close to that of the piezoelectric is usually employed. The most commonly used backings consist of epoxy resin with high ultrasonic absorption filled with tungsten powder to increase the density and therefore the acoustic impedance. Figure 2a shows the ultrasonic velocities in a tungsten/Araldite composite computed (8) as $k_1 a \to 0$ using the self consistent scattering theory of Sayers and Smith (8-10). This theory treats material of both type 1 and 2 as deviations from the effective properties of the medium, and therefore considers two types of scatterers to be present for two-phase materials. The addition of tungsten to Araldite results initially in a sharp fall in ultrasonic velocity, the velocity only recovering to a value equivalent to that of pure Araldite at a tungsten volume concentration of 50%. The density of the tungsten/Araldite mixture rises linearly with the addition of tungsten and the net result is a monotonic rise in acoustic impedance as shown in figure 2b. The acoustic impedance of PZT lies in the region of $3 \times 10^6 \mathrm{g\ cm^{-2} s^{-1}}$ and it follows that high volume fractions of tungsten are required to form an acoustic match with PZT. The transition from high to low impedance values takes place over a narrow range of volume fraction, and therefore accurate control of the volume fraction of tungsten is required.

When the radius a of the tungsten particles in the composite is comparable to the wavelength, wave scattering by the particles becomes important and the zero frequency results of figure 2 need to be modified. Figure 3a shows the variation of v_1'/v_{11} with $k_1 a = \omega a/v_{11}$ for 10% volume fraction of tungsten in araldite using the single scattering theory of equation (2). All particles are assumed to have the same radius a. It is seen that the rapid decrease of ultrasonic velocity with increasing volume fraction of tungsten in tungsten/Araldite composites shown in figure 2a is purely a low frequency phenomenon. Acoustic matching with

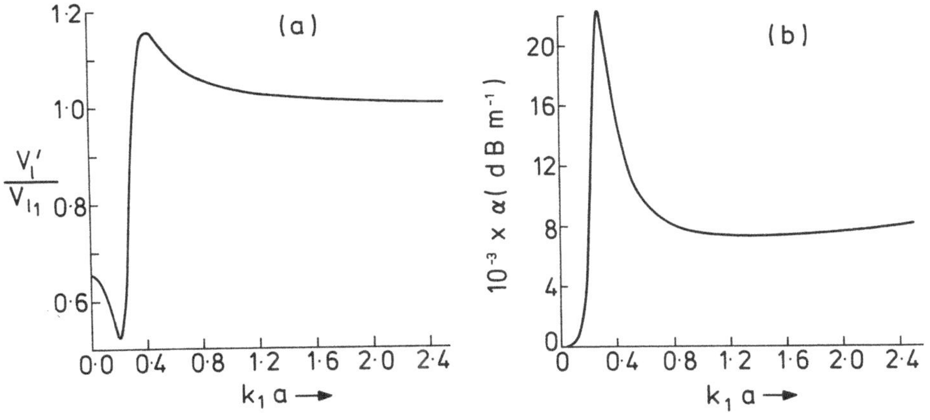

Figure 3. Variation of (a) normalised longitudinal velocity v_1'/v_{11}; (b) ultrasonic attenuation α with k_1a for 10% tungsten in Araldite (8). v_1' is the longitudinal velocity of the composite and v_{11} that of Araldite. $k_1 a = \omega a/v_{11}$. A radius a = 100 μm was used.

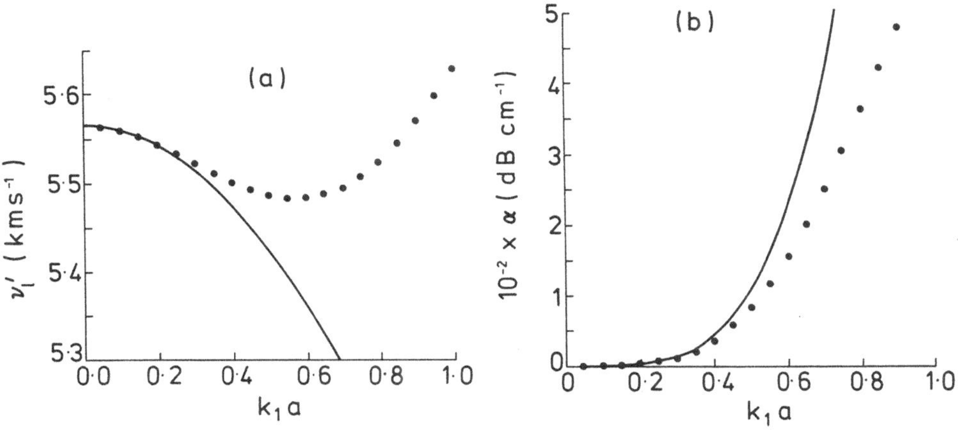

Figure 4(a) Variation of the ultrasonic velocity v_1' of the porous medium with k_1a; (b) variation of attenuation α with k_1a for the same case. The longitudinal velocity v_{11} of the matrix is assumed to be 6×10^3 m s^{-1}. A porosity $\delta = \Delta V/V = 0.1$ and a value $(K_1/k_1) = 2$ are assumed. Solid curve - single scatterer result of equation (3) valid for small k_1a; • - numerical evaluation of equation (2).

PZT could therefore be achieved at much lower volume fractions of tungsten if larger particles could be used. Figure 3b shows the variation of ultrasonic attenuation due to scattering with k_1a. The attenuation is seen to be dominated by a strong peak at $k_1a = 0.28$. A choice of particle size such that the centre frequency of the transducer lies in this region would lead to a significant increase in the attenuation of the backing.

The theory has also been applied to backings made from tungsten powder and a plastic metal such as Al, Cu, Pb and Sn (8).

3.2 Porous Materials

For a porous material the coefficients A, B and C in equations (3-5) are functions only of K_1/k_1 given by

$$\frac{K_1}{k_1} = \frac{v_{11}}{v_{t1}} = \left(\frac{\lambda_1 + 2\mu_1}{\mu_1}\right)^{\frac{1}{2}} = \left(\frac{2(1-\sigma_1)}{(1-2\sigma_1)}\right)^{\frac{1}{2}} \tag{7}$$

where σ_1 is the Poisson's ratio of the matrix and $K_1/k_1 > \sqrt{2}$ for an isotropic medium. Explicit expressions for A, B and C are given in reference (6).

In the limit $K_1/k_1 >> 1$ equations (4) and (5) give

$$\frac{v_1'}{v_{11}} = 1 - \frac{3}{8}\left(\frac{K_1}{k_1}\right)^2 \frac{\Delta V}{V}\left[1 + \left(\frac{K_1}{k_1}\right)^2 \frac{k_1^2\, a^2}{4}\right] \tag{8}$$

$$\alpha = \frac{\pi\, n_o}{8} K_1^4\, a^6 \tag{9}$$

For rubber, for example, $K_1/k_1 = 40$. The velocity in this case therefore drops off very rapidly with increasing porosity and has a very large frequency dependent component.

The value of K_1/k_1 considered above is not typical. Most isotropic solids have a value of K_1/k_1 of about 2. For $K_1/k_1 = 2$, $A = -13/8$, $B = -883/360$. Equation (4) then gives

$$\frac{v_1'}{v_{11}} = 1 - \frac{13}{16}\left(1 + \frac{883}{585} k_1^2\, a^2\right)\frac{\Delta V}{V} \tag{10}$$

For a 10% volume fraction of pores

$$\frac{v_1' - v_{11}}{v_{11}} = -0.081 - 0.123\,(k_1 a)^2 \tag{11}$$

Figure 4 compares the low frequency expansion of equation (3) with the numerical evaluation of equation (2) for the case of 10% porosity. The low frequency expansion is seen to be valid for $k_1\, a < 0.25$. For a spherical void of radius 10μm, this limits the ultrasonic frequency to be less than 40 MHz in Al_2O_3 for example. The porosity is most easily evaluated from the zero frequency limit of the velocity, whilst the size of the pores can be determined from the frequency dependence of either the velocity or the attenuation.

Winkler (11) has measured the frequency dependence of the longitudinal phase velocity and attenuation in several high-porosity (20-26%)

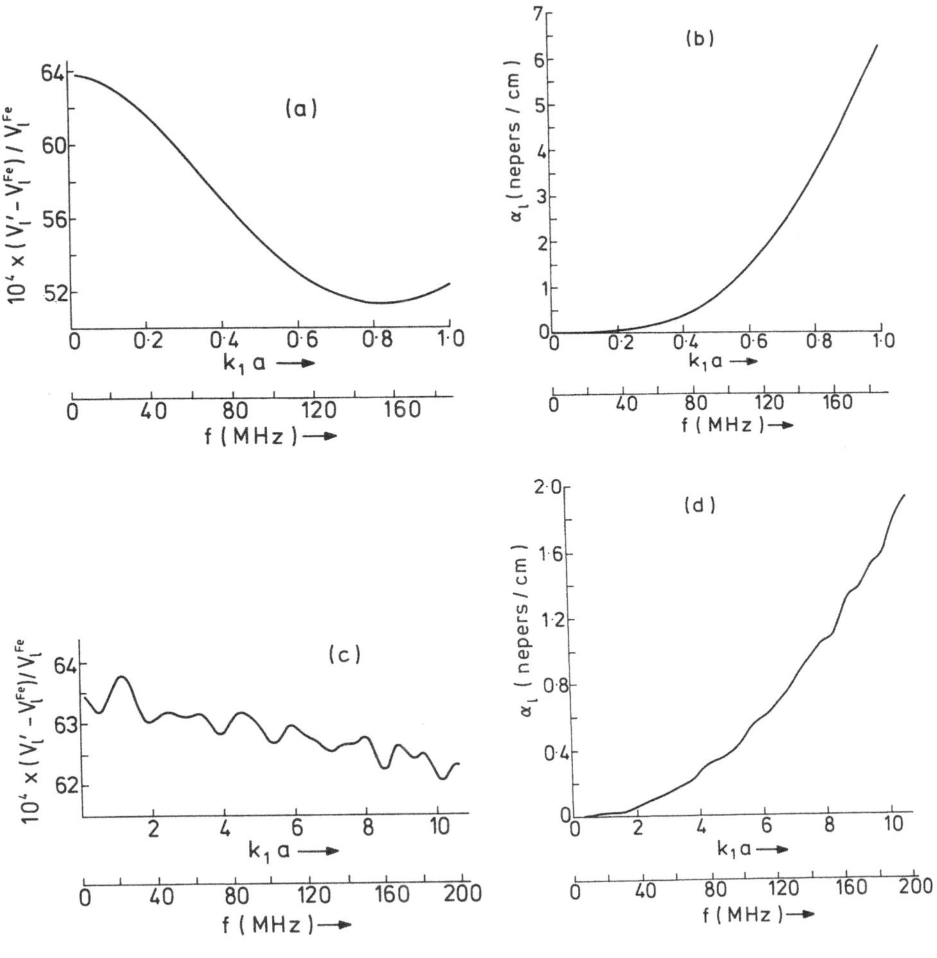

Figure 5. Variation of (a) the ultrasonic velocity v_1' and (b) the attenuation α with $k_1 a$ for 5% volume of MnS spheres of radius a = 5 μm in iron. v_1^{Fe} is the longitudinal wave velocity in the iron matrix. The same quantities are shown in (c) and (d) for a 5% volume fraction of MnS forming a shell around the iron crystallites which are assumed to be spheres of radius a = 50 μm.

sandstones from 0.4 to 2 MHz. The dry samples all show negative velocity dispersion (velocity decreasing with increasing frequency) and an attenuation increasing as the third to fourth power of the frequency. These observations are in qualitative agreement with the predictions of equations (4) and (5) and suggest a scattering mechanism as being important in these samples (11). Brine saturated rocks show positive velocity dispersion and an attenuation greater than that in the dry samples which increases with a first to second power dependence (11). Since the acoustic mismatch between the pores and the grains would be expected to be reduced upon saturation, these results suggest that a local fluid-flow loss mechanism is present in the saturated samples (11).

4. Scattering by a Spherical Shell

The treatment of the scattering by the second phase as that from spherical scatterers in an isotropic medium is not generally valid. In many cases, a second phase forms by segregation at a grain boundary. For example, in the face centered cubic to body centered cubic transformation which occurs upon cooling steel from a temperature above the transformation point, a ferrite layer forms at the boundaries of the original austenitic grains, the interiors of the grains transforming to pearlite, a layered structure consisting of crystallites of ferrite and cementite. It was found by Kamagaki (12) that this thick layer of ferrite reflects the ultrasonic waves and increases the attenuation. Another example is the attenuation of ultrasound in polycrystalline aluminium (13). In impure aluminium, impurities are found to segregate to grain boundaries and result in a higher attenuation than found in pure aluminium.

The scattering by a sphere of medium 2 and radius a, surrounded by a spherical shell of medium 3 with inner radius a and outer radius b and embedded in a matrix of medium 1 has been considered (14). Application to free-machining steel containing MnS inclusions was made. Figure 5 compares the velocity and dispersion due to scattering from a 5% volume fraction of MnS spheres of radii 5μm in iron with the case when the same volume fraction of MnS is assumed to form a shell around the iron crystallites which are assumed to be spheres with 50μm radius. The ultrasonic velocity and attenuation are clearly sensitive to the geometrical arrangement of the second phase as well as its volume fraction.

REFERENCES

1. Vary, A, in 'Research Techniques in Non-Destructive Testing' Vol IV, ed. R S Sharpe, Academic Press, New York (1980) 159.
2. Klinman, R, Webster, G R, Marsh, F J and Stephenson, E T, Mat. Eval. 38 (1980) 26.
3. Sayers, C M and Smith, R L, J Phys D 16 (1983) 1189.
4. Kinra, V K, Ker, E and Datta, S K, Mech. Res. Commun. 9 (1982) 109.
5. Ying, C F and Truell, R, J. Appl. Phys. 27 (1956) 1086.
6. Sayers, C M, J Phys D 14 (1981) 413.
7. Willis, J R, J Mech. Phys. Solids 28 (1980) 307.
8. Sayers, C M and Tait, C E, Ultrasonics 22 (1984) 57.
9. Sayers, C M, J Phys. D. 13 (1980) 179.
10. Sayers, C M and Smith R L, Ultrasonics 20 (1982) 201.
11. Winkler, K W, J Geophys. Res. B 88 (1983) 9493.

12. Kamigaki, K. Sci. Rep. RITU A9 (1957) 48.
13. Hirone, T and Kamigaki, K. Sci. Rep. RITU A7 (1955) 455.
14. Sayers, C M, Wave Motion 7 (1985) 95.

DISCUSSION

Comment: Schmitt
In one of your graphs you showed the impedance of the backing as function of the tungsten volume up to 60% and higher. Could you, please, comment how these values are calculated in spite of the assumption of low volume fraction limit?

Reply: Sayers
The results shown in Fig. 2 were obtained using the self consistent scattering theory of Sayers and Smith (Ref. 8-10). In a two-phase material this theory considers both types of materials and deviations from the effective properties of the medium, and therefore considers two types of scatterers to be present. If the scatterers are surrounded by an effective medium of longitudinal wavenumber k, the wavenumber β of the coherent longitudinal wave passing through the composite is given by

$$(\beta/k)^2 = 1 + 4\pi n_0 \langle f(0) \rangle /k^2 ,$$

with $\langle f(0) \rangle$ the forward scattered amplitude averaged over the two types of scatterers. Self consistency requires $\beta = k$ allowing this equation to be solved for k.

Comment: Alippi
Am I correct in thinking that in your last slide the material parameters inside and outside the shell were the same? Is your programme capable of dealing with general problems of three different materials?

Reply: Sayers
The first and second answers are both positive.
1. The material inside and outside the shell was taken as iron with $\rho = 7.87$ g/cm^3, $v_L = 5.9$ km/s, $v_s = 3.23$ km/s and the shell was assumed to be MnS with $\rho = 3.99$ g/cm^3, $v_L = 7.395$ km/s, $v_s = 4.347$ km/s.
2. The program can treat the case of three different materials, but the shell has to be spherical.

Comment: Bonnet
What is the exponent term (if there is one) of the power expansion of the attenuation given by multiple scattering as a function of frequency?

Reply: Sayers
At low frequencies, the scattering contribution to the attenuation varies as the fourth power of the frequency in both single scattering theories. At higher frequencies there is a departure from Rayleigh scattering, but this cannot be written as a single power law. The precise frequency dependence depends on the elastic properties of the scatterers and reflects the individual particle resonances and their interactions.

Comment: Hutchins
Would a measurement of the attenuation add in the determination of residual stress in metallic samples with texture?

Reply: Sayers
In a textured polycrystal both the attenuation and the velocity depend on the propagation and polarization directions with respect to the principal texture axes. Therefore, it is possible, in principle, to use this dependence to separate the effects of texture and stress. This has been discussed by Allen et al. in Research Techniques in NDT, Ed.R.S. Sharpe, Vol. 6 Academic Press, London (1982) and is being pursued by Göbbels , Hirsekorn and Schneider at the Iz fP.

Comment: Zarembowitch
In order to separate the effect of stress on the acoustical birefringence it is necessary to obtain additional information and I agree with your Rayleigh wave approach. Some people have used the frequency dependence of the shear wave as mentioned in Mahedevan's paper, that I personally find wrong. What is your opinion on this approach?

Reply: Sayers
We have measured the freqùency dependence of the shear wave birefringence and found an effect much smaller than that reported by Mahedevan. Nevertheless, the frequency dependence of the shear wave birefringence is being used by Göbbels and Hirsekorn (NDT International 17, 337 (1984)) as an alternative for separating the effects of texture and stress.

STUDY OF THE DISLOCATION DYNAMICS BY ULTRASONIC CYCLIC BIAS STRESS EXPERIMENTS

G. Gremaud

1. INTRODUCTION

Recent progress in dislocation dynamics study has been done using cyclic bias stress experiments. In this new technique attenuation α and velocity v of ultrasonic waves are measured in a sample subjected to a low frequency cyclic stress σ. Closed curves $\Delta\alpha$ (σ) and $\Delta v/v(\sigma)$ are measured during each cycle of the applied stress. The shapes of these curves and their evolution are characteristic for each mechanism controlling the dislocation motion. For this reason, the shapes of these curves have been called the "signatures" of the mechanisms controlling the dislocation dynamics.

In this paper, it will be shown why the closed curves $\Delta\alpha$ (σ) and $\Delta v/v$ (σ) can be called the "signatures" of a particular dislocation motion mechanism and how powerful this technique is in finding the mechanisms controlling the dislocation dynamics.

2. PLASTIC AND ANELASTIC DEFORMATION DUE TO DISLOCATION MOBILITY

Dislocations are linear topological defects of the lattice in crystalline materials (Fig. 1). Plastic and anelastic properties of crystalline materials are generally controlled by the motion of these dislocations, according to the fact that the total strain of a crystal is given by

$$\varepsilon = \varepsilon_{el} + \varepsilon_d \quad (1)$$

where ε_{el} is the elastic strain, related to the applied stress σ by the Hookerelation:

$$\varepsilon_{el} = J_{el}\,\sigma \quad (2)$$

and where ε_d is the plastic or anelastic strain, related to the average displacement u of the dislocation by the Orowan relation :

$$\varepsilon_d = \Lambda b u \quad (3)$$

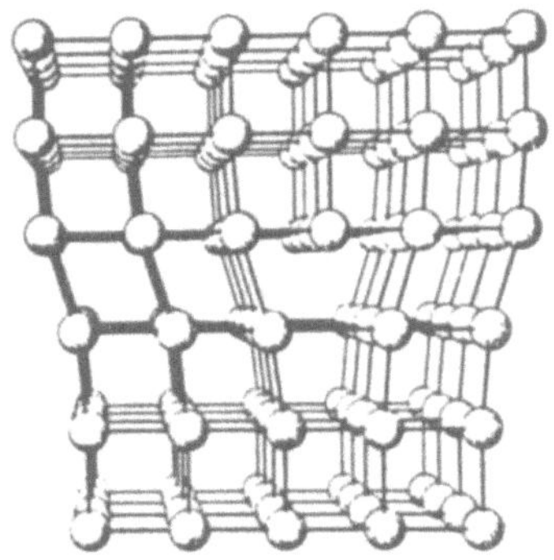

FIGURE 1. Schematic representation of an edge dislocation in a cubic lattice.

with Λ the dislocation density (in cm of dislocation per cm^3) and b the Bürgers vector (charge of deformation transported by one dislocation).

For the understanding of the plastic and the anelastic deformation, the study of the dislocation dynamics is then very important. This study is generally made by tensile tests in the case of plastic deformation (when there is no restoring force associated with the dislocation displacement u), or by internal friction measurements in the case of anelastic deformation (when there is an equilibrium position for the dislocations, due to a restoring force associated with the dislocation displacement u) /1/.

The dislocation mobility can be controlled by different interactions:

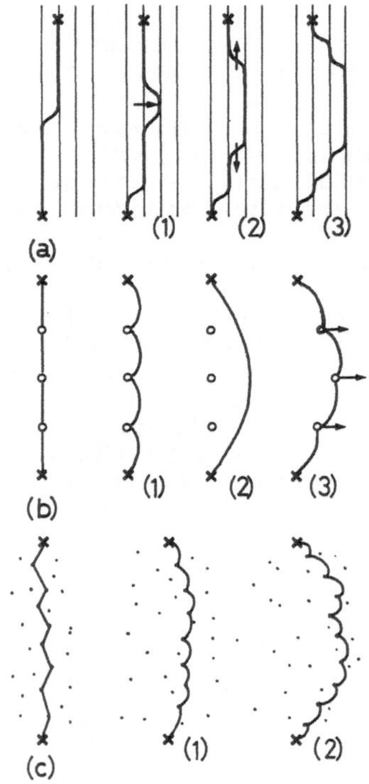

(i) the dislocation-phonon interaction at high velocity of the dislocations, which leads to a viscous friction force proportional to the dislocation velocity /2, 3/.

(ii) the dislocation-lattice interaction at low temperature, which leads to linear obstacles (Peierls hills) due to the lattice periodicity /4/. These obstacles can be crossed by a mechanism of thermally activated kink pair formation (generation of solitons on the dislocation lines), followed by a lateral diffusion of the kinks (migration of the solitons along the dislocation lines), as shown in figure 2.a.

(iii) the interactions of the dislocations with localized obstacles (substitutional impurities, interstitials, vacancies, precipitates, other dislocations, grain boundaries, etc).

These interactions lead to a lot of different dislocation motion mechanisms /5/. Some examples are given in figures 2 in the case of interaction with point obstacles: the rigid pinning on a row of immobile point obstacles (fig. $2.b_1$), the catastrophical

FIGURE 2. Some examples of interaction mechanisms represented in the case of anelastic deformation (existence of a restoring force due to the line tension in the case of a dislocation bowing between two hard pinning points).

breakaway from a row of immobile point obstacles (fig. $2.b_2$), the dragging of a row of mobile point obstacles (fig. $2.b_3$) and the hysteretic motion in a random distribution of immobile point obstacles by successive pinning and depinning of the dislocation (fig. 2.c).

3. BIAS STRESS EXPERIMENTS

In this paper, measurements of the dislocation dynamics will be considered only in the case of the anelastic deformation. In this deformation range, the most commonly used measurement technique is the internal friction. Unfortunately, the mechanisms which control the dislocation mobility are responsible for internal friction spectrums presenting a lot of similarities: similar internal friction peaks as a function of the temperature in the case of thermally activated relaxational interaction mechanisms, similar dependence of the internal friction on the measurement amplitude in the case of hysteretic interaction mechanisms /1/. This means that it can be very difficult to find the interaction mechanism from internal friction measurements. This difficulty can be avoided by doing bias stress experiments instead of internal friction experiments. Bias stress experiments have been recognized by several authors as a powerful method to study the dislocation dynamics /6 to 11/. The principle of this measurement technique is simple.

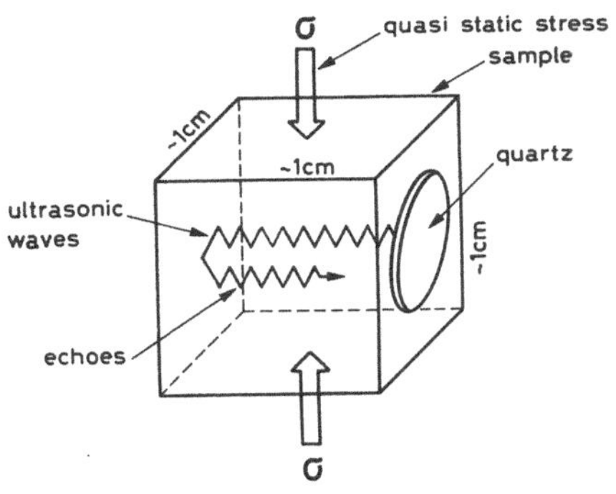

FIGURE 3. Principle of bias stress experiments.

In this technique attenuation α and velocity v of ultrasonic waves are measured in a sample subjected to a quasi-static stress σ (fig. 3) and reported as a function of this quasi-static stress σ in order to obtain the plots $\Delta\alpha(\sigma)$ and $\Delta v/v(\sigma)$. In our case, the measurement device is composed of four distinct parts /12/:

(i) a high precision echometer able to measure automatically the attenuation and the relative change of propagation time with a sensitivity of 10^{-4} and 10^{-6} dB/μs respectively, and with a measurement repetition rate of 10 kHz.

(ii) a mechanical system allowing to generate a compression stress of any waveform to the sample: sinusoidal stresses can be generated between 10^{-3} and 2 Hz with amplitudes from 2 to 500 kg/cm^2.

(iii) a cryogenic system allowing to regulate the temperature of the sample between 4 K and 300 K.

(iv) a computer system which controls the measurement system, gets the experimental results, treats them on-line by a fast Fourier transform in order to eliminate the electronic noise, and finally draws the calibrated curves $\Delta\alpha(\sigma)$ and $\Delta v/v(\sigma)$.

4. PHYSICAL INTERPRETATION OF BIAS STRESS EXPERIMENTS

The physical interpretation of bias stress experiments is the following. The quasi-static stress σ is used to induce a motion of dislocations interacting with some obstacles, as illustrated by two examples in figure 4: the depinning from a row of point obstacles by a catastrophic breakaway mechanism (a) and the jumping over linear obstacles (Peierls hills) by a kink

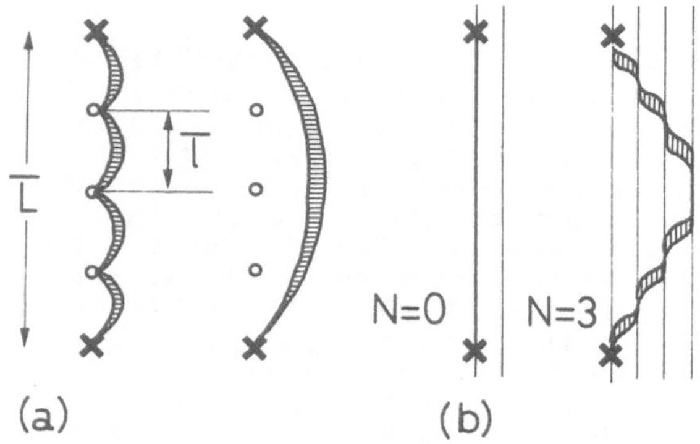

FIGURE 4. Principle of the physical interpretation of bias stress experiments in the cases of a breakaway mechanism (a) and of a kink-pair formation (b).

pair formation mechanism (b). On the other hand, the ultrasonic field, which is of much higher frequency and much lower amplitude than the quasi-static applied stress, cannot activate the preceding mechanisms. It can only induce a small vibration (represented by the shaded area in figure 4) of the dislocation segments (fig. 4.a) or of the kinks (fig. 4.b). The strong attenuation of these vibrations by the phonon field (dislocation-phonon or kink-phonon interaction) leads to an attenuation of the ultrasonic wave, which depends essentially on the average length $\bar{l}$ of the dislocation segments in the case of fig. 4.a, or on the average number $\bar{N}$ of created kink pairs on the dislocation lines in the case of fig. 4.b:

$$\Delta\alpha = \Delta\alpha\{\bar{l}, \sigma(t), T\} \text{ or } \Delta\alpha = \Delta\alpha\{\bar{N}, \sigma(t), T\} \tag{4}$$

The dependence of $\Delta\alpha$ on $\sigma(t)$ is due to the softening or the hardening of the bowed dislocation, and the dependence on T is due to the temperature dependence of the phonon field.

As $\bar{l}$ or $\bar{N}$ depends on the mechanism which is activated by the quasi-static applied stress $\sigma(t)$, $\bar{l}$ or $\bar{N}$ depends also on $\sigma(t)$:

$$\bar{l}=\bar{l}[\sigma(t),t,T,...] \text{ or } \bar{N}=\bar{N}[\sigma(t),t,T,...] \tag{5}$$

It is clear that this dependence on $\sigma(t)$ is strongly correlated with the interaction mechanism controlling the dislocation motion. This means that the plot of $\Delta\alpha$ as a function of σ, given by

$$\begin{aligned} \Delta\alpha &= \Delta\alpha\ \{\bar{l}[\sigma(t), t, T, ...], \sigma(t), T\} \text{ or} \\ \Delta\alpha &= \Delta\alpha\ \{\bar{N}[\sigma(t), t, T, ...], \sigma(t), T\} \end{aligned} \tag{6}$$

depends also strongly on the interaction mechanism. This clearly shows also that two models are needed to interpret correctly a $\Delta\alpha(\sigma)$ curve. The first model is concerned with the calculation of the high-frequency ultrasonic attenuation due to the dislocation segment vibration $\Delta\alpha\ \{\bar{l}, \sigma, T\}$ or due to the kink vibrations $\Delta\alpha\ \{\bar{N}, \sigma, T\}$.

The second model is concerned with the calculation of the low frequency dislocation-obstacle interaction mechanism, which allows an estimate of the expressions $\bar{I}[\sigma(t), t, T, ...]$ or $\bar{N}[\sigma(t), t, T, ...]$ that have to be introduced in the expressions for $\Delta\alpha$.

5. CYCLIC BIAS STRESS EXPERIMENTS AND THE CONCEPT OF SIGNATURE

Cyclic bias stress experiments have been recently developed /13, 14/. They consist in applying a cyclic low frequency stress on the sample and measuring the closed curves $\Delta\alpha(\sigma)$ and $\Delta v/v(\sigma)$ on each cycle of the applied stress. Such experiments are directly connected with low frequency internal friction measurements when the cyclic applied stress amplitude remains in the anelastic range of the material /13/. But they can also be correlated with fatigue experiments if the cyclic applied stress amplitude becomes high enough to reach the plastic domain of the material /14/.

From the equation (6) giving the ultrasonic attenuation $\Delta\alpha$ as a function of $\sigma(t)$, it is clear that the shape of the closed curves $\Delta\alpha(\sigma)$ depends strongly on the interaction mechanism which controls the dislocation motion. For this reason, the closed curves $\Delta\alpha(\sigma)$ have been called the "signatures" of the interaction mechanism. Now several different signatures have been experimentally obtained and theoretically related with a specific dislocation motion mechanism. Some of these signatures are reported in figure 5: the "bow signature" (fig. 5.a) due to breakaway from a row of immobile point defects /11, 15, 16/, the "moustache signature" (fig. 5.b) due to pinning-depinning in two clouds of slowly mobile point defects /11, 15, 16/, the "eight signature" (fig. 5.c) due to dragging of a cloud of mobile point defects /11, 15, 16/ the "Bordoni signature" (fig. 5.d) due to kink pair formation mechanism /17, 18/ and the "flip-flop signature" (fig. 5.e) due to dragging of point defects situated between two kinks or between the two partials of a dislocation /18/.

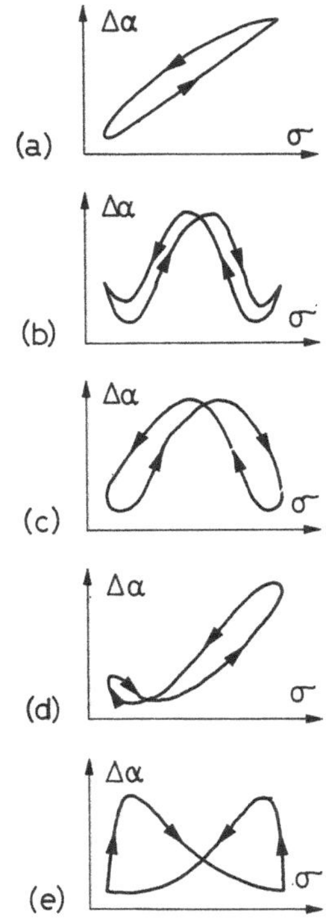

FIGURE 5. Some experimentally obtained and theoretically explained signature shapes, for different interaction mechanisms.

6. ONE EXAMPLE OF PHENOMENOLOGICAL EXPLANATION

In this section, a phenomenological explanation will be given as an example, for the signatures due to interaction mechanisms between dislocations and more or less mobile impurities. A more complete theoretical description has been published elsewhere /15, 16/.

The impurities situated around dislocations can act as soft pinning points for the dislocations. The mean length $\bar{l}$ between these pinning points (figure 6) can be measured with an ultrasonic wave, according to the fact that the changes of the attenuation α and of the velocity v of the ultrasounds depend directly on the mean length $\bar{l}$ of the vibrating dislocation segments as

$$\Delta\alpha \simeq \bar{l}^4 \text{ and } \Delta v/v \simeq \bar{l}^2 \qquad (7)$$

Under the effect of the cyclic applied stress σ, the dislocations can also be displaced inside the atmosphere of impurities by a mechanism of successive pinning and depinning. The measurement of the ultrasonic attenuation allows then to follow the changes of $\bar{l}$ during the motion of the dislocation inside the cloud of impurities. It is then possible to do an indirect measurement of the local density of impurities around the dislocations.

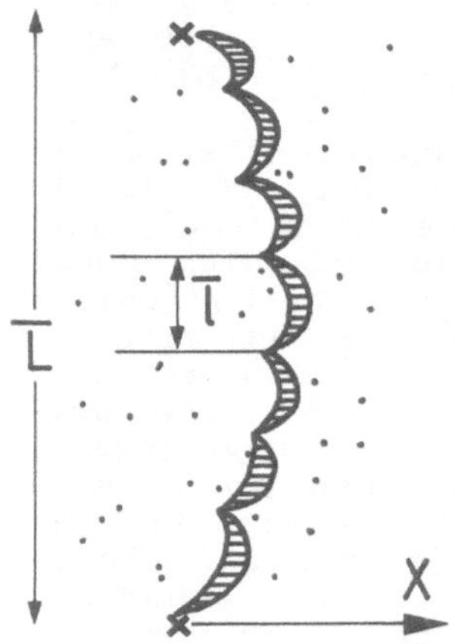

FIGURE 6. Motion of a dislocation inside a cloud of impurities.

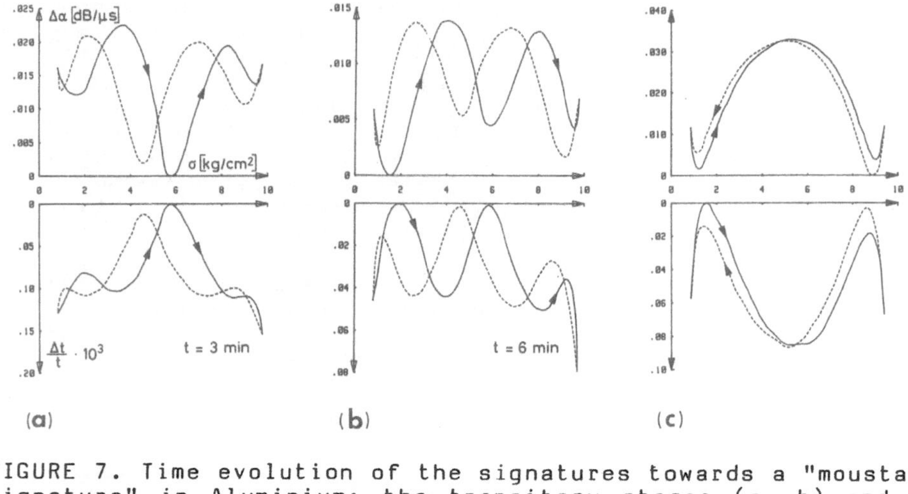

FIGURE 7. Time evolution of the signatures towards a "moustache signature" in Aluminium: the transitory stages (a, b) and the stationary state (c) of the signatures.

Such a measurement is presented in figure 7. A static stress of 5 kg/cm^2 has been applied to a sample of pure Aluminium (99.999 %) for one hour at room teemperature before applying a sinusoidal compression stress between 0.5 and 9.5 kg/cm^2 with a frequency of 0.1 Hz. After three minutes of cycling (fig. 7.a), one can see a minimum of attenuation at about 5 kg/cm^2, which is due to a cloud of impurities formed during the application of the static stress of 5 kg/cm^2. After 120 minutes (fig. 7.c), the impurities have migrated and a "moustache signature" is then observed. This typical signature can simply be explained. The stress field around a dislocation is schematically represented in figure 8.a. The oscillation of the dislocation under the sinusoidal applied stress displaces cyclically this stress field between two extremal positions along the x axis of the glide plane of the dislocation (fig. 8.b). The time average stress field can be easily calculated on one cycle (fig. 8.c). The impurities migrate slowly in this average stress field and, after a more or less long time, finish to form two clouds of impurities with a maximum density near the extremities of the zone swept by the dislocation (fig. 8.d). This density distribution of the impurities induces a typical shape of the attenuation curve as a function of the position x of the dislocation (fig. 8.e), because the average length $\bar{l}$ between impurities depends on their density. The dislocation crosses these two clouds with a phase lag with regard to the stress. This phase lag depends on the friction due to the successive pinning-depinning processes of the dislocation in the cloud of impurities, and appears as the value $\Delta\alpha$ on the attenuation curve plotted as a function of the applied stress σ (fig. 9.a).

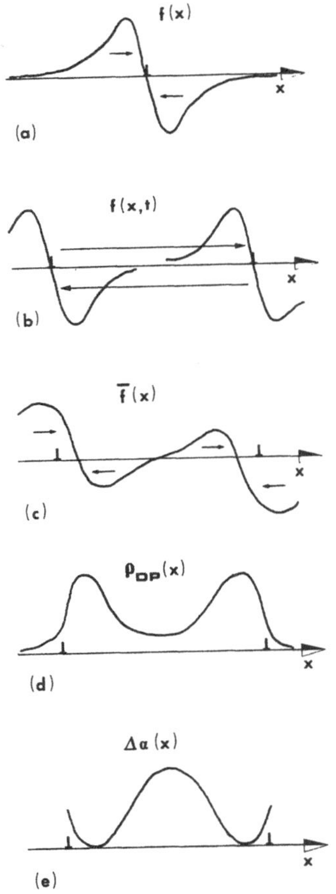

FIGURE 8. Phenomenological explanation of the "moustache signature".

If the mobility of the impurities is not too low, each cloud can partially vanish when the dislocation is at the opposite side of the glide plane. This appears as the value $\Delta\alpha$ in figure 9.b. If the mobility of the impurities becomes high enough, each cloud can completely disappear when the dislocation is at the opposite side of the glide plane. In this case the mechanism can be explained as one cloud which migrates in the stress field of the dislocation. It follows the dislocation, but always with a phase lag due to the impurity diffusion as shown in figure 10.a. Such a mechanism must have characteristic $\Delta\alpha(\sigma)$ and $\Delta v/v(\sigma)$ signatures due to the fact that the dislocation crosses the cloud twice in each cycle of the applied stress (fig. 10.b).

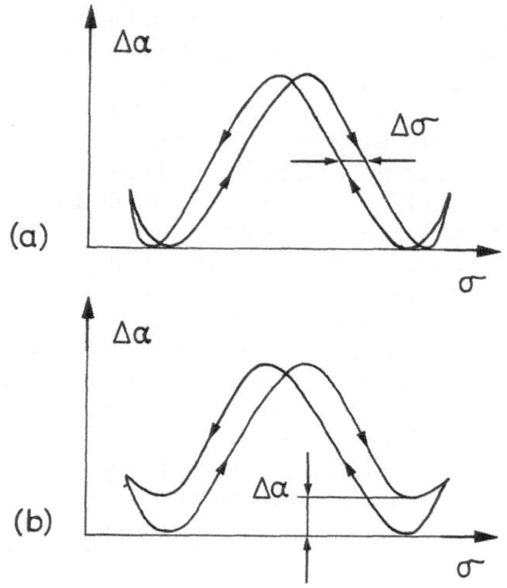

FIGURE 9. Theoretical "moustache signature".

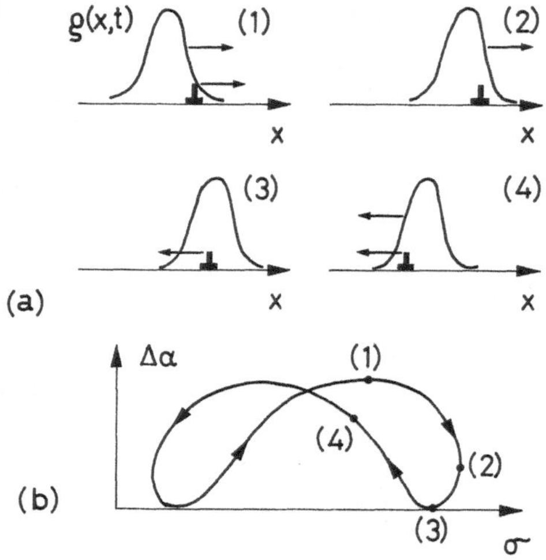

FIGURE 10. The mechanism of perpendicular dragging of a cloud of impurities (a) and its characteristic signature shape (b).

This kind of signature has been called the "eight signature". It has been observed in the same sample of Aluminium that was used for measurements of fig. 7, with substitutionnal impurities at room temperature, when the frequency of the applied stress is decreased at about 10^{-4} Hz (fig. 11).

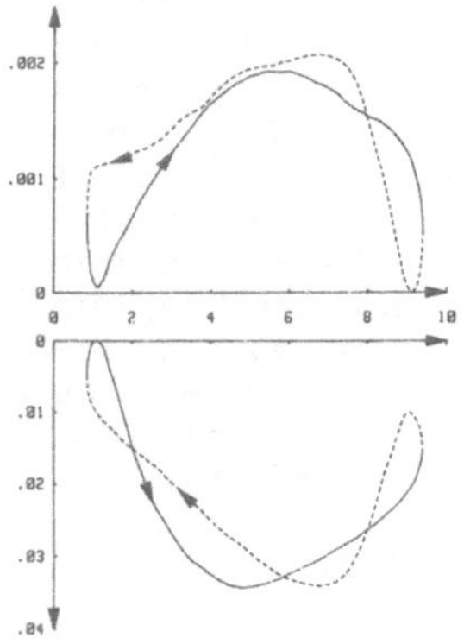

FIGURE 11. "Eight signature" experimentally obtained in Aluminium.

7. CONCLUSION

In this paper, it has been shown that important progress can be made in dislocation dynamics studies by using cyclic bias stress experiments. This technique allows for the first time "to see" the mechanisms controlling the dislocation dynamics simply by analyzing their "signature", the characteristic shape of the closed curve $\Delta\alpha(\sigma)$.

It is also interesting to note that time and temperature dependences of the signatures can be measured. Such measurements lead to very interesting results /19, 20/, which can be summarized by the following points:

(i) the time dependence measurements permit transitory stages and stationary states to be identified, which give interesting informations about the mobility of the obstacles that interact with the dislocation.

(ii) the temperature dependence of the signatures can be easily correlated with the low frequency internal friction spectrum when using a harmonic cyclic bias stress, which is very useful for the interpretation of internal friction spectrums.

REFERENCES

1. W. Benoit, G. Gremaud, R. Schaller, in "Plastic Deformation of Amorphous and semi-crystalline Materials" (Les Editions de Physique, Les Ulis, France, 1982), p. 65
2. A. V. Granato, Scripta Met. 18 (1984) 663
3. C. Elbaum, Scripta Met. 18 (1984) 657
4. A. Seeger, J. de Phys. 42 (1981) C5-201
5. D. Lenz, K. Lücke, in "Internal Friction and Ultrasonic Attenuation in Crystalline Solids" (Springer Verlag, Berlin, vol. II, 1968), p. 48
6. A. Hikata, R. Truell, A. Granato, B. Chick, K. Lücke, J. Appl. Phys. 27 (1956) 396
7. A. Hikata, B. Chick, C. Elbaum, R. Truell, Acta Met. 10 (1962) 423
8. A. Hikata, R. A. Johnson, C. Elbaum, Phys. Rev. B 2 (1970) 4856
9. D. Lenz, Echenhofer, K. Lücke, Scripta Met. 5 (1971) 387
10. A. Vincent, J. Perez, P. F. Gobin, J. Phys. Chem. Solids 35 (1974) 1253
11. G. Gremaud, W. Benoit, J. de Phys. 42 (1981) C5-369
12. G. Gremaud, J. de Phys. 42 (1981) C5-1141
13. G. Gremaud, thesis, EPF-Lausanne (1981)
14. M. Omry, thesis, INSA-Lyon (1984)
15. G. Gremaud, J. de Phys. 44 (1983) C9-607
16. G. Gremaud, Li Ping Ho, W. Benoit, J. de Phys. 44 (1983) C9-581
17. M. Bujard, G. Gremaud, J. Baur, W. Benoit, proceedings of the 8th International Conference on Internal Friction and Ultrasonic Attenuation in Solids, to be published in J. de Physique
18. M. Bujard, thesis, EPF-Lausanne (1985)
19. G. Gremaud, in "Dislocations 1984" (Ed. du CNRS, Paris, 1984), p. 191
20. G. Gremaud, M. Bujard, invited paper at the 8th International Conference on Internal Friction and Ultrasonic Attenuation in Solids, to be published in J. de Physique

ANALYSIS OF SIGNALS GENERATED BY MULTIPLY REFLECTED ULTRASONIC WAVES IN PLATES

H.KARAGÜLLE

ABSTRACT

This study deals with the mechanics of the output signals generated by multiply reflected ultrasonic waves in plates where the transmitting and receiving transducers are coupled to the same face.

The theoretical analysis considers successive generalized reflections of the waves radiated by the transmitting transducer. The output signal is superposition of successive reflections.

Digital signal processing techniques are applied for obtaining spectral information on reflections in experimental output signals. Short-time steady-state, short-time Fourier, and short-time cepstrum analyses are used for the plates where the output signals contain nonoverlapping tone burst reflections, nonoverlapping pulse reflections, and overlapping pulse reflections, respectively.

Experiments are conducted in the 0.75 to 2.25 MHz frequency range in aluminum plates having thicknesses of 1.25 to 10 cm. Good agreement is observed between the experimental and theoretical magnitudes of individual nonoverlapping tone burst reflections. It is observed theoretically and experimentally that the magnitude spectra of individual nonoverlapping pulse reflections have approximately the same characteristics as the magnitude spectra of the corresponding individual overlapping pulse reflections, where the two corresponding reflections are generated from two multiply reflected waves having the same ray angles in two plates of different thicknesses.

The potential use of this study for nondestructive evaluation applications is discussed.

1. INTRODUCTION

Conventional ultrasonic testing (UT) techniques are used to detect crack-like defects in a structure, and analysis of the output signals in the time domain or in the frequency domain are confined to the output signals with nonoverlapping echoes from the boundaries of the structure or from defects. There is growing recognition of the need for nondestructive evaluation (NDE) techniques which go beyond flaw detection. Metallic, ceramic and composite components in structures may be free of critical-size defects while still being susceptible to failure under design loads due to inadequate or degraded distributed mechanical properties. UT is becoming an increasingly, important NDE technique in the characterization of microstructural defect states of materials because stress waves interact with such defects dynamically.

Recently, Vary et al.(1) introduced an ultrasonic NDE parameter called the "stress wave factor" (SWF). Separate transmitting and receiving transducers are coupled to the same face of the test structure. An input pulse having a broadband frequency spectrum is applied to the transmitting transducer and the number of oscillations (or some modifications thereof) in the output signal at the receiving transducer exceeding a preselected voltage threshold is defined as the SWF. Unlike conventional UT which is limited to the analysis of nonoverlapping echoes, the SWF is also valid for the analysis of overlapping echoes.

The SWF analysis has shown encouraging laboratory results in several NDE applications (1-3). The purpose of this study is to develop a quantitative understanding of the mechanics of the SWF waveforms in isotropic-elastic plates.

2. THEORETICAL ANALYSIS

In the theoretical analysis, the stress waves radiated into the plate by the transmitting transducer are studied first. It is assumed that the transmitting transducer transforms an electrical voltage into a uniform normal stress on the plate.

The schematic of the plate with axially symmetric steady-state excitation is shown in Fig. 1a. h is the thickness of the plate and a_1 is the radius of the transmitting transducer. In Fig. 1b, the cartesian coordinates x,y,z; the cylindrical coordinates r,φ,z; and the spherical coordinates R,θ,φ used in the analysis are shown. The spatial and time behavior of the σ_{zz} component of the stress tensor at x=0 boundary plane is given in Fig. 1a; where ω is radian frequency, t is time, and $i=\sqrt{-1}$. The amplitude of σ_{zz} is taken as the unity since only the frequency response of the plate is of interest.

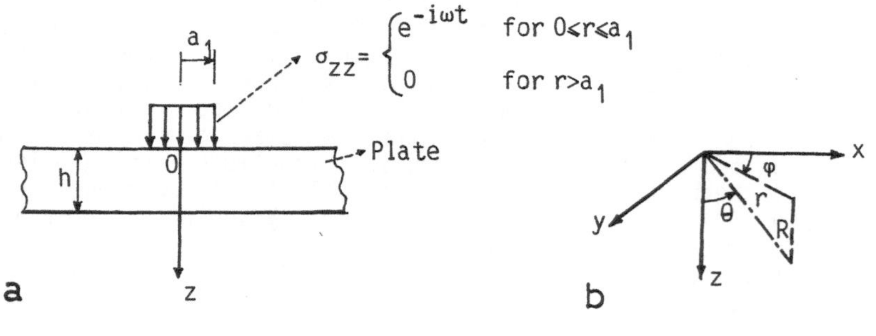

FIGURE 1. Schematics of a) plate with axially symmetric excitation and b) coordinate systems.

The boundary conditions at x=0 plane also require that the σ_{rz} and $\sigma_{z\varphi}$ stress components should vanish. On the other hand, the stress-free conditions at the bottom face of the plate (x=h plane) require that the σ_{zz}, σ_{rz} and $\sigma_{z\varphi}$ stress components should vanish.

One way to address the boundary conditions stated above is to expand the radially propagating plate waves which are derived by the nonlinear Rayleigh-Lamb frequency spectrum (4). Such an approach would be especially useful at low frequencies where the thickness of the plate is of the order of the longitudinal wavelength because a small number of radially propagating waves would be excited.

Another way to approach the problem, which is more practical at high frequencies, is to consider the radiated waves from the transmitting transducer into the plate and then to introduce successive generalized reflections of those waves at the bottom and top faces of the plate (5,6).

Axially symmetric longitudinal (P) and shear (S) waves are radiated into the plate by the transmitting transducer. Each reflection produces both longitudinal and shear waves for each incident longitudinal and shear wave. Thus the P and S waves radiated by the transmitting transducer reflect as PP, PS, SP, and SS waves at z=h plane. For example, the SP wave is the reflection of the S wave as a longitudinal (P) wave. Then, these reflected waves reflect again as PPP, PPS, PSP, ... (a total of 8) at z=0 plane. For example, the PSP wave is the reflection of the PS wave as a longitudinal (P) wave. In this way, increasingly number of successive reflected waves at the bottom and top faces of the plate are introduced.

The waves which travel p times as a longitudinal wave and s times as a shear wave through the plate are called the waves with p,s. For example the waves with p=2,s=1 are the PPS, PSP, and SPP waves. The total of p and s is designated as n.

It is assumed that the receiving transducer produces an electrical voltage proportional to the normal stress, σ_{zz}, generated from an incident wave and averaged over its contact area. The receiving transducer therefore observes only the reflected waves at z=h for which n is even.

The points on the receiving transducer contact area are called the receiving points. After dropping the steady-state term $\exp(-i\omega t)$, σ_{zz} at a receiving point due to incident waves with p,s may be called the frequency response at the receiving point due to the waves with p,s; and it is designated as $H_{p,s}(\omega)$. The average of $H_{p,s}(\omega)$ over the receiving transducer contact area is called the frequency response at the receiving transducer due to the waves with p,s; and it is designated as $\bar{H}_{p,s}(\omega)$.

The general equations for $H_{p,s}(\omega)$ and $\bar{H}_{p,s}(\omega)$ and their detailed derivation and evaluation are given in Ref. 6. $H_{p,s}(\omega)$ is given by an integral which may be evaluated numerically and asimptotically. The asimptotic evaluation by the method of the stationary phase gives

$$H_{p,0}(\omega) \approx -ia_1 D_1(\xi_0,\omega)\,[\,Q_{pp}(\xi_0)\,]^{(p-1)}\,\frac{\exp(i\omega t_{p,0})}{R_1} \tag{1}$$

where only longitudinal waves (s=0) are considered. In Eq. 1 ξ_0 is the stationary point given by $\xi_0/\sqrt{1-\xi_0^2} = r/(ph)$ where r is the distance between the center of the transmitting transducer and the receiving point; $R_1 = ph/\sqrt{1-\xi_0^2}$; $t_{p,0} = R_1/c_1$ where c_1 is the longitudinal wave velocity; $D_1(\xi_0,\omega)$ is called the directivity function for the P wave and is given by

$$D_1(\xi_0,\omega) = 2\pi\,\frac{\sqrt{1-\xi_0^2}\,(2\xi_0^2 - k^2)^2}{\xi_0\,G(\xi_0)}\,J_1\!\left(2\pi\xi_0\,\frac{a_1}{\lambda_1}\right) \tag{2}$$

$k=c_1/c_2$ where c_2 is the shear wave velocity; J_1 denotes a Bessel function of the first kind and of order zero;

$$G(\xi_0) = (2\xi_0^2 - k^2)^2 + 4\xi_0^2\sqrt{1-\xi_0^2}\,\sqrt{k^2-\xi_0^2} \tag{3}$$

$Q_{pp}(\xi_0)$ is called the P to P reflection coefficient function and is given by

$$Q_{pp}(\xi_0) = \frac{4\xi_0^2\,\sqrt{1-\xi_0^2}\,\sqrt{k^2-\xi_0^2} - (2\xi_0^2 - k^2)^2}{G(\xi_0)} \tag{4}$$

For the interpretation of the equations above, the multiply reflected ray corresponding to the wave with p=4,s=0 is constructed geometrically in Fig. 2.

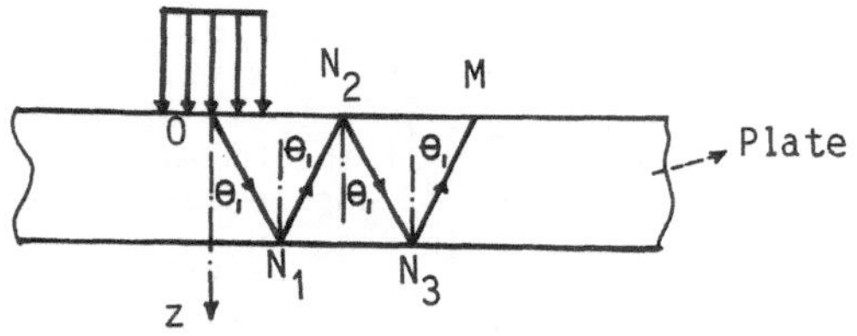

FIGURE 2. Plate and the ray of the wave with p=4,s=0.

A P ray emanating from the center of the transmitting transducer, 0, reflects as P rays at N_1, N_2 and N_3 and propagates to the receiving point M. The reflection angles are given by Snell's law (7) and they are all θ_1 as shown in Fig. 2.

Using the ray geometry it can be shown that $\xi_0=\sin\theta_1$. Further it can be shown that R_1 is the total distance travelled by the ray and $t_{p,o}$ is the time delay required to travel this distance with the speed of c_1. Also, it can be shown that $Q_{PP}(\xi_0)$ is the P to P plane wave reflection coefficient for $\xi_0=\sin\theta_1$.

For the asimptotic results given above to be valid, the dimensionless quantity $R_1\lambda_1/a_f^2$ must be adequately large (6).

3. EXPERIMENTS AND ANALYSIS OF OUTPUT SIGNALS

A schematic of the experimental system used in this study is shown in Fig. 3. Aluminum (6061-T6) plates are considered. Typical input signals used in the experiments are shown in Fig. 4. Typical output signals are shown in Fig. 5. The plate thicknesses (h) and the distances between the transmitting and receiving transducers (ℓ) are indicated in the title of Fig. 5. The time delays for some possible waves to arrive at the receiving transducer center are calculated using asimptotic ray analysis, and their locations are indicated above the time signals in Fig. 5 with the values of p and s.

It is observed from Fig. 5a that the output signal is dominated by reflections of the P waves only. These reflections can be observed as unoverlapped short-time steady-state (finite duration sinusoidal or tone burst) signals. This is because the thickness of the plate is sufficiently large and thus the difference between the time delays for the successive significant reflections is adequately large. In this case it is possible to obtain spectral information on individual reflections by changing the center frequency (f) of the tone burst and then by reading the amplitudes of short-time steady-state reflections. Such an analysis is called short-time steady-state analysis.

It is observed from Fig. 5b that the output signal is dominated by reflections of the P waves only and of the P waves containing a single S. These reflections can be observed as pulses and most of them are not overlapped significantly. Here, it is no longer possible to observe unoverlapped short-time steady-state reflections because the thickness of the plate is not large enough. In this case, where significant reflections as pulses are not overlapped significantly, short-time Fourier analysis can be used to obtain spectral information on individual reflections.

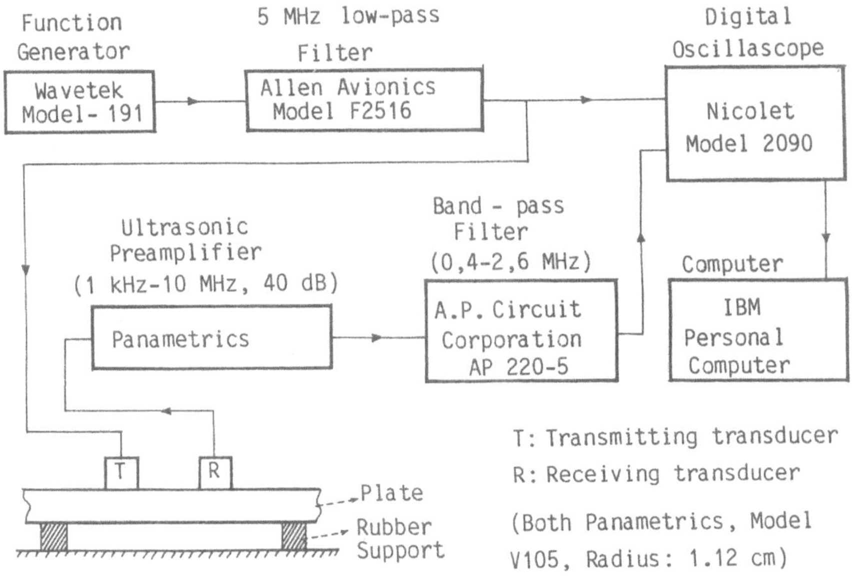

FIGURE 3. Schematic of experimental system.

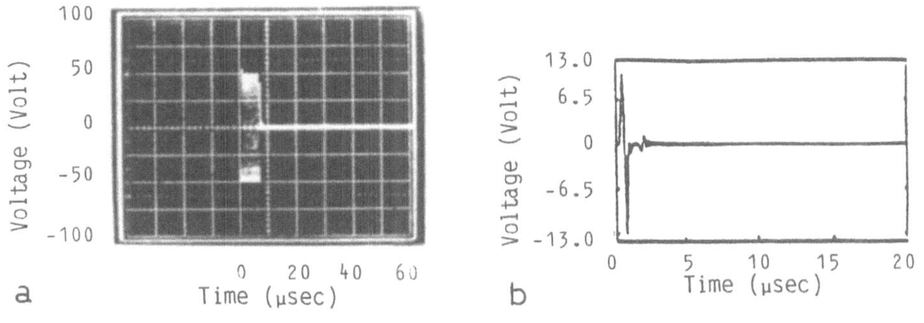

FIGURE 4. Input signals a) tone burst with f=1.5 MHz center frequency, b) pulse with -6 dB frequency band 0.4 to 3 MHz.

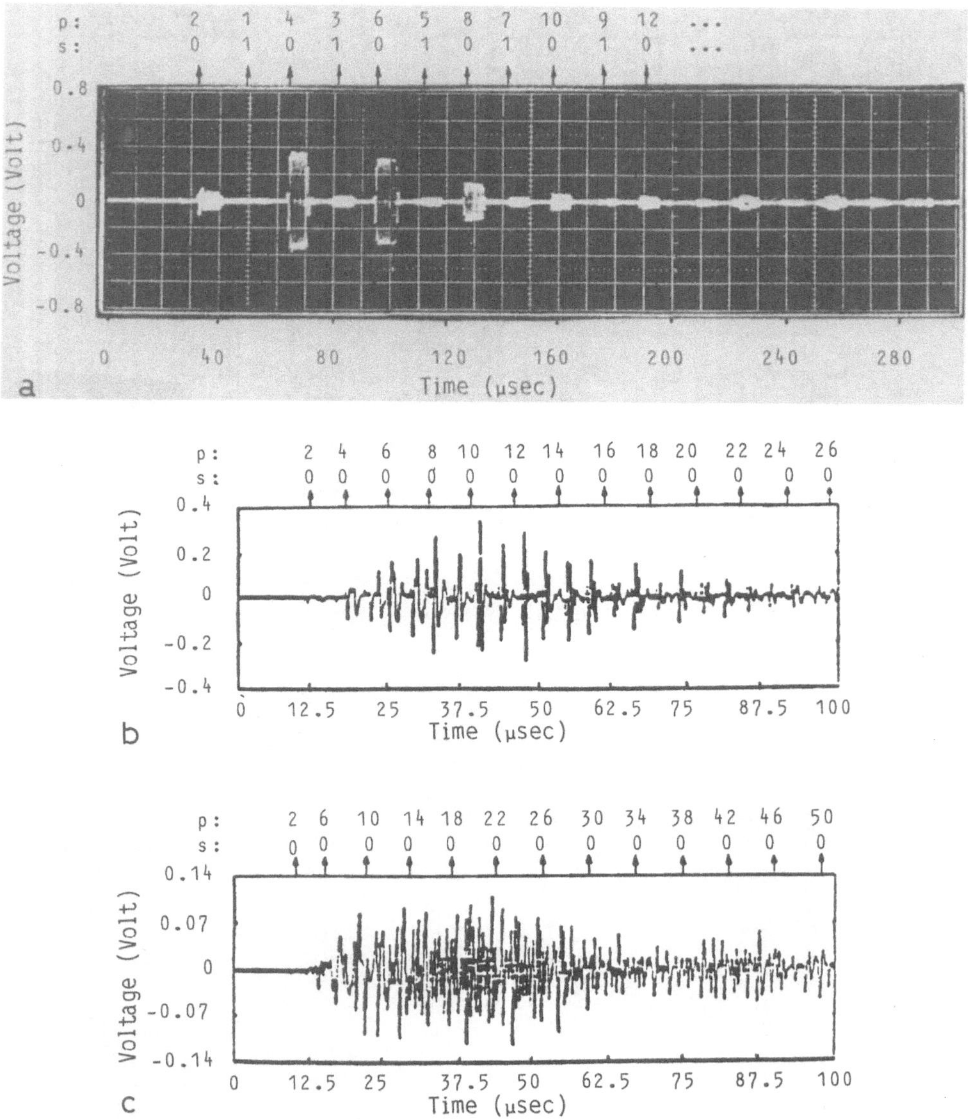

FIGURE 5. Output signals for a) h=10 cm, ℓ=5.5 cm; b) h=2.4 cm, ℓ=6 cm; c) h=1.25 cm, ℓ=6.25 cm.

In the short-time Fourier analysis, a rectangular data window is used and its time interval is chosen such that the short-time signal contains only an individual reflection and excludes all the other reflections completely. Then, by taking the fast Fourier transform (FFT) of the short-time signal the magnitude spectrum of the individual reflection can be found directly.

It is observed from Fig. 5c that the output signal contains overlapped pulse reflections and they cannot be identified separately. This is because the thickness of the plate is so small that the difference between the time delays of the successive significant reflections is smaller than the length of a single reflection. Here, the short-time Fourier analysis does not give a direct access to spectral information of an individual reflection. However, the short-time cepstrum analysis (6,8) may give an estimate of the magnitude spectrum of a single reflection although it is overlapped with other adjacent reflections.

A schematic of the short-time cepstrum analysis is shown in Fig. 6.

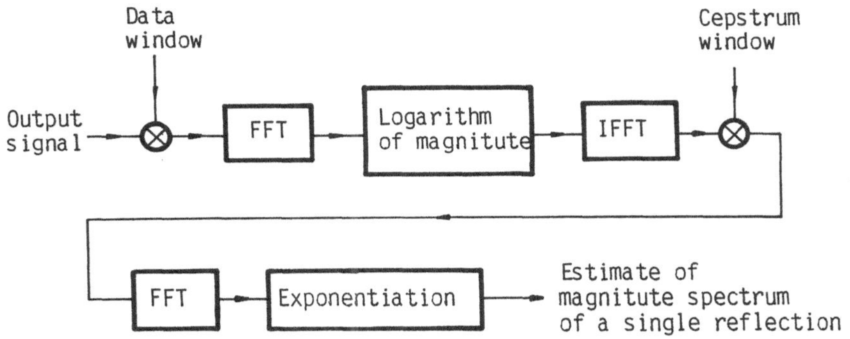

FIGURE 6. Schematic of short-time cepstrum analysis.

A suitable data window is used so that the short-time signal contains only several reflections and it is assumed that the reflections in the short-time signal have approximately the same waveform but they differ only by a scale. Then the short-time signal can be modelled as the convolution of an impulse train with the individual reflection having the maximum scale in the short-time signal. As shown in Fig. 6, the FFT, the logarithm of the magnitude spectrum, and the inverse fast Fourier transform (IFFT) operations are applied to the short-time signal, respectively. The resultant transformed signal is called the cepstrum. The cepstrum is multiplied by a suitable window (6) and then the FFT and exponentiation operations are applied to obtain an estimate of the magnitude spectrum of the individual reflection.

4. THEORETICAL AND EXPERIMENTAL RESULTS

Theoretical and experimental steady-state amplitudes of some possible reflections for the plate with h=10 cm are given in Fig. 7. The time delays for the reflections considered are indicated above the figures with the values of p and s. Different frequencies (f) and distances between the centers of the transmitting and receiving transducers (ℓ) are considered.

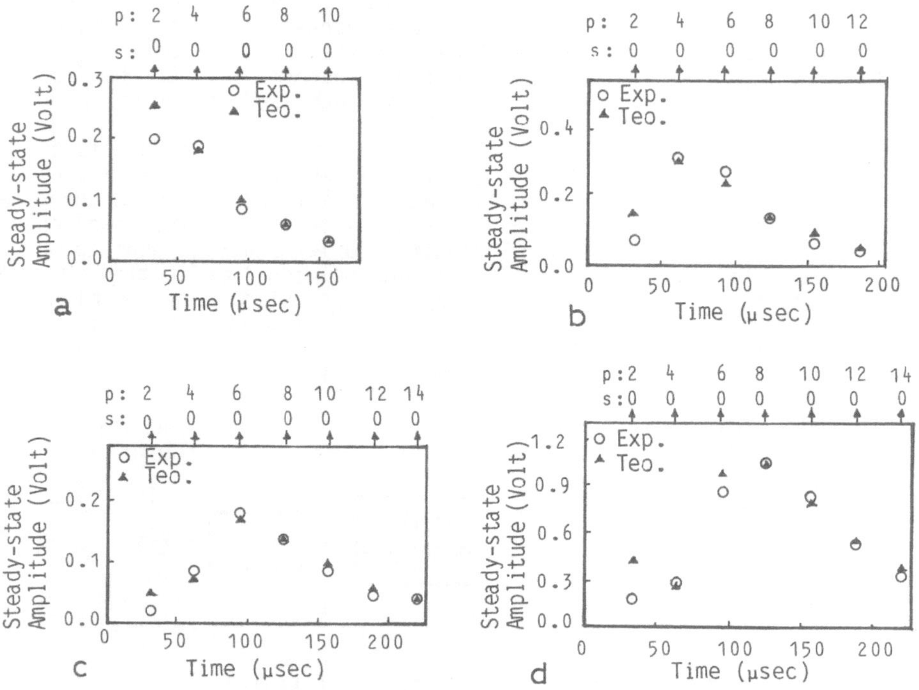

FIGURE 7. Theoretical and experimental steady-state amplitudes of multiply reflected longitudinal waves in an aluminum plate of h=10 cm thickness for a) f=0.75 MHz, ℓ=5.5 cm; b) f=1.50 MHz, ℓ=5.5 cm; c) f=2.25 MHz, ℓ=5.5 cm; and d) f=1.50 MHz, ℓ=10 cm.

As observed from Fig. 7, as f or ℓ increases the reflection with a larger value of p has maximum amplitude compared with other reflections. This can be explained by the angle and frequency dependent behavior of the directivity function given in Eq. 2 (6). The directivity function has a maximum for the ray angle θ_1=0 and decreases to zero for smaller ray angles at higher frequencies. The reflections with larger p have smaller ray angles but they have larger total ray path lengths. Also larger ℓ means larger ray angles.

The magnitude spectra of some reflections in the outputs given in Figs. 5b and 5c are found by short-time Fourier and cepstrum analyses, respectively, and their characteristics are given in Table 1. Rectangular and Hamming data windows (8) are used for Figs. 5b and 5c, respectively. In the short-time cepstrum analysis for Fig. 5c, a cepstrum window having duration from -1.25 μsec to 1.25 μsec is used for all the reflections considered.

As observed from Eq. 1, only the directivity function determines the frequency dependence of the steady-state amplitude of a reflection. Thus, the theoretical results suggest that the reflections with the same ray angles have the same waveform because all the frequency components of the input signal are transmitted to the receiving transducer in the same direction. However, the reflections with the same ray angles may have different scales because they may travel different distances and their value of p may be different.

TABLE 1. Characteristics of some reflections with only P waves in experimental output signals.

Output Signal (Figure)	Reflection (Value of p)	Ray Angle (Degree)	Average Resonant Frequency* (MHz)	-3 dB Frequency Bandwidth (MHz)
	8	17.35	1	0.74
5b	10	14.04	1.19	0.73
	12	11.77	1.49	0.83
	16	17.35	1.17	1.04
5c	20	14.04	1.3	0.84
	24	11.77	1.44	0.98

* The average resonant frequency is the average of the frequencies over the -3 dB frequency band.

The reflections with $p=p_1$ for Fig. 5b have the same ray angles as the reflections with $p=2p_1$ for Fig. 5c as observed in Table 1.The experimental results given in Table 1 show that the reflections with the same ray angles have approximately the same spectral characteristics, which agrees with the theoretical analysis.

As observed from Table 1, the average resonant frequency moves to higher frequencies for larger values of p. Again, this can be explained by the frequency and angle dependent behavior of the directivity function.

5. CONCLUSIONS AND RECOMMENDATIONS

In this study, stress wave transmission characteristics of an isotropic-elastic plate are analyzed theoretically and experimentally. Separate transmitting and receiving transducers coupled to the same face of the plate are considered.

The output signal is the superposition of signals (reflections) generated by stress waves which multiply reflect between the bottom and top faces of the plate. The magnitude spectra of individual reflections contain information about the frequency and angle dependent directivity function of the transmitting transducer, and attenuation of the plate material due to losses or microstructural defects. Since ray angles and total ray path lengths are different for different reflections, the effects of the directivity function and material attenuation on the magnitude spectra of different reflections are also different.

The frequency components of an individual reflection can be obtained experimentally by short-time steady-state, short-time Fourier, or short-time cepstrum analyses if the output signal contains nonoverlapping tone burst reflections, nonoverlapping pulse reflections, or overlapping pulse reflections, respectively.

The frequency components of a reflection can be obtained theoretically by the asimptotic ray analysis if the total ray path length is sufficiently large. Thus, although the thickness of the plate may be small, in general, there will be reflections in the output signal for which total ray path lengths are sufficiently large. Then asimptotic ray analysis which can be quantifiably and easily interpreted for NDE applications will be valid for such reflections.

Efforts in developing NDE techniques to characterize microstructural defect states of structures should benefit from this study. Microstructural defects may affect the directivity function of the transmitting transducer and the material attenuation. These both effects may be observed on the magnitude spectra of reflections.

The results of this study may be also used to extend the applicability of conventional ultrasonic testing (UT) techniques which are currently limited to the analysis of nonoverlapping reflections.

Acoustic emission (AE) studies should also benefit from this study. In AE, the source is a defect rather than a transmitting transducer. Different defects may have different directivity functions and these defects may be characterized by the analysis of the magnitude spectra of reflections.

Finally, the results in this study should be extended to structures other than plates. Structures made of composite materials should be considered. The effects of microstructural and macrostructural defect states should be analyzed theoretically and experimentally.

ACKNOWLEDGMENTS

I am grateful to Professor James H.Williams, Jr. and Dr. Samson S. Lee at the Massachusetts Institute of Technology for advice and help concerning the work reviewed here. NATO generously supported my attendance at the Advanced Study Institute.Finally the Materials and Structures Division at the NASA Lewis Research Center (Project Monitor, Alex Vary) is gratefully acknowledged for its support of this research.

REFERENCES

1. Vary A and Bowles KJ: An Ultrasonic-Acoustic Technique for Nondestructive Evaluation of Fiber Composite Quality: Polymer Engineering and Science, Vol 19, No 5, April 1979, pp 185-191.
2. Williams JH,Jr and Lampert NR: Ultrasonic Evaluation of Impact-Damaged Graphite Fiber Composite: Materials Evaluation, Vol 38, No 12, pp 68-72.
3. Wehrenberg RH,II: New NDE Technique Finds Subtle Defects: Materials Engineering, Vol 92, No 3, September 1980, pp 59-63.
4. Pardee WJ: Radially Propagating Surface and Plate Waves: The Journal of the Acoustical Society of America, Vol 71, No 1, January 1982, pp 1-4.
5. Ceranoglu AN and Pao YH: Propagation of Elastic Pulses and Acoustic Emission in a Plate, Parts 1-3: Journal of Applied Mechanics, Vol 48, No 1, March 1981, pp 125-147.
6. Karagülle H: PhD Thesis: Department of Mechanical Engineering, MIT, Cambridge, MA, USA, May 1984.
7. Graff KF: Wave Propagation in Elastic Solids, North Holland Publishing Company, NY, 1973.
8. Rabiner LR and Schafer RW: Digital Processing of Speech Signals, Prentice-Hall, Inc, Englewood Cliffs, NJ, 1978.

FLAW CLASSIFICATION IN COMPOSITES USING ULTRASONIC ECHOGRAPHY

J.P. DUMOULIN , J.F. de BELLEVAL

Composites are becoming more and more popular in industrial applications and are tending to replace metals in a certain number of cases either for cost or weight advantages.

The testing of these materials gives rise to new problems and it is necessary to adapt and even completely remodel the methods used for classical materials.

At the University of Compiegne an investigation has been undertaken into the use of ultrasonic testing for epoxy carbon composites which are used, amongst other applications, for aircraft structures.

I. THE MATERIAL, EXISTING TEST TECHNIQUES, DISADVANTAGES

The materials under investigation are epoxy carbon composites composed of layers of carbon fibres in an epoxy resin matrix. Each unidirectional layers are approximately 0.13mm thick. The material is therefore heterogenic and anisotropic leading to difficulties of modeling the propagation of ultrasonic waves even in the absence of flaws. In order to understand the problems better, the following points may be noted :

- The velocity of the ultrasonic wave in the fibre axis direction is three times greater than in the direction normal to the layers,
- The wavelength in the direction perpendicular to the fibres corresponding to the centre frequency of the transducers used (5MHz) is of the order of five times the layer thickness,
- The attenuation of the ultrasonic waves at 5MHz is approximately 16dB for unflawed, 1cm thick, material and reaches 30dB for porous material (1% porosity), in usual measurement conditions.

The principal flaws encountered on relatively thin plates (a few millimeters) are delamination which are classed according to their dimensions as complete delamination or micro delamination, and porosity between layers. This has led to the use of ultrasonic transparency testing for this type of plate. The attenuation of an ultrasonic wavetrain is measured as it traverses the material, either with two transducers and a waterjet method which helps the coupling between the transducers and the material (Fig. 1) or in a tank, by the so-called double transmission method, where the wave emitted by the transducer is reflected by an acoustic mirror and crosses the material twice before being detected by the same transducer (Fig. 2). In the presence of delaminations, the wave is practically not transmitted by the plate (the amplitude of the detected wave is very small). In the presence of porosity the ultrasonic absorption increases considerably. The measurement of the wave attenuation and the application of an empirical formula taking into account the number of layers and some of the material characteristics permits an average porosity density per layer to be calculated, which is a criterion of the strength of the structure. This method gives satisfactory results for plates with fewer than 40 layers.

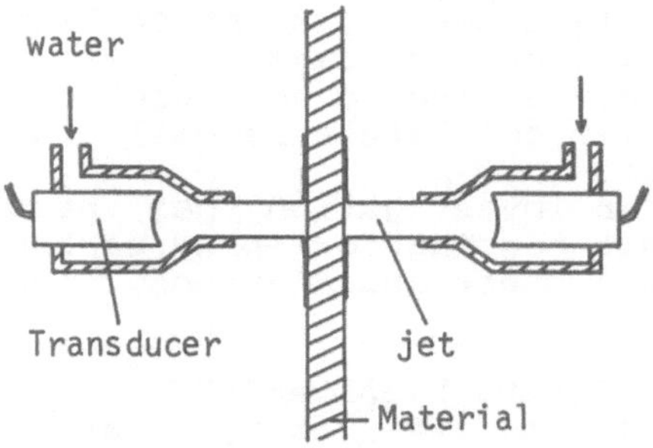

FIGURE 1 Waterjet Method

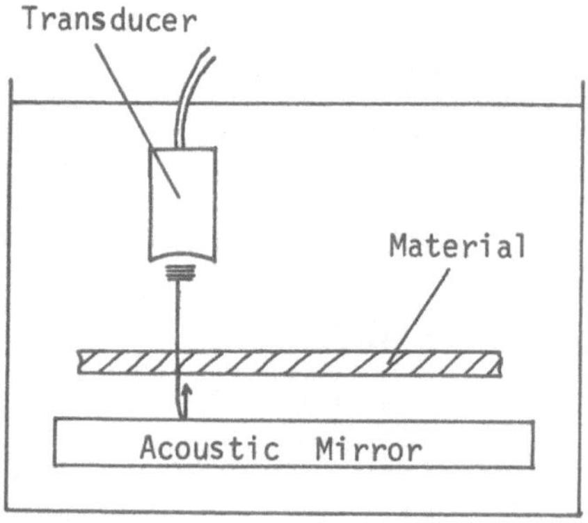

FIGURE 2 Double transmission Method

The use of this type of material for structures of increasing complexity (a complete aircraft wing for example) introduces new problems as indicated below.

For thicker plates an evaluation of the average porosity density is no longer sufficient. A rate of porosity distributed over a few layers giving the same attenuation will not have the same effects on the strength of the material.

Differents flaws could be present in places where the structure has more complex geometrical forms (ribs etc ...)

To solve these problems the transparency measurements which can only give results averaged over the total thickness of the structure are insufficient, which necessitates the use of echographic methods which can give additional information.

The interpretation of the results of such methods is however sometimes difficult. The problems encountered are due to the high absorption of the ultrasonic waves in the material, the presence of a high level of so-called structural noise (this noise is due to the reflections at the inhomogeneities of the material) and the difficulty of identifying the different flaws which give similar detected signals.

These difficulties have led us to try to extract the necessary information by more sophisticated methods than these used at present and in particular, signal processing techniques.

The present investigations have been orientated towards three problems :

- the identification of two types of flaw giving similar echographic signals, first a microdelamination and secondly a porosity concentrated over several layers,
- the testing of thin plates (of about 1mm thickness) for which the above method is too approximate,
- the measurement of the local absorption in thick plates (up to a centimetre).

The work presented in this paper corresponds to the first problem.

II. PROBLEM DESCRIPTION - CHOICE OF METHOD

The goal is to discriminate in the composite structure described above between two types of flaw, the first called microdelamination, which is delamination between layers over an area of several mm^2 , the other corresponding to porisity concentrated over several layers. The microdelaminations were simulated by inserting teflon film and the porisity by modifying the polymerisation cycles.

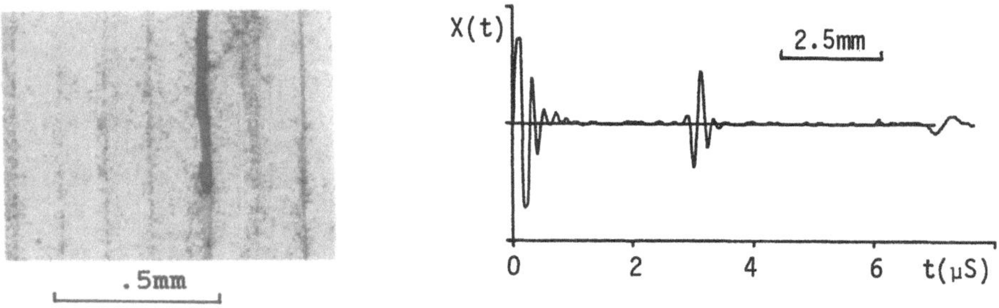

FIGURE 3 Microdelamination

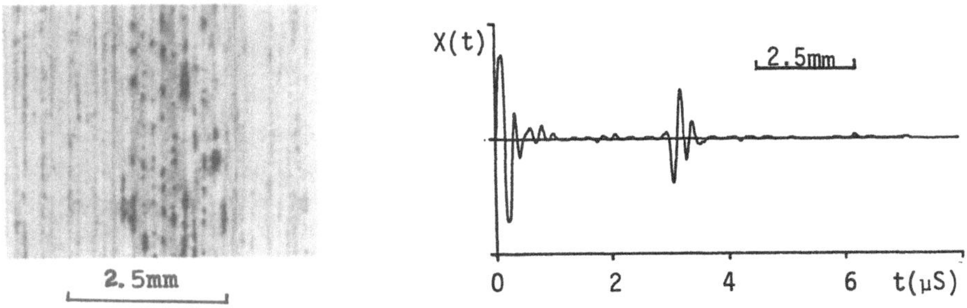

FIGURE 4 Concentrated Porosity

Figures 3 and 4 illustrate the problem using two echographic results one for the microdelamination, the other for a porosity concentrated over 12 layers. The signals detected in each case are similar, they are not of the same form but there are no characteristics aiding discrimination. The background noise from the structure is also considerable. The time signal being unable to provide satisfactory results, it seems logical as others have done /2/ to examine the frequency analysis of the signal. In order to have more information on the flaws a method developed in our laboratories was used (N. MERCIER /4,5/) permitting not only the amplitude but also the phase of signal to be exploited. The results were insufficient to obtain a satisfactory discrimination /3/ between the two types of flaw for the following reasons :

- when the frequency analysis is carried out on the complete echographic signal (corresponding to the total thickness of the sample) this is too much global information and that corresponding to the flaws are masked.
- when the frequency analysis is carried out on the part of the signal corresponding to a flaw, a "time window" is used to select this part. The result in this case is very dependent on the position of the window and it is impossible to completely eliminate the effects, principally because of the background noise in the neighbourhood of the useful part of the signal.
- finally the material has a high ultrasonic absorption which varies with frequency and the echographic response of a flaw is therefore dependent on the depth of the flaw, which is particularly important for the frequency domain.

Globally these results are due to the fact that frequency analysis is well adapted to relatively stable or stationary signals but less to signals that evolve as in this case.

Stationary signals are characterised by second order functions of a single variable (correlation function depending on time or spectral density depending on frequency). Non stationary signals are characterised by functions of two variables, two time variables, two frequency variables or one time and one frequency variable. This permits a joint representation in time and frequency which has a more significant physical interpretation. These representations try to define an energy distribution in the time frequency plane /5 to 10/ or in a more realistic way using the uncertainty relation of GABOR on the elementary cells of area $\Delta t \, \Delta \nu$ of this plane /11/.

The echographic signals which interest us being non stationary or evolutionary, it is this type of representation which will be used.

III. THE BASE OF THE METHOD

A certain number of time frequency representations are proposed by different authors. Some of which have been referred to in references 5 to 11. Our choice has been guided by the following constraints :

- retain a localisation of the information permitting the interface echos and structural noise effects to be eliminated,
- estimate a frequency content of relatively simple interpretation,
- try to eliminate the frequency shift of the signal as a function of flaw depth linked to the ultrasonic absorption,
- use relatively rapid methods of calculation.

The representation of FLANAGAN /9/ or the "sliding windows" overcomes these constraints. It concerns the spectral content of a troncated time signal using a sliding time window. This representation better known as "short term spectrum" is often

used with a window adapted to the signal investigated.
It is defined by :

$$\rho(t,\nu) = \left| \int_{-\infty}^{+\infty} X(\tau)W(t-\tau)\, e^{-2\pi i\nu\tau}\, d\tau \right|^2 \qquad (1)$$

$X(t)$ is the time signal (an echographic signal in this case)
$W(t)$ is the time window used.

This representation has the advantage of being a positive quantity which can be associated with the notion of energy in the (t,ν) plane, and the spectrum only takes into account the signal around the instant considered.

Figure 5 shows the results corresponding to the signal of Figure 3, the classical echographic signal and the joint representation (t,ν). A better resolution is observed in the classical representation than in the joint representation where a compromise related to the uncertainty relation of GABOR has to be made between the time and frequency resolution.

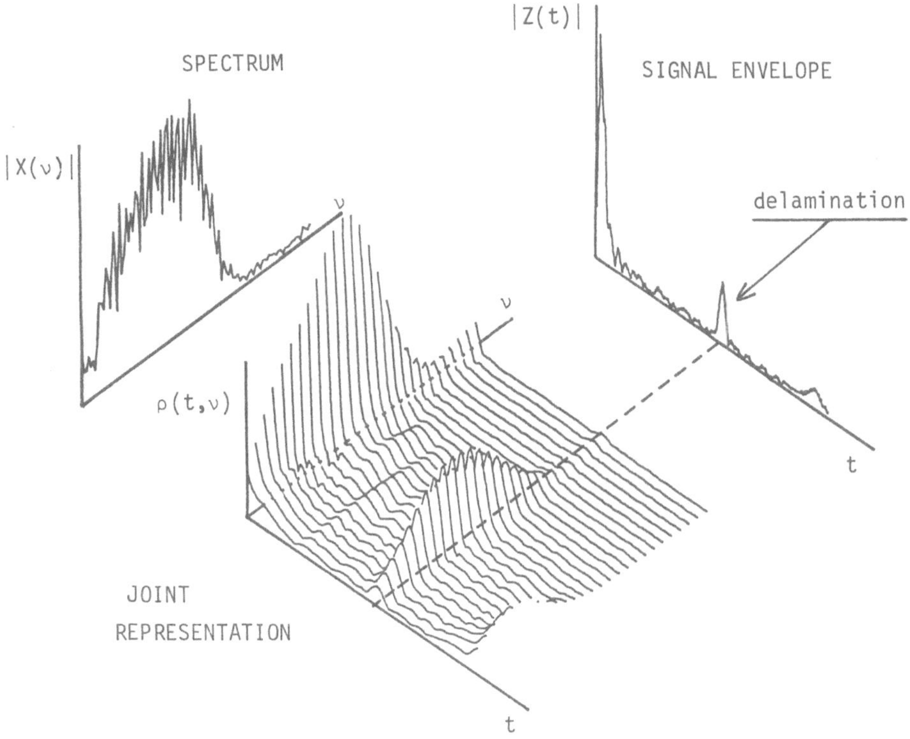

FIGURE 5 Classical and joint Representation

The use of this type of representation complicates the interpretation at first sight because it is a function of two variables rather than one. Thus it is necessary to extract simpler parameters from this representation to permit the discrimination of flaws, with criteria which are as independent as possible of the attenuation in the material and, therefore, of the frequency shift as a function of depth. For this reason the parameters chosen are strictly related to the instantaneous frequency. These parameters are thus defined.

3.1. Instantaneous Frequency.

The instantaneous frequency introduced by Ville /7/ is defined as the derivative of the argument of the analytical signal Z(t) at $(2\pi)^{-1}$:

$$\nu_i(t) = \frac{1}{2\pi}\frac{d\phi(t)}{dt} \tag{2}$$

where $\phi(t)$ is defined by $Z(t) = A(t)\exp(i\phi(t))$ and Z(t), the analytical signal associated with the real signal X(t), is a complex signal with a real part equal to X(t) and the imaginary part is the HILBERT transform of X(t).

It can be approximated by the first spectral moment of the representation $\rho(t,\nu)$

$$\tilde{\nu}_i(t) = \frac{\int_0^\infty \nu\, \rho(t,\nu)d\nu}{\int_0^\infty \rho(t,\nu)d\nu} \tag{3}$$

3.2. Instantaneous Bandwidth.

The instantaneous bandwidth is defined as the second order moment of $\rho(t,\nu)$. It is expressed :

$$B_i(t)=2\left[\frac{\int_0^\infty (\nu-\tilde{\nu}_i(t))^2\, \rho(t,\nu)d\nu}{\int_0^\infty \rho(t,\nu)d\nu}\right]^{1/2} \tag{4}$$

or

$$B_i(t)=2\left[\frac{\int_0^\infty \nu^2\, \rho(t,\nu)d\nu}{\int_0^\infty \rho(t,\nu)d\nu} - \tilde{\nu}_i(t))^2\right]^{1/2}$$

This is an equivalent bandwidth around the instantanous

frequency.

3.3. Time Distribution of the Instantaneous Frequency.

This name is given to the second order moment in time corresponding to the instantaneous frequency at time t. It is expressed :

$$E_{\tilde{\nu}_i}(t) = \left[\frac{\int_{t_{min}}^{t_{max}} (\tau - t)^2 \rho(\tau, \tilde{\nu}_i(t)) d\tau}{\int_{t_{min}}^{t_{max}} \rho(\tau, \tilde{\nu}_i(t)) d\tau} \right]^{1/2} \qquad (5)$$

To have a parameter localised around the instant t, the integration is effected between the limits tmin and tmax of the window $W(\tau - t)$ which defined $\rho(t, \nu)$ (1).

3.4. Product Bandwidth-Time Distribution.

This parameter provides no additional information because it is the product of the two preceding parameters. It is a non dimensional value (a surface in the time-frequency plane) related to the energy concentration around the point $(t, \tilde{\nu}_i(t))$, for the total time history of the signal :

$$P_i(t) = B_i(t) \times E_{\tilde{\nu}_i}(t)$$

Note that, in the definition of these parameters whether in frequency or time, the time variable has been privileged to obtain localised information and, in the frequency domain, the parameters have been defined with respect to the instantaneous frequency to follow the frequency evolution of the signal as a function of the penetration of the wave in the material.

Figure 6 shows the echographic signal of a composite with a delamination and the distribution $\rho(t, \nu)$ associated. On the projection of this representation are the parameters $\tilde{\nu}_i(t)$, $B_i(t)$ and $E_{\tilde{\nu}_i}(t)$ corresponding to the echo related to the delamination.

IV. EXPERIMENTAL RESULTS

The results presented were obtained on specimens in epoxy carbon of 78 layers (10mm thick), two types of flaws were simulated in these specimens :

- microdelaminations simulated by 2/100mm thick teflon film placed between layers.
- regions of concentrated porosity obtained by premature aging of the layers before compression.

The echographic signals were digitized at 100 MHz by a transient digital recorder Biomation 8100 which was piloted by a PDP 11/23 minicomputer which also processed the data. The

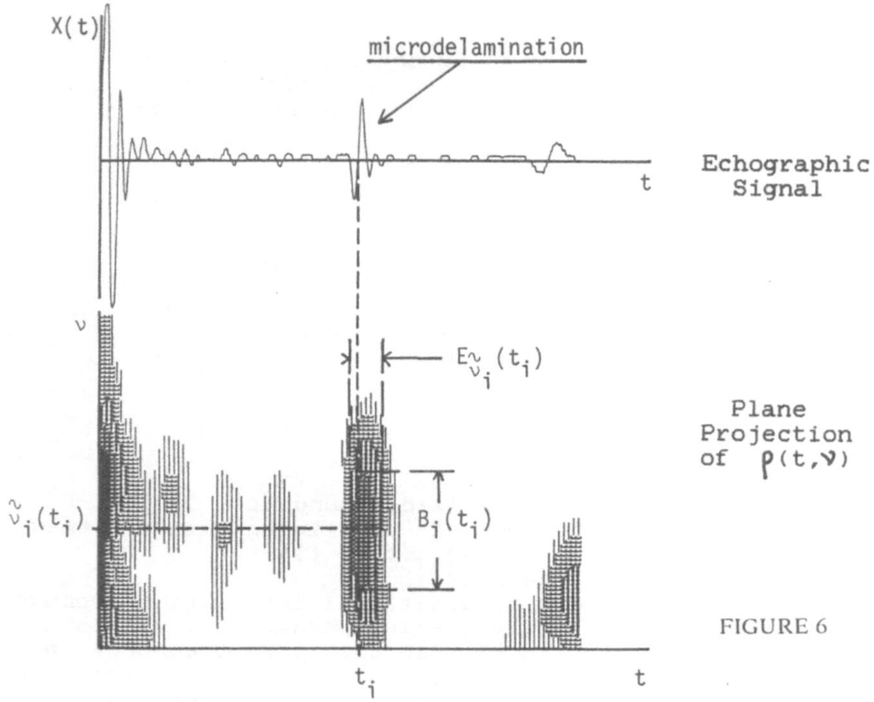

FIGURE 6

window used was a Hamming of 64 points :

$$w(t) = 0.54 + 0.46 \cos (2\pi t/T)$$

The transducer was a focussed wideband transducer of 5MHz central frequency.

Figure 7 shows the echographic signal, a micrographic section, and the parameters $E_{\tilde{\nu}_i}(t)$ and $P_i(t)$ for a microdelaminage. Figures 8 and 9 are same representations for a porosity concentrated in a zone of 12 layers of carbon in the case of figure 8 and in two zones of six layers at two different depths for figure 9. The parameters $E_{\tilde{\nu}_i}(t)$ and $P_i(t)$ have only been presented in the neighbourhood of the echos corresponding to the flaws studied, previously detected with the time signal.

We can observe on **these** three figures that the parameters $E_{\tilde{\nu}_i}(t)$ and $P_i(t)$ both present local minima when the flaw is a microdelamination or local maxima in the case of concentrated porosity. Systematic tests have been done on a great number of flaws (about 50) and this phenomenon has always been observed. Hence this is a criterion for discrimination between the two kinds of flaws.

Some simple physical considerations permit an intuitive

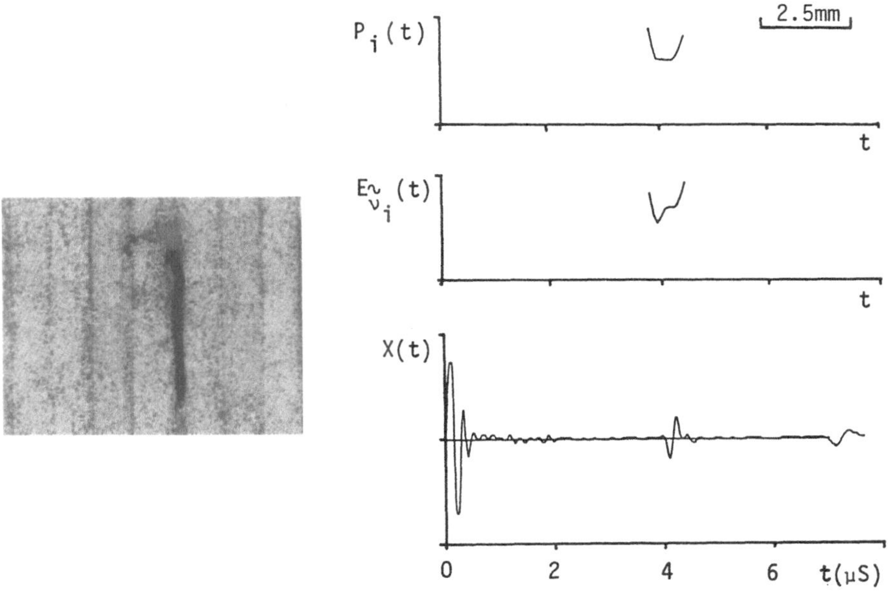

FIGURE 7 Microdelamination

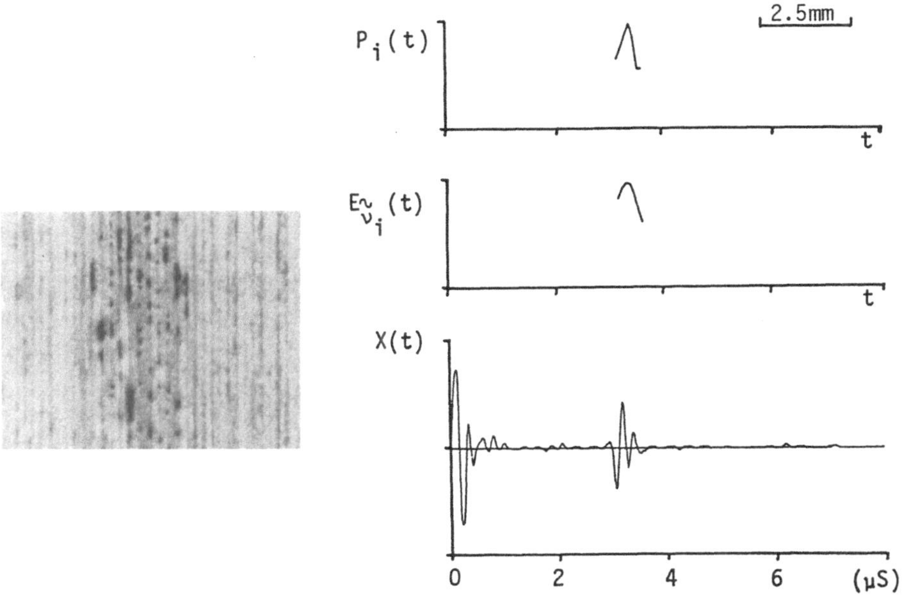

FIGURE 8 Concentrated Porosity

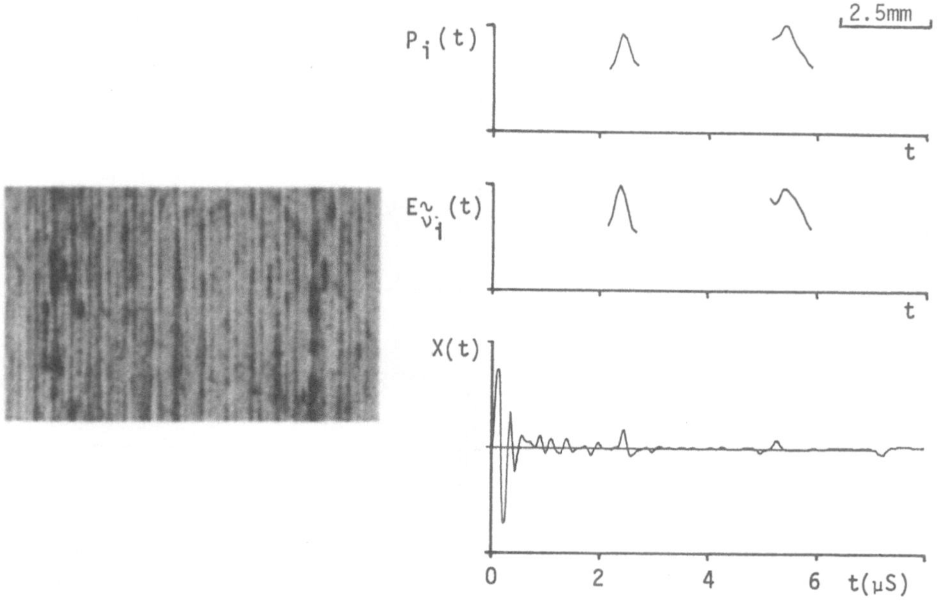

FIGURE 9. Concentrated Porosities at two Depths

interpretation of this result. A microdelamination behaves as a small isolated plane reflector whereas the porosities, which are concentrated on few layers (6 or 12 for our examples), are spread reflectors on a more important depth and tnen scatter waves during longer time. This can be noticed by means of a careful comparison between time signals corresponding to each type of flaw at a constant depth . However a change in depth, by the frequency shift hence induced, causes a variation of echo duration which prevents this discrimination.

V. CONCLUSION

We developed a method of processing ultrasonic echographic signal based upon time-frequency representation which has permitted the discrimination of two types of flaws present in an epoxy-carbon composite. This discrimination has been achieved because the extraction, by means of this kind of representation, of several local frequency parameters, become free from frequency shift linked to material absorption. The two kinds of flaws have been discriminated because they can be assimilated, one (microdelamination) with a partial reflector and the other (concentrated porisity) with a scatterers concentration, and these physical characteristics have been put forward by means of the chosen parameters.

This method can be certainly applied to other cases, which we studied, such as the detection of flaws partially hidden in structural noise, the discrimination of adhesive bound defects

For the further developements of this study and moreover, by the use of a simple modeling of signals and ultrasonic propagation, we have shown /1/ that it is possible to deduce the local absorption of ultrasound in the medium and then to better quantify the state of stucture.

Other studies on echographic signal processing are more adapted to other particular cases, as the use of spectral amplitude and phasis of the signal /12,13/, and are led at the same time in our laboratory. All these methods introduce new parameters which could be exploited on a more global way than that is currently done, by mean of pattern recognition methods which are otherwise developped at the University of Compiegne.

Aknowledgement: This study has obtain a financial support from the DRET (Ministry of Defence) and has been led in collaboration with the Laboratoire Central of the Société Nationale Industrielle Aérospatiale (S.N.I.A.S)

REFERENCES

1. J.P. Dumoulin : Traitement du signal adapté au contrôle de matériau composite par ultrasons: Représentation conjointe temps-fréquence. Thèse de Doctorat, Université de Compiègne, to be presented Jan 86.

2. D.W. Fitting, L. Adler : Ultrasonic Spectral Analysis for non Destructive Evaluation, Plenum Press 1981.

3. J.P. Dumoulin : Caractérisation des défauts d'une structure composite carbone-epoxy par ultrasons, Rapport final UTC/DAVI N° 85-09

4. N. Mercier, J.F. de Belleval : Use of the phase of the signal in ultrasonic spectral analysis to evaluate flaws, Ultrasonics International 1981, Brighton G.B.

5. B. Escudie : Représentation en temps et fréquence des signaux d'énergie finie : analyse et observation des signaux Ann. Telecomm., 34, N° 3-4, 1979.

6. B. Escudie , J. Grea : Sur une formulation générale de la représentation en temps et en fréquence dans l'analyse des signaux d'énergie finie, C.R. Acad. Sc. Paris, t. 283, N° 15, Série A, 1976, pp 1049-1051.

7. J. Ville : Théorie et applications de la notion de signal analytique, Cables et Transmissions, 2, N° 1, 1948, pp.61-74.

8. A.W. Rihaczek : Signal Energy distribution in time and frequency, I.E.E.E. IT-14, 1968, pp. 369-374.

9. J.F. Flanagan : Speech analysis synthesis and perception, Germany : Spinger Verlag, 1965.

10. M. Fink, F. Hottier, J.F. Cardoso : Ultrasonic signal processing for in vivo attenuation measurement : short time Fourier analysis, Ultrasonic Imaging 5, 1983, pp.117-135.

11. J.L. Lacoume, W. Kofman : Description des processus non stationnaires par la représentation temps-fréquence, Applications, Colloque GRESTI 1975, pp. 96-101.

12. N. Mercier, J.F. de Belleval : Phase information in ultrasonic spectroscopy, I.E.E.E. Ultrasonics Symposium 1983, Atlanta, U.S.A.

13. N. Mercier : Transfer function characteristics of an echographic system ; Application to a parallel face target, Ultrasonics International 1985, London.

APPLICATION OF LIGHT DIFFRACTION BY A SYSTEM CONSISTING OF AN ULTRASONIC WAVE AND A PHASE OPTICAL DIFFRACTION GRATING

P. KWIEK, A. MARKIEWICZ

1. INTRODUCTION

During the last ten years much interest has been concentrated on the phenomenon of light diffraction by complex systems consisting of two ultrasonic beams or of a combination of an ultrasonic beam and an optical grating [7.8]. Investigation of light diffraction by a system combined of an ultrasonic wave and an optical grating was firstly started by Calligaris, Ciuti and Gabrielli [1,2,3] and then continued by Patorski [4.5] and Kwiek [6].

Besides the investigations carried on in the field of acoustooptics there is research in optics concerned with the phenomenon of light diffraction by two optical gratings. However there is a difference between these two phenomena: in acoustooptics, indeed, because of Doppler effect, the time modulation may occur in the diffraction image.

In the present paper the authors discuss the possibilities of applying the phenomenon of light diffraction by a system consisting of an ultrasonic wave and an optical phase grating to measure ultrasonic velocity. In this method, the spatial period of the optical phase grating is taken as a standard for ultrasonic wavelength and the ultrasonic velocity can be evaluated by matching the spatial periods of both gratings (ultrasonic and optical).

2. THEORY

Let us consider two adjacent diffraction gratings, the former being a plane progressive ultrasonic wave, the latter an optical phase grating placed along x-axis, as shown in Fig. 1.

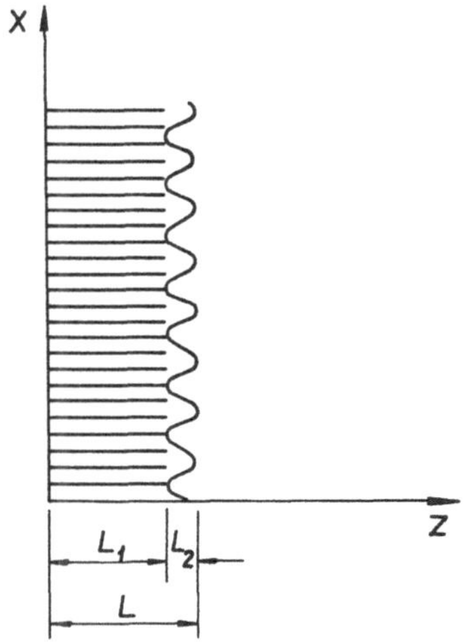

FIGURE 1. The geometry of the diffraction system.

A plane light wave of frequancy ω and complex amplitude E

$$E = E_0 \cdot e^{i(\omega t - kz)} \tag{1}$$

illuminates the diffraction system. Assuming to be in the Raman-Nath region, the amplitude of light after passing through the considered system has the form

$$E\,(x,z = L,t) = E_0 \cdot e^{i[\omega t - kL - a_1 \sin(\Omega t - K_u x) - a_2 \sin K_0 x]} \tag{2}$$

where

Ω = radial frequency of ultrasonic wave

$K_u = \frac{2\pi}{\Omega}$ and $K_0 = \frac{2\pi}{\omega}$ wave numbers of ultrasonic wave and optical phase grating, respectively respectively

$a_1 = \frac{2\pi\Delta n_u L_1}{\lambda}$ — Raman-Nath parameter of ultrasonic wave

$a_2 = \frac{2\pi\Delta n_0 L_2}{\lambda}$ — phase change of light caused by optical phase grating

$\Delta n_u, \Delta n_0$ amplitudes of refractive index changes caused by ultrasonic wave and optical phase grating, respectively.

The expression (2) after developing into Bessel function series becomes

$$E\,(x,z = L,t) = E_0 \cdot \sum_{m,r=-\infty}^{+\infty} J_{m,r}(a_1) \cdot J_r(a_2)\, e^{ir\Omega t} \cdot e^{-i[(\omega - m\Omega)t - k(n_0 L - m\frac{\lambda}{\Lambda}x)]} \tag{3}$$

According to Eq. (3) the diffracted light is a set of plane waves propagating at angles $\alpha_m = \arctan m\frac{\lambda}{\Lambda}$ with respect to the z axis. The diffraction angles are the same as in the case of light diffraction by ultrasonic waves, while the amplitudes are a combination of the light amplitudes caused by first and second gratings and modified by the time dependent factor $e^{i\Omega t}$. This latter effect results from the interference of light components diffracted into the same direction with different frequencies, precisely frequency-shifted components are caused by ultrasonic wave (Doppler effect), while stationary optical phase grating does not affect light frequency. The interference procedure is schematically shown in Fig. 2.

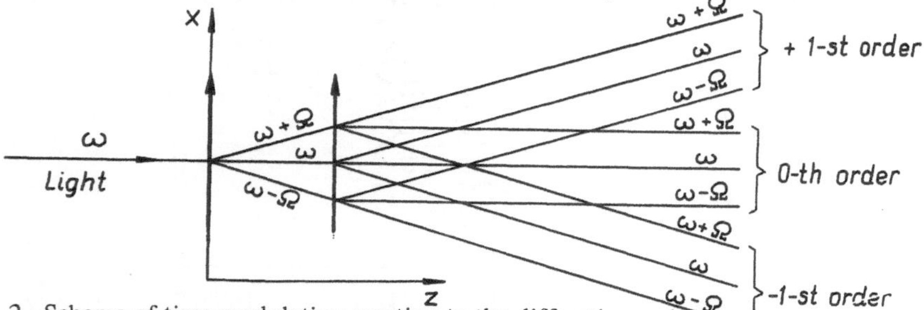

FIGURE 2. Scheme of time modulation creation in the diffraction orders.

When the phase shifts introduced by the ultrasonic wave and the optical phase grating are equal, i.e. $a_1 = a_2$, then the espression (3) simplifies and the diffracted light becomes

$$I_r = \mathcal{J}_r^2 \left(2a_1 \cos \frac{\Omega t}{2}\right) \tag{4}$$

The phenomenon of the light modulation caused by the combination of moving and stationary gratings was discovered by Gabrielli [1] in the diffraction system combined of an ultrasonic wave and an amplitude optical grating. The effect was later confirmed by Kwiek and Markiewicz [6] in the slightly different system where the amplitude optical grating was replaced by a phase grating. In the latter system the light intensity in the diffraction orders is expressed by a simpler formula (4) than in the former one.

If we assume that the spatial periods of both diffracting structures are not exactly the same, the diffraction orders produced by each structure do not overlap in space and this effect decreases the modulation factor of the diffracted light intensity. The extreme case, for which the modulation still occurs, depends upon the geometrical profile of the diffraction orders, the aperture of the illuminating light and the properties of the lens producing diffraction image. In order to evaluate the influence of the aperture and of the lens upon the modulation factor, one should introduce the factors of the optical system into Eq. (2) and it would appear that the analytical solution of the problem does not exist: however numerical evaluation could be found.

The authors suggest measurement of the ultrasonic velocity by looking for the maximum value of the modulation factor which is achieved when the spatial periods of the gratings match. Thus the measuring procedure can be limited to searching the maximum of the modulation factor (for $\Lambda_u = \Lambda_o$), provided that we know accurate Λ_o value of the phase grating.

3. EXPERIMENT

In the experimental setup, shown in Fig. 3, an argon laser of 0.488 μm wavelength was used (without longitudinal modes selection). The laser light of angular divergence 0.5 mm passes through the collimating system C of 10 cm diameter aperture, and illuminates the ultrasonic wave. The tank has 6 liters capacity, is filled with distilled water and it has a special shape to guarantee progressive ultrasonic wave only (reflected waves are damped 45 dB).

After crossing the ultrasonic progressive wave, the light wave illuminates the optical phase grating (OPhG) whose angular position could be changed by tilting. Tilting of the optical phase grating causes that the actual spatial period is a product of the real spatial period times cos β, where β is the angle of tilt. In that way the spatial period could be precisely changed. The light diffracted by both diffraction structures is focused by a lens of 10 cm diameter and focal length 0.6 m and then directed to the photomultiplier via a slit that separates different diffraction orders. The light intensity is observed on the osciloscope. The change in the modulation factor is caused by the sloping of the optical grating. When the modulation factor is a maximum one can evaluate the ultrasonic wavelength through the spatial period of the optical phase grating, used as a standard.

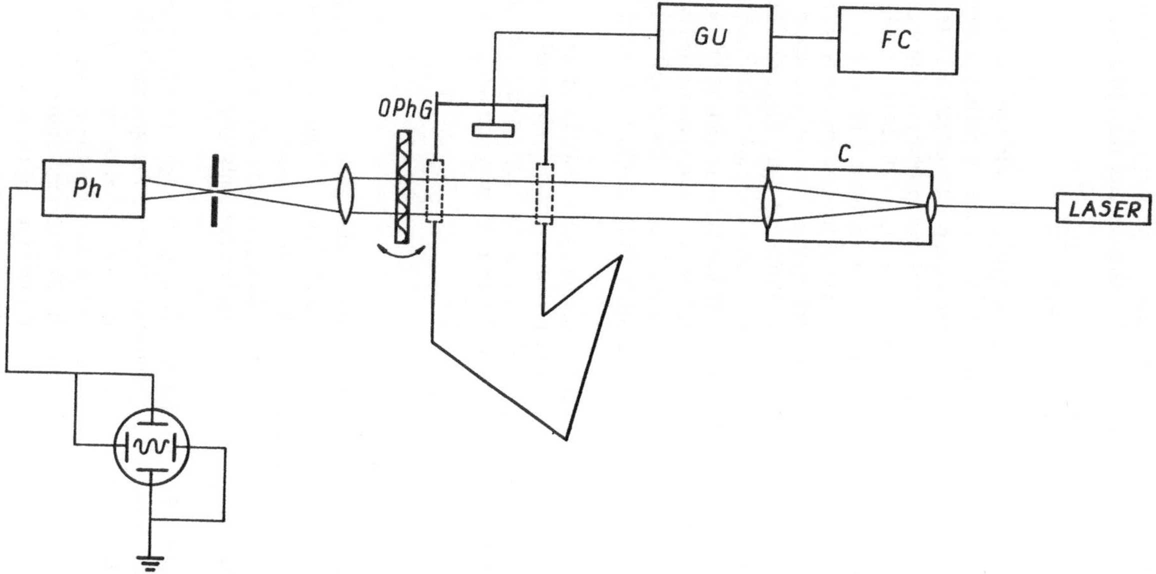

FIGURE 3. The experimental setup
C - collimating system, GU - generator of ultrasonic wave, FC - frequency counter, OPhG - optical phase grating, Ph - photomultiplier.

4. RESULT AND DISCUSSION

The examination of the accuracy of the method presented was performed in the expetimental setup with the $BaTiO_3$ transducer of 3 cm diameter and resonant frequency 1.66 MHz. Raman-Nath parameter of ultrasonic wave was established to be 0.4 and it was equal to the maximum phase shift in the light wave resulting from the optical phase grating. Temperature of water was kept constant within 0.05°C accuracy during the series of measurements.

Figs. 4 a, b, c, d are photographs of the osciloscope signals recorded for the 1-st diffraction order when the spatial periods of both structures varied with respect one to the other from $\Lambda_u - \Lambda_o = 0$ (Fig. 4a) up to $\Lambda_u - \Lambda_o = 5.5 \times 10^{-3}\ \Lambda_o$ (Figl. 4d). The difference between the periods was caused by tilting the optical phase grating within angles 0° and 6°.

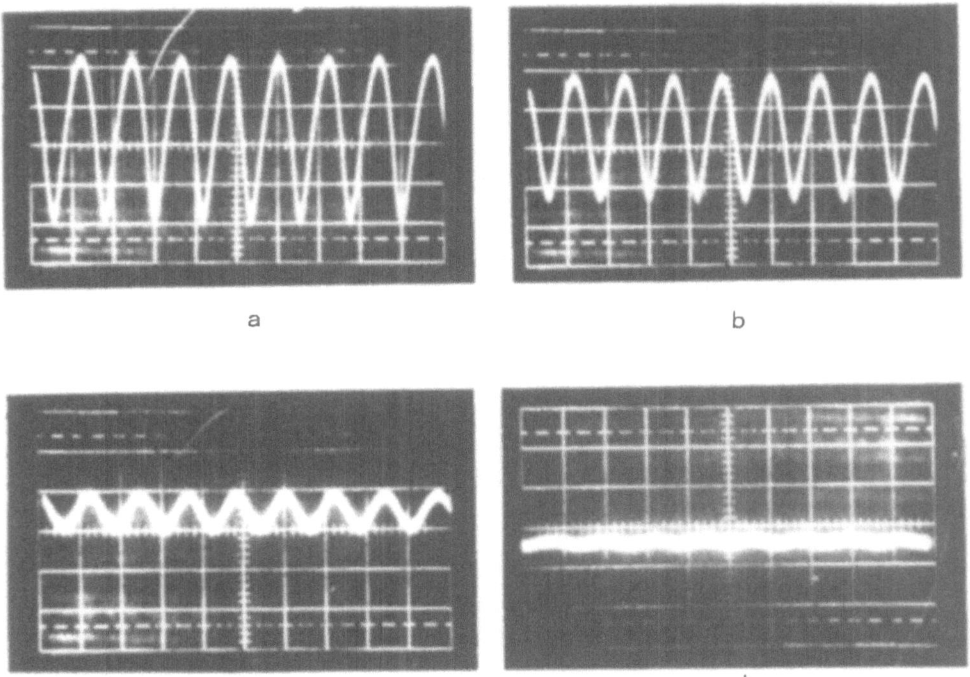

a b c d

FIGURE 4. Modulation of 1st order light intensity in several cases of spatial periods mismatch.

a) $\Lambda_u - \Lambda_o = 0$ ($\beta = 0°$)
b) $\Lambda_u - \Lambda_o = 6.1 \cdot 10^{-4}\ \Lambda_o$ ($\beta = 2°$)
c) $\Lambda_u - \Lambda_o = 2.4 \cdot 10^{-3}\ \Lambda o$ ($\beta = 4°$)
d) $\Lambda_u - \Lambda_o = 5.5 \cdot 10^{-3}\ \Lambda_o$ ($\beta = 6°$)

Maximum of modulation factor was observed on the osciloscope for the series of the successive measurements.

Using the Student-Fischer distribution at 0.95 level of confidence, the confidence region was estimated to be ± $1.7. 10^{-2}$ m/s for the ultrasonic velocity.

The measurements were carried out in two ways in order to change the spatial periods of both diffraction gratings. One way was obtained by tilting the optical phase grating, the other by tuning the ultrasonic frequency to achieve the maximum modulation factor and in both the confidence region for the frequency was the same and equal to ± 290 Hz.

The authors believe that the accuracy could be increased by one order of magnitude if the indicator of the maximum of the modulation factor were replaced by a voltmeter and stabilization of the light and ultrasonic sources were properly accomplished. The proposed method of the ultrasonic velocity measurement after some developments could become more effective in liquid crystals and solid state and for surface acoustic waves.

REFERENCE

1. Gabrielli, I., Acustica 21 (1969), 97.
2. Calligaris, F., Ciuti, P., Gabrielli, I., Acustica 23 (1976), 74.
3. Calligaris, F., Ciuti, P., Gabrielli, I., J.A.S.A. 61 (1977), 959.
4. Patorski, K., Acustica 52, 4 (1983), 246.
5. Patorski, K., Acustica 53, 1 (1953), 1.
6. Kwiek, P., Markiewicz, A., Submitted to Acoustics Letters.
7. Proc. of the First Spring School on Acoustoopics and Applications, 1980.
8. Proc. of the Second Spring School on Acoustoopics and Applications, 1983.

III. MATERIALS

THREE-DIMENSIONAL COMPOSITES

B.A. AULD

1. INTRODUCTION

The 1980's are truly becoming a decade for artificial materials realized by combining material phases that do not appear together in nature. In this synthesis procedure it is possible to fabricate composite materials with more useful properties than those of the individual phases by themselves. Most composites now considered for practical applications have only two constitutive phases, and this paper will consider only this class of composite materials; but improved fabrication technology will undoubtedly lead to applications for more complicated materials in the future. The rapid development of composite material science has been driven both by the need for improved performance and by new technology suitable for fabrication of sophisticated two-phase materials. Composites are now being studied and used for structural applications (metal matrix and plastic matrix materials imbedded with strengthening fibers), layered semiconductor structures used to realize material properties not otherwise available (a procedure now termed "bandgap engineering"), and artificial piezoelectric materials with improved properties for sonar hydrophones and ultrasonic transducers [1-10].

Composites used in the above applications may have either a random or a periodic distribution of the two phases. For example, in the structural domain all of the fiber-matrix materials have random composition. (There does, however, exist a single-phase artificial structural material that is periodic. This is the class of so-called carbon-carbon composites used for some rocket motor parts. It is a sort of high tech basket weaving, consisting of a three-dimensional pattern of interwoven carbon fibers. There is no matrix, although one could in principle be added to form a two-phase three-dimensional periodic composite.) The microscopic semiconductor composites (heterostructures), fabricated by molecular beam epitaxy, are one-dimensional. They may be strictly periodic, but are more often fabricated with properties that are tapered spatially according to some specified law. Macroscopic composites used for piezoelectric applications are of both random and periodic types, and have been investigated in one-, two- or three-dimensional forms. They have also been made with both two and three component phases.

This paper deals with elastic waves and vibrations in macroscopic piezoelectric composites having the form of strictly periodically arrayed piezoceramic inclusions in a polymer matrix--i.e., elastic superlattices. In some applications it is more economical to use randomly-structured piezoceramic-polymer composites, but with less accurate control over the material properties. For ultrasonic medical and nondestructive transducer applications two-dimensional piezoceramic-polymer superlattices have been shown to be superior to single phase transducer materials in terms of impedance match to water, reduction of lateral modes and crosstalk, and piezoelectric coupling constant. Although the materials under active consideration for practical applications at the present time have only two-dimensional periodicity, the analysis presented here will begin with the general problem of elastic wave propagation in an infinite three-dimensional elastic superlattice. This will then be specialized to infinite two-dimensional superlattices corresponding to the materials now being used in composite transducers. Finally, the paper will conclude with a brief discussion of elastic vibrations in a two-dimensional superlattice of finite dimensions.

Elastic wave interactions with periodic gratings have already been treated in connection with surface wave (SAW) devices, by both equivalent circuit [11] and perturbation [12] methods. This paper deals with elastic wave propagation in materials with periodically varying density and stiffness. In previous analyses of wave propagation in elastic superlattices only those cases have been treated where the wavelength is large compared with the superlattice period [13,14]. The earlier studies used constant strain and constant stress models for calculating the spatial average elastic constants by explicitly matching boundary conditions across interfaces. A more recent publication treats the same long-wavelength limit but avoids the need for matching interface boundary conditions by means of a first-principles hamiltonian approach [15]. In this paper the calculation presented by Brillouin for scalar wave propagation on a periodic lattice [16] is followed and is applied <u>without</u> restriction to the long-wavelength case.

2. WAVE PROPAGATION IN THREE-DIMENSIONAL ELASTIC SUPERLATTICES

This section considers a general cubic superlattice of elastic inclusions imbedded in a matrix with different elastic properties. Each elastic inclusion (or motif) is of arbitrary shape and is isotropic, as is the matrix itself. The cube axes of the unit cell are aligned along the x,y,z axes and the periodic length along each axis is d. In this analysis the concepts of scalar wave propagation in a periodic medium are extended to the vector problem of elastic wave propagation in a periodic medium. What is involved is an extension of the Floquet method [16-18] to this more general problem. Although the formulation presented here is specifically limited to cases where the two phases of the superlattice are individually isotropic (the overall superlattice has, of course, anisotropic properties), it is easily extended to superlattices with individually anisotropic phases.

For a <u>spatially uniform</u> isotropic medium, the wave equation governing time harmonic ($\exp i\omega t$) elastic wave propagation is conventionally written, in full subscript notation as

$$c_{ijk\ell}\partial_j \partial_k u_\ell = -\rho\omega^2 u_i \tag{1}$$

where ∂_j and ∂_k denote partial derivatives. For a nonuniform medium, where the $c_{ijk\ell}$'s are functions of the spatial coordinates, some of those derivatives operate directly on the displacement components u_k and others operate on products of the c's and derivatives of the u_k's. To separate these terms, it is convenient to write Eq. (1) in reduced subscript notation as [19]

$$\nabla_{iI} c_{IJ} \nabla_{Jj} u_j = -\rho\omega^2 u_i \tag{2}$$

Here, c_{IJ} and ρ are periodic functions of x, y, and z. It is convenient to express them as

$$c_{IJ}(x,y) = \overline{c}_{IJ} + \delta c_{IJ}(x,y,z) \tag{3}$$

and

$$\rho(x,y) = \overline{\rho} + \delta\rho(x,y,z) \tag{4}$$

where $\overline{c}_{IJ}$ and $\overline{\rho}$ are spatial averages. It is apparent from this that the spatial average values of the functions of x,y,z on the right are zero, a fact that will later become important. Application of the derivative operators ∇_{iI} and ∇_{Jj} in Eq. (2) converts it to the general form

$$\underline{\underline{\overline{W}}} \cdot \underline{u} = -\underline{\underline{\delta W}}\,(x,y,z) \cdot \underline{u} \tag{5}$$

where all effects of spatial periodicity have been transferred to the right-hand side of the equation. The matrix-differential operators in Eq. (5) are defined as

$$\underline{\underline{\overline{W}}} = \underline{\underline{\Gamma}}\,(\overline{c}_{IJ}) + \overline{\rho}\omega^2\underline{\underline{I}} \tag{6}$$

and

$$\underline{\underline{\delta W}}(x,y,z) = \underline{\underline{\Gamma}}\,[\delta c_{IJ}(x,y,z)] + \underline{\underline{\delta\Gamma}}\,(x,y,z) + \delta\rho(x,y,z)\omega^2\underline{\underline{I}} \tag{7}$$

where $\underline{\underline{\Gamma}}(\overline{c}_{IJ})$ is the spatial average elastic wave matrix (obtained by replacing wave vector components in the spatial average Christoffel matrix with spatial derivatives), $\underline{\underline{I}}$ is the identity matrix, and $\underline{\underline{\delta\Gamma}}(x,y,z)$ takes account of the spatial derivatives of the elastic constants. It is a three-by-three matrix with the following form:

(i) The top diagonal element is

$$(\partial_x c_{11})\partial_x + (\partial_y c_{44})\partial_y + (\partial_z c_{44})\partial_z \tag{8(a)}$$

and the other elements down the diagonal are obtained by cyclically permuting x,y,z.

(ii) The upper off-diagonal elements, starting with the second element in the top row and moving clockwise are:

$$\begin{aligned}&(\partial_x c_{12})\partial_y + (\partial_y c_{44})\partial_x\\&(\partial_x c_{12})\partial_z + (\partial_z c_{44})\partial_y\\&(\partial_y c_{12})\partial_z + (\partial_z c_{44})\partial_y\end{aligned} \tag{8(b)}$$

(iii) The lower off-diagonal elements, across from those in (ii), are obtained by interchanging c_{12} and c_{44}. It can be seen from Eqs. (3) and (4), and the fact that the elastic wave matrix is a linear function of the elastic constants, that the spatial average of Eq. (7) is zero.

Equation (5) is a generalization to three-dimensional vectors of the problem of scalar wave propagation in three-dimensional periodic media, treated by Brillouin [16], and can be analyzed in the same way. Following Brillouin's procedure the problem is first addressed by perturbation theory, to determine the general nature of the solutions. In this case $\underline{u}$ is taken as some solution for the problem of the spatial average medium, plus a first-order perturbation $\varepsilon\underline{u}^{(1)}$. Here ε is an order parameter assigned to the right-hand side of Eq. (5), so that $\varepsilon = 0$ corresponds to a uniform medium. The assumed solution is therefore

$$\underline{u} = \underline{A} \exp(-i\underline{\beta_0} \cdot r) + \varepsilon \underline{u}^{(1)} \tag{9}$$

where $\underline{A}$ is a polarization vector for one of the three possible elastic wave solutions in a spatially uniform medium, and

$$\underline{r} = x\,\underline{\hat{x}} + y\,\underline{\hat{y}} + z\,\underline{\hat{z}}$$

In what follows, the two phases of the elastic superlattice are assumed to be isotropic.

Because the operator $\underline{\underline{\delta W}}$ in Eq. (5) contains material parameters that are triply-periodic functions of x,y and z, it can be expressed as a triple Fourier series. For the assumed cubic elastic superlattice, with period d along the x,y,z axes,

$$\underline{\underline{\delta W}}(x,y,z) = \sum_{m,n,\ell \neq 0} \underline{\underline{\delta W}}_{mn\ell} \exp\{-i[(2\pi m x/d) + (2\pi n y/d) + (2\pi \ell z/d)]\} \tag{10}$$

The absence of the terms $m,n,\ell = 0$ is due to the fact, noted above, that the spatial average of Eq. (7) is zero. Substituting Eqs. (9) and (10) into Eq. (5), retaining only first-order terms in ε, and noting that $\underline{\underline{\delta W}}$ is defined to be of order ε reduces Eq. (5) to the form

$$\underline{\underline{\overline{W}}} \cdot \underline{u}^{(1)} = - \sum_{m,n,\ell \neq 0} \underline{\underline{\delta W}}_{mn} \cdot \underline{A} \exp(-i_{\underline{mn}} \cdot \underline{r}) \tag{11}$$

where

$$\underline{\beta}_{mn\ell} = \left(\beta_{0x} + \frac{2\pi m}{d}\right) \underline{\hat{x}} + \left(\beta_{0y} + \frac{2\pi n}{d}\right) \underline{\hat{y}} + \left(\beta_{0z} + \frac{2\pi \ell}{d}\right) \underline{\hat{z}} \tag{12}$$

Since $\underline{\underline{\overline{W}}}$ in Eq. (11) is a linear operator, the solution for $u^{(1)}$ is obtained by considering independently the partial solutions excited by the individual exponential driving terms on the right of Eq. (11). That is,

$$\underline{u}^{(1)} = \sum_{m,n,\ell \neq 0} a_{mn\ell} \underline{\hat{u}}_{mn\ell} \exp(-i\underline{\beta}_{mn\ell} \cdot \underline{r}) \tag{13}$$

where $a_{mn\ell}$ are the partial solution amplitudes, and $\underline{\hat{u}}_{mn\ell}$ are their unit polarizations. The polarizations $\underline{\hat{u}}_{mn\ell}$ of the partial solutions are, as yet, unknown; but they can be expanded in an orthogonal set of basis functions consisting of the three unit polarizations $\underline{\hat{u}}_{mn\ell\eta}$ of elastic waves propagating with wave vectors $\underline{\beta}_{mn\ell}$ (Fig. 1). In defining these unit polarizations, the subscript η runs over the values L, S, S′ corresponding to one longitudinal and two shear waves. For uniqueness, the shear polarizations are defined as shown in Fig. 1. With these definitions, Eq. (13) can be rewritten as

$$\underline{u}^{(1)} = \sum_{m,n,\ell \neq 0} \; \sum_{\eta = L,S,S'} a_{mn\ell\eta} \underline{\hat{u}}_{mn\ell\eta} \exp(-i\beta_{mn\ell} \cdot \underline{r}) \tag{14}$$

where $a_{am\ell\eta}$ are the amplitudes of the basis polarizations in the $mn\ell$ partial solution.

The advantage of the above decomposition of $\underline{u}^{(1)}$ is that the unit polarization vectors in Eq. (14) are all eigenvectors of $\underline{\underline{W}}$. This means that the left-hand side of Eq. (11) becomes a sum of terms of the form

$$\underline{T}_{mn\ell\eta} = \bar{\rho}(\omega^2 - \omega^2_{mn\ell\eta})a_{mn\ell\eta}\hat{\underline{u}}_{mn\ell\eta}\exp(-i\underline{\beta}_{mn\ell}\cdot\underline{r}) \tag{15}$$

where the radian frequency with subscripts is the natural frequency of an elastic wave of polarization η with the wavenumber equal to $\beta_{mn\ell}$. Since the three polarizations for each partial solution are orthogonal, and each partial solution is driven by its corresponding forcing term, the coefficients $a_{mn\ell\eta}$ can be expressed in terms of A just as in the scalar problem treated by Brillouin.

Continuation of the above process to second and higher orders in ε shows that no new exponential terms are introduced. The general solution of Eq. (5) can therefore be taken in the form of Eq. (14), but including the terms $m,n,\ell = 0$ to account for the first term in Eq. (9). On the left-hand side of Eq. (5) this generates a summation over terms of the form of Eq. (15). Therefore, Eq. (5) reduces to a set of coupled linear equations in the amplitudes $a_{mn\ell\eta}$.

$$\bar{\rho}\sum_{mn\ell\eta}(\omega^2 - \omega^2_{mn\ell\eta})a_{mn\ell\eta}\hat{\underline{u}}_{mn\ell\eta}\exp(-i\underline{\beta}_{mn\ell}\cdot\underline{r}) \tag{16}$$

$$= -\sum_{pqs\zeta}\underline{\underline{\delta W}}(x,y,z)\cdot a_{pqs\zeta}\hat{\underline{u}}_{pqs\zeta}\exp(-\underline{\beta}_{pqs}\cdot\underline{r})$$

where the subscripts $m,n,\ell = 0$ and $p,q,s = 0$ are now included in the summations. Since the exponential terms in Eq. (15) can be shown to satisfy the orthogonality relation

$$\iiint_{\substack{\text{unit}\\\text{cell}}}\left(\exp(-i\underline{\beta}_{tuv}\cdot\underline{r})\right)^* \exp(-i\underline{\beta}_{mn\ell}\cdot r)\,dxdydz = d^3\delta_{pqs,mn\ell} \tag{17}$$

single terms can be extracted from the summation on the left-hand side of Eq. (16) by multiplying the equation with the complex conjugate of the tuv exponential, integrating over the unit cell, and taking the scalar product with the ξ-th tuv unit polarization vector. Because the tuv polarization vectors are orthogonal, this procedure selects one term on the left-hand side of Eq. (16). Repetition of the procedure for all possible space harmonics and polarizations gives an infinite set of linear equations for the a's. That is,

$$d^3\bar{\rho}(\omega^2 - \omega^2_{tuv\xi})a_{tuv\xi} = -\sum_{pqs\zeta}K_{tuv\xi,pqs\zeta}a_{pqs\zeta} \tag{18}$$

where

$$K_{tuv\xi,pqs\zeta} = \iiint_{\substack{unit\\cell}} \exp(i\underline{\beta}_{tuv} \cdot \underline{r})\,\underline{\hat{u}_{tuv\xi}} \cdot \underline{\underline{\delta W}}(x,y,z) \cdot \underline{\hat{u}_{pqs\zeta}} \; \exp(-i\underline{\beta}_{pqs} \cdot \underline{r})\; dxdydz \qquad (19)$$

with t,u,v,p,q,s summed over all positive and negative integers, including zero, and ξ,ζ summed over L,S,S′ in Fig. 1. Because the spatial average of $\underline{\underline{\delta W}}$ is zero, the K's are zero for t = p, u = q, v = s.

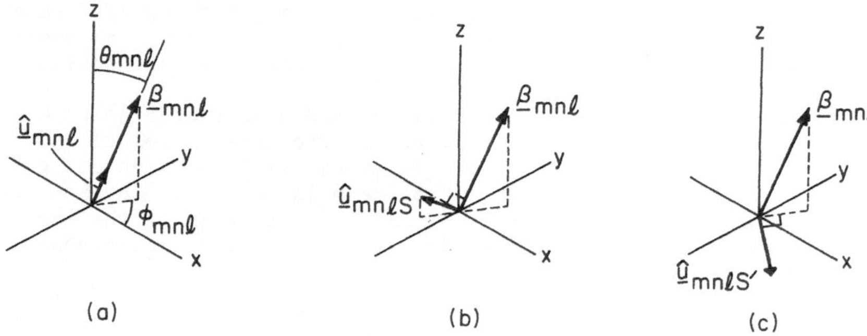

Figure 1. Definitions of the unit polarization vectors $\hat{\underline{u}}_{mn\ell\eta}$

Equations (18) and (19) extend to the problem defined by Eq. (5) the method presented by Brillouin in Reference 16 for scalar wave propagation in a periodic medium. This medium is now commonly called the Floquet Method [18]. (In electrical device terminology the terms with different exponentials in Eq. (14) are called space harmonics.) It is seen from the equation that each space harmonic has three independent amplitude coefficients, required by the vector nature of the space harmonic. In the scalar problem, by contrast, each space harmonic has only one amplitude coefficient, since the space harmonic is a scalar.

Equation (18) is an infinite set of linear equations for the space harmonic amplitudes of elastic waves propagating in a cubic superlattice. Values of the space harmonic coupling constants in Eq. (19) depend on the shape of the inclusions and on the material properties of the two phases. A dispersion relation (ω as a function of β_0 in Eq. (12)) is defined by setting the determinant of Eq. (18) to zero. The nature of this relation is fully described by the literature on scalar wave propagation in periodic media [16-18]. An infinite number of branches is found for the ω versus β_0 function. These correspond to "dispersion relations" for the various space harmonics, plotted against β_0. The complete plot of these curves gives the Brillouin diagram for the medium. If the periodic inclusions are all allowed to become vanishingly small, or their material properties are allowed to approach those of the imbedding matrix, the higher space harmonic amplitudes go to zero, and the Floquet solution reduces to a spatially uniform plane wave, conveniently defined as the zero space harmonic. As can be seen from Eq. (12) the three components of β_0 (vector) define the direction of propagation of this limiting spatially-uniform plane wave. In solving for the Brillouin diagram in a three-dimensional problem it is necessary to specify the direction of β_0 (vector) in order to solve the characteristic equation of Eq. (18) for ω versus β_0. For elastic wave propagation there are three possible uniform plane wave solutions for each propagation direction (one longitudinal, and two shear) in the limiting spatially-uniform medium. This means that there will be three Brillouin diagrams for each direction of propagation.

Just as in the spatially-periodic scalar wave propagation problems, propagation in an elastic superlattice is characterized by the existence of stopbands (or forbidden bands) in which the waves cannot propagate, and decay exponentially with distance. These are associated with the phenomenon of Bragg reflection from periodically spaced planes of inclusions, oriented at various angles relative to the superlattice. A three-dimensional elastic superlattice therefore has three sets of stopbands (one longitudinal and two shear) associated with the various sets of Bragg planes for the lattice. In a medium with weak periodicity these stopbands are narrow, and all but a few of the lowest space harmonics in the Floquet expansion are negligibly small. Under these conditions, a reasonable approximation to the dispersion curve in the vicinity of a stopband can be obtained by including in Eq. (18) only the zero space harmonic and the space harmonic of the same species (longitudinal or shear) generated from it by Bragg reflection. (Recall that the zero space harmonic may be either longitudinal or shear.) The technique which was first developed for the scalar problem, can also be described (and implemented more accurately) by selecting the two space harmonics for which the resonance terms on the left-hand side of Eq. (18) vanish simultaneously at the Bragg frequency. This reduces Eq. (18) to two coupled linear equations, which can be solved algebraically. In the literature this truncation method is sometimes called the coupled-mode approximation, although the terminology is not completely standardized [18].

3. TWO-DIMENSIONAL ELASTIC SUPERLATTICES

As was noted in the Introduction the periodic piezoceramic-polymer composites of greatest practical interest at the present time are two-dimensional structures. Figure 2 shows commonly used unit cell geometries for these elastic superlattices. The square geometries at the left and the right are fabricated by slotting the piezoceramic and filling the slots with polymer, while the circular geometry is made by aligning the circular rods in a jig and imbedding them in a polymer matrix.

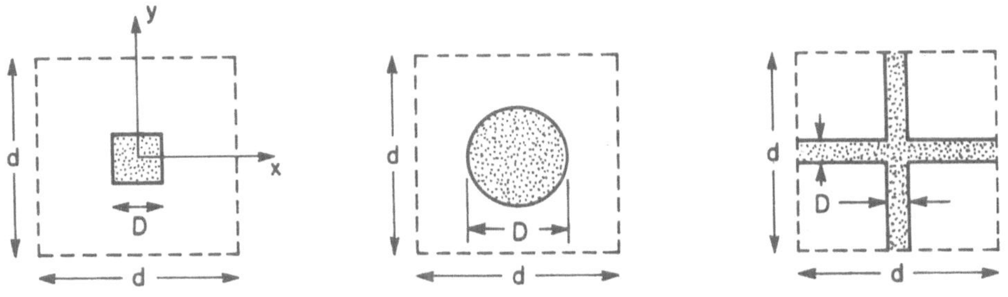

Figure 2. Unit cell geometries for two-dimensional elastic superlattices. In a piezoceramic-polymer composite the stippled areas are ceramic.

The analysis of Section 2 is adapted to this problem by noting that the material properties are now functions of x and y only, so that the expansion in Eq. (10) reduces to a double-Fourier series. This means that the third term in the exponent is suppressed and that the summation is over m,n only. After the substitutions noted, Eqs. (11) and (12) appear with subscript ℓ suppressed. Also, the final term in Eq. (12) contains only β_{0z}. Apart from these changes, the analysis proceeds as before. It now leads to Eqs. (18) and (19), with subscripts v and s suppressed and with d squared on the left of Eq. (18), instead of d^3. The integrand of Eq. (19) is a function of x,y only and the integration over the unit cell is carried out over these variables. In the integrand the unit polarization vectors are defined in Fig. 1, but with subscript ℓ suppressed. For the two-dimensional superlattice the z component of β_{mn} vector is β_{0z}, and is arbitrary [20,21].

In the two-dimensional version of Eq. (19) the integration can be carried out analytically for the geometries of Fig. 1 [20,21]. For the square rod geometry at the left of the figure this gives, with R the radius vector in the x,y plane,

$$K_{\ell m\eta,pq\zeta} = \int_{cell} (\underline{\hat{u}}_{\ell m\eta} \cdot \underline{\underline{\delta W}}(x,y) \cdot \underline{\hat{u}}_{ps\zeta}) \times \exp[-i\overline{\underline{\beta}_{\ell m} - \underline{\beta}_{pq}} \cdot \underline{R}]\, dxdy$$
$$\approx (D/d)^2 \operatorname{sinc} \frac{(\ell - p)\pi D}{d} \operatorname{sinc} \frac{(m - q)\pi D}{d} \tag{20}$$

For the circular rod geometry the sinc functions are replaced by jinc functions. The philosophy and implementation of the coupled-wave approximation is the same as described for the three-dimensional superlattice in Section 3 above, namely two space harmonics are selected for which the terms on the left-hand side of Eq. (18) are simultaneously resonant at the Bragg frequency.

4. PIEZOCERAMIC-POLYMER TRANSDUCER RESONANCES

Piezoceramic-polymer transducers now being used consist of plates of a two-dimensional superlattice composite material with faces parallel to the x,y plane. Although the two-dimensional version of Section 3 is capable of treating wave propagation in any direction, attention will be focused on wave propagation either normal to the plane of the periodicity (xy-plane) or parallel to that plane. These propagation directions correspond, in the first case, to the longitudinal thickness resonances of a composite transducer and, in the second case, to lateral resonances of the transducer. Lateral resonances occur in all transducers [22,23]. However, certain resonances in composite transducers are enhanced by Bragg-scattering from the lateral periodicity and by phase-matched coupling of the electrical excitation to selected lateral resonances through the periodically distributed piezoelectric coupling. Other lateral resonances are suppressed.

Figure 3 gives a typical transducer input impedance curve, showing three strong resonances. Here f_ℓ is the longitudinal thickness resonance and f_{t1}, f_{t2} are the two dominant lateral standing wave resonances--along the unit cell edge and the unit cell diagonal, respectively (Fig. 4). Since the sample is uniformly electroded, the piezoelectric elements (shown stippled in Fig. 2) are driven in phase, so that the elastic vibrations excited must also be in phase at these points. This means that the lateral resonances observed in Fig. 3 should consist of standing waves with one full-wavelength, rather than one half-wavelength, spacing between the piezoelectric elements--along the unit cell edge for f_{t1} and along the unit cell diagonal for f_{t2}. Laser probe measurements, performed by the technique of Reference 24, confirm this hypothesis [9,20,25].

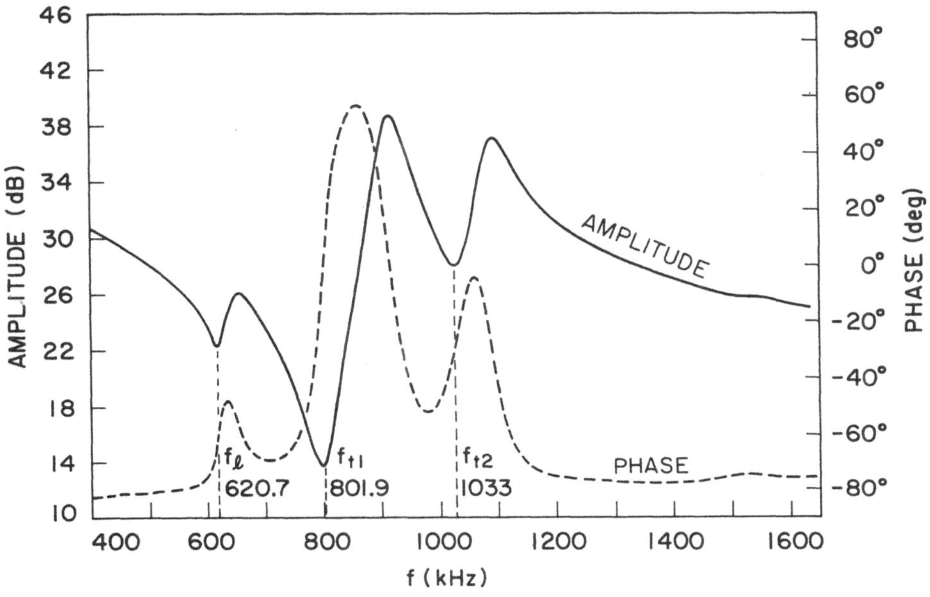

Figure 3. Typical resonance spectrum for a longitudinal piezoceramic-polymer composite transducer.

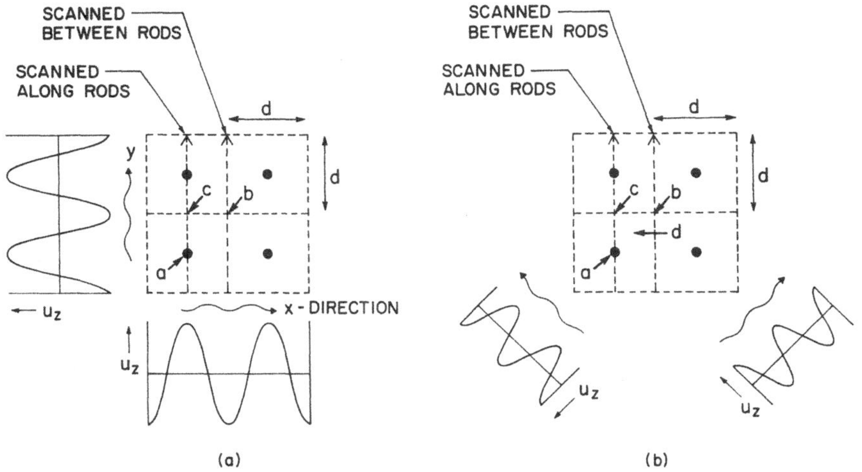

Figure 4. Standing wave patterns for lateral wave resonances along the unit cell edge and diagonal directions of a square two-dimensional super-lattice.

Modeling of lateral resonances in composite transducers is first approached by assuming wave propagation in the xy-plane of an infinite composite. The model can be further simplified by supposing that only a z-polarized elastic displacement exists, reducing the elastic wave function to a scalar form similar to that appearing in electronic band theory. In a first approximation, coupled-mode theory is used to find the edges of the stopbands. These are located near frequencies where Bragg-scattering resonances occur between certain planes of piezoceramic rods (Bragg planes) in the periodic array of rods. (Resonances occur at frequencies for which the Bragg planes are spaced by integral multiples of $\lambda/2$ for the z-polarized shear wave assumed in the model.) These Bragg resonances are just the enhanced lateral resonances referred to above.

At $\lambda/2$ spacing between the Bragg planes, the resonant wave functions are standing waves along the unit cell edge, the unit cell diagonal, and perpendicular to the higher Bragg planes. These resonances cannot, however, be excited piezoelectrically in a uniformly electroded piezoceramic-polymer because neighboring ceramic rods vibrate 180° out of phase. The transducer cannot be driven in this mode of resonance because the driving electric field is everywhere in phase. The lowest lateral mode that can be excited piezoelectrically in this geometry is that which occurs when the edge of the unit cell equals one full wavelength (f_{t1} in Fig. 3 and the diagram at the left in Fig. 4). At this resonant frequency several allowed Bragg-scattering resonances occur, but only the mode shown in Fig. 3 has the piezoceramic rods all vibrating in phase. For the longitudinal thickness resonance (f_ℓ in Fig. 3), Bragg-scattering resonances do not appear, because there is no periodicity of the medium along the z-propagation direction. In this case the simplest model is a z-propagation longitudinal wave with properties determined by $\bar{\rho}$ and $\bar{c}_{33}$ (spatial averages over the unit cell), and with weakly-excited space harmonics, because no resonance of the space harmonics can occur. In other words, the Bragg-scattering condition cannot be satisfied.

The above model for the transducer resonances given good agreement for measurements of the longitudinal thickness mode but only fair agreement for the lateral modes. One reason for this is that the theory uses a spatial average of the density and the elastic stiffness constants c_{IJ} to model the effective properties of the medium at low frequencies, where scattering from the periodicity is unimportant. This corresponds to the <u>constant strain model</u> for the static (or low frequency) behavior of a composite. If a particular strain component (the J-th) has the same value (S_J) in the polymer matrix and in the piezoceramic rod, the average I-th component of stress over the unit cell is (see Fig. 2)

$$\left(\frac{(A - a)((c_{IJ})_{plastic} + a(c_{IJ})_{PZT}}{A}\right) S_J = \bar{c}_{IJ} S_J \tag{21}$$

corresponding to the spatial averaging of the stiffness c_{IJ}. For the longitudinal thickness resonance (f_ℓ) the plastic matrix and the piezoceramic rods vibrate like springs in parallel, and the model reduces to this constant-strain description at low frequencies. For the lateral resonances in Fig. 4, the plastic and the piezoceramic are elastic members connected in series, so that the stress (not the strain) is common to both; and the low frequency behavior of the medium is best represented by the <u>constant stress model</u>. In this case the counterpart of the previous equation is

$$\left(\frac{(A - a)(s_{IJ})_{plastic} + a(s_{IJ})_{PZT}}{A}\right)T_J = \bar{s}_{IJ}T_J \qquad (22)$$

for the average strain over the unit cell. Vibrations of the composite material at low frequencies are now most accurately calculated from the spatially averaged compliances $\bar{s}_{IJ}$, rather than $\bar{c}_{IJ}$. A theory producing this result is outlined in the next section.

5. COMPARISON OF THEORY AND EXPERIMENT

Experiments on the resonances of finite piezoceramic-polymer resonators (Fig. 3) can be explained qualitatively, as in Section 4, by the theory of an infinite two-dimensional superlattice. However, a more accurate comparison requires that the additional boundary conditions at the surfaces of the resonator be taken into account. The infinite superlattice was treated by expressing the acoustic wave solutions in a series of plane wave harmonics. Each order of space harmonic (for example the mn-th space harmonic) includes plane waves with the three possible plane wave polarizations corresponding to the space harmonic propagation factor $\underline{\beta}_{mn}$ in the two-dimensional superlattice version of Fig. 1. There are therefore three space harmonic amplitudes $a_{mn\eta}(\eta = L,S,S')$ in each order (mn). Substitution of the space harmonic expansion into Eq. (5), and use of the orthogonality properties of the space harmonic wave functions, led to an infinite set of coupled equations for the space harmonic amplitudes, Eq. (18). In the infinite superlattice, standard methods can be used to solve these equations [16-18], including the coupled-wave approximation in which only space harmonics that are near "resonance" are retained. However, problems arise in treating the composite plate structure by coupled-mode theory, because the space harmonics are coupled by both the spatial inhomogeneities (δ_ρ, c_{IJ}) and the plate boundary conditions. These boundary conditions couple not only space harmonics of different orders, but also different polarizations in the same order of space harmonic, so that the coupling terms in the two-dimensional Eq. (18) become much more complicated. Selection of the important terms in the Floquet solution therefore becomes very difficult. This problem can be avoided by taking the space harmonic terms in the Floquet solution for the plate to be Lamb mode and SH mode wave functions, rather than combining plane wave space harmonics to match the plate boundary conditions. By choosing Lamb and SH mode wave functions as a basis [19], one pre-matches the space harmonics to the plate boundary conditions. Since the Lamb and SH modes are decoupled by symmetry in a superlattice with isotropic component phases, the problem separates immediately into Lamb-type and SH-type superlattice waves.

Consider the Lamb-type waves, which are relevant to lateral resonances in a composite longitudinal transducer. The superlattice waves are first expanded in a space harmonic series. Each harmonic is then decomposed into an infinite series of Lamb wave terms by using the mode orthogonality relations [19]. This is in contrast to the infinite superlattice problem (Sections 2 and 3), where each space harmonic was decomposed into three different plane wave polarizations. The Lamb wave formulation eventually leads to equations similar to Eq. (18); but there are now only two space harmonic subscripts (tu,pq) and the subscripts η,ζ run over the infinite series of Lamb modes. Because of these additional degrees of freedom, the stopband spectrum becomes extremely dense and complicated. However, since the plate boundary conditions are automatically taken into account by the Lamb waves, the basic equations are no more involved than for the infinite superlattice. This means that the coupled wave approximation can still be used, for a weakly-periodic plate.

Two-dimensional Lamb wave space harmonics are, nevertheless, still quite inconvenient to use, and a further approximation beyond the coupled-mode theory is usefully applied. This can be illustrated by the lateral edge resonance at the left of Fig. 4. In this case, the Bragg reflection is between lines of piezoceramic rods parallel to edges of the unit cell. If these lines of rods are approximated by slabs of piezoceramic, the problem is converted to a one-dimensional superlattice. One-dimensional Lamb space harmonics may then be used. Figure 5 compares measured lateral resonances in two-dimensions with theoretical predictions of the Bragg frequencies, obtained from a coupled-mode approximation in the one-dimensional Lamb wave model. The theoretical points are intercepts on the lowest dotted curve at the same abscissas as the corresponding experimental points. (Tabulations of the physical parameters of the measured composite resonators are given in Reference 2.) Use of Lamb wave space harmonics can be shown to give a constant stress model (Eq. (22)) for the average properties of the composite material, explaining the good agreement found in the figure. The infinite superlattice model, which gives a constant strain model (Eq. (21)), would be expected to show poor agreement for the lateral resonances; and this has, in fact, been observed [26]. Good agreement with the longitudinal thickness resonances is, however, obtained with the constant strain model.

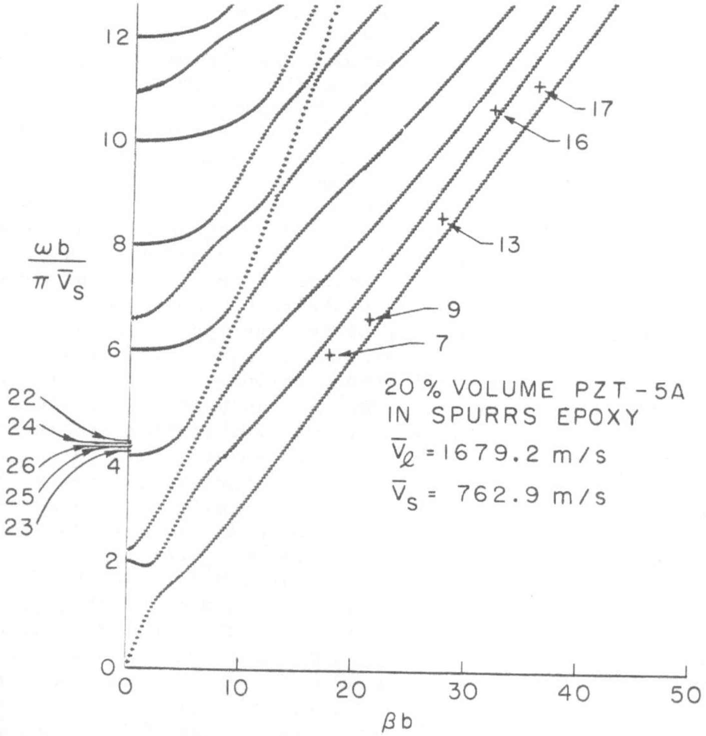

Figure 5. Comparison of measured lateral resonant frequencies with Bragg frequencies predicted by the one-dimensional Lamb wave model for piezoceramic-polymer transducer.

6. SUMMARY

A brief review has been given of elastic superlattice theory and its application to resonances in piezoceramic-polymer transducers. Some important aspects of the problem could not be reported here for lack of space; others have not, indeed, yet been analyzed. All of the theory/experiment comparisons discussed above were based on coupled-mode theory, where only two space harmonics are retained in the elastic wavefunction. For piezoceramic-polymer superlattices, the coupling constants in Eq. (20) are large, since the material components are very different for the two component phases, and volume fractions of ceramic as high as 28.6% are used. Furthermore, the coupling constants decrease slowly with increasing order of the space harmonics. Numerical calculations of the dispersion relation using increasing numbers of harmonics (2,4,8, etc.) show that reasonable convergence to a solution is obtained with 12 harmonics, even with 28.6% volume fraction of piezoceramic [27]. The effect of adding more harmonics is to decrease the width of the stopband and to shift its midpoint down from the Bragg frequency. Another refinement that has been studied is to include the effect of the side boundaries on the lateral resonances of a transducer. This analysis shows that the resonances observed in Fig. 3 are at one edge of the stopband, not at the Bragg frequency. Finally, the treatment of piezoceramic-polymer superlattices presented here completely ignores the effect of piezoelectricity. Examination of this question has only just begun. In Reference 28 the thickness mode-coupling strength of a piezoceramic-polymer transducer has been modeled in the long wavelength limit, showing that there exists an optimum volume fraction of piezoceramic. A similar analysis for lateral modes remains to be done for the lateral resonances.

REFERENCES

1. Newnham RE, Bowen LJ, Klicker KA, and Cross LE: Materials in Engineering 2, 93, 1980.
2. Gururaja TR, Schulze WA, Shrout TR, Safari A, Webster L, and Cross LE: Ferroelectrics 9, 1245, 1981.
3. Safari A, Newnham RE, Cross LE, and Schulze WA: Ferroelectrics 41, 197, 1982.
4. Newnham RE, Safari A, Sa-gong G, and Giniewicz J: 1984 Ultrasonics Symposium Proceedings, McAvoy B(ed.) (IEEE, Piscataway, NJ, 1984), 501.
5. Gururaja TR, Schulze WA, Cross LE, and Newnham RE: 1984 Ultrasonic Symposium Proceedings, McAvoy B(ed.) (IEEE, Piscataway, JN, 1984), 533.
6. Smith WA, Shaulov AA, and Singer BM: 1984 Ultrasonic Symposium Proceedings, McAvoy B(ed.) (IEEE, Piscataway, NJ, 1984), 539.
7. Shaulov AA, Smith WA, and Singer BM: 1984 Ultrasonic Symposium Proceedings, McAvoy B(ed.) (IEEE, Piscataway, NJ, 1984), 545.
8. Takeuchi H. Nakaya C, and Katakura K: 1984 Ultrasonic Symposium Proceedings, McAvoy B(ed.) (IEEE, Piscataway, NJ, 1984), 507.
9. Gururaja TR, Schulze WA, Newnham RE, Auld BA, and Wang Y: IEEE Trans. on Sonics and Ultrasonics, SU-32, 481, 1985.
10. Takeuchi H and Nakaya C: Ferroelectrics (in press, 1985).
11. Li RC, Melngailis J: IEEE Trans. Sonics and Ultrasonics SU-22, 189, 1975.
12. Datta S and Hunsinger BJ: 1979 Ultrasonic Symposium Proceedings, McAvoy B(ed.) (IEEE, Piscataway, NJ, 1984), 677.
13. Lees S and Davidson CL: IEEE Trans on Sonics and Ultrasonics, SU-24, 222, 1977.

14. Nayfeh AN, Crane RL, and Hoppe WC: J. Appl. Phys. 55, 685, 1984.
15. Tao R and Sheng P: J. Acoust. Soc. Am. 77, 1651, 1985.
16. Brillouin L: Wave Propagation in Periodic Structures (New York: McGraw-Hill, 1946).
17. Elachi C: IEEE Proceedings, 64, 1666, 1976.
18. Gaylord TK and Moharam MG: IEEE Proceedings, 73, 894, 1985.
19. Auld BA, Acoustic Fields and Waves in Solids, Vols. I and II (New York: Wiley-Interscience, 1973).
20. Auld BA, Kunkel HA, Shui Y, and Wang Y: 1983 Ultrasonics Symposium Proceedings, McAvoy B(ed.) (IEEE Piscataway, NJ 1984), 554.
21. Auld BA, Shui YA, and Wang Y: Journal de Physique 45, 159, 1984.
22. Ikegami S, Ueda I, and Kobayashi S: J. Acoust. Soc. Am. 55, 339, 1974.
23. Ueha S, Sakuma S, and Mori E: J. Acoust. Soc. Am. 73, 1842, 1983.
24. Kunkel HA and Auld BA: 1981 Ultrasonics Symposium Proceedings, McAvoy B (ed.) (IEEE, Piscataway, NJ, 1981), 438.
25. Gururaja TR, Schulze WA, Cross E, Auld BA, Shui YA, and Wang Y: Ferroelectrics 54, 183, 1984.
26. Auld BA and Wang Y: 1984 Ultrasonics Symposium Proceedings, McAvoy B (ed.) (IEEE, Piscataway, NJ, 1984), 528.
27. Wang Y and Auld BA: "Acoustic wave propagation in one-dimensional periodic composites," presented at the 1985 IEEE Ultrasonics Symposium, San Francisco, CA. October 1985.
28. Smith WA, Shaulov A, and Auld BA, "Tailoring the properties of composite piezoelectric materials for medical ultrasonic transducers," presented at the 1985 IEEE Ultrasonics Symposium, San Francisco, CA. October 1985.

DISCUSSION

Comment: Sayers
In this approach, are there material constants of the two component phases allowed to have discontinuity at their interfaces, or must the changes be smooth across the interface?

Reply: Auld
The Floquet (Brillouin) - or space harmonic - method does, indeed, allow for discontinuities of the material properties. An explicit proof of this is given in the reference by Tao and Sheng cited in the paper.

Comment: Adler
Is this method applicable to a superlattice of empty pores?

Reply: Auld
Yes. It has been applied to a porous superlattice, in the long wavelength limit by Tao and Sheng, in the paper referred to.

Comment: Cheeke
You mentioned that losses may be included in the analysis. Doesn't this decrease the accuracy drastically when the losses are large?

Reply: Auld
If you mean that the coupled mode approximation (including only resonant space harmonics) is no longer valid when the losses are large, then, that is certainly true. When the losses are large, more space harmonics must be included, until convergence is obtained. We are using 12 space harmonics in some of our lossless calculations, and this should take care of any practical losses.

Comment: Mayer
In the measurement of finite thickness plates shown on one of your figures, the experimental points lie very far out of the wavenumber axis of the Lamb wave dispersion curves. In this case, isn't likely that the waves are Rayleigh, rather than Lamb, waves?

Reply: Auld
It is very likely that these lateral resonances involve predominantly Rayleigh waves. However, these appear first at moderately large wave number in the two lowest order Lamb modes as symmetric and antisymmetric combinations of Rayleigh waves on the top and bottom of the plate.

PIEZOELECTRIC MATERIALS

J.J. GAGNEPAIN

1.INTRODUCTION

The propagation of acoustic waves in elastic solids can be described by linear laws only in the case of infinitelysmall amplitudes. In fact the conditions of propagation are nonlinear. This is due to the structure of the medium. These nonlinearities, which will be called intrinsic nonlinearities, are characterized by higher order material constants ; Hooke's law is the linear approximation of the stress-strain relations. In addition real waves have a finite amplitude, and during the propagation nonlinearities are induced by the deformation of the medium. These induced nonlinearities are comparable to the previous ones, and both must be taken into account.

Because of these nonlinearities the solid becomes inhomogeneous ; and must be considered as such. As a consequence harmonic generation, velocity-amplitude dependence, intermodulation can take place. On the other hand, if the solid is submitted to a bias, which can be induced by external perturbation (temperature, forces, acceleration field, electric field, ...), the nonlinearities will be at the origin of a coupling between the static or quasistatic deformation and the high frequency wave. The resultsis modification of the wave characteristics, and in particular of its velocity.

These nonlinear phenomena are a limitation of the performances of piezoelectric devices like resonators, oscillators, delay lines and filters, since they prevent to operate them at higher powers ; they are causes of instabilities and noises, and they are responsible for their sensitivities to the surrounding world. A large part of the efforts, along the history of piezoelectricity, consisted in trying to reduce such effects, and this was achieved mainly by using the anisotropy of these piezoelectric crystals. However, as by products, nonlinearities are also used for realizing functions adapted to signal processing (convolutors, correlators, ...) or for measuring physical quantities like temperature, forces, acceleration fields, etc, by means of sensors which transform the variation of the physical quantity into a frequency variation.

2. NONLINEAR ELECTRO-ACOUSTIC EQUATIONS

2.1. Initial and final state

Let a_j be the coordinates of a material point M of the solid at rest, which correspond to the initial state. After deformation the material point undergoes a displacement and takes a new position M' of coordinates x_j, denoted as the final state. The a_j's are called Lagrangian or material coordinates, and the x_j's the eulerians or spatial coordinates.

The components of the mechanical displacement are

$$u_j = x_j - a_j \tag{1}$$

and the instantaneous velocity of a particle is $v_j = \frac{dx_j}{dt} = \dot{x}_j$.

2.2. Strains

A length element $d\ell_0$ of the solid at rest becomes $d\ell$ after deformation, such that

$$d\ell_0^2 = da_j \, da_j = \alpha_{ij} \, dx_i \, dx_j \tag{2}$$

$$d\ell^2 = dx_j \, dx_j = \beta_{ij} \, da_i \, da_j \tag{3}$$

α_{ij} and β_{ij} are the Cauchy's and Green's strain tensors. Since they are not zero for rigid deformations, it is prefered to use the η_{ij} tensor defined by

$$d\ell^2 - d\ell_0^2 = 2\eta_{ij} \, da_i \, da_j \tag{4}$$

with

$$\eta_{ij} = \frac{1}{2} \left(\frac{\partial x_k}{\partial a_i} \frac{\partial x_k}{\partial a_j} - \delta_{ij}\right) \tag{5}$$

which by using (1) into (5) takes the form

$$\boxed{\eta_{ij} = \frac{1}{2} \left(\frac{\partial u_j}{\partial a_i} + \frac{\partial u_i}{\partial a_j} + \frac{\partial u_k}{\partial a_i} \frac{\partial u_k}{\partial a_j}\right)} \tag{6}$$

In this definition, the strain tensor is given with respect to the initial coordinates a_j. A similar definition could be given with respect to the final state. This would yield the tensor

$$\varepsilon_{ij} = \frac{1}{2} \left(\frac{\partial u_j}{\partial x_i} + \frac{\partial u_i}{\partial x_j} - \frac{\partial u_k}{\partial x_i} \frac{\partial u_k}{\partial x_j}\right) \tag{7}$$

These relations contain a linear part, which corresponds to the usual strain definition in the approximation of infinitesimal deformations, and a quadratic term, which corresponds to nonlinearities of the "second order" (in terms of energy definition as it will be shown further).

2.3. Stresses

Considering in the usual manner an elementary tetrahedron submitted to external surface forces Q_j, the induced internal stresses T_{ij} are given by

$$Q_j = n_i \, T_{ij} \tag{8}$$

where n_i is the component of the unit normal to the surface element, the surface forces Q_j are acting on.

2.4. Electrical forces

Tiersten [1] developed a macroscopic model of the electroelastic solid, which consists in an electronic charge continuum coupled to the elastic continuum. The lattice continuum is assumed to have a positive charge density, and the electronic continuum to have a negative charge density. During a motion the electronic continuum is permitted to displace, and the opposite charge displacement gives to the structure a distribution of dipoles. Each elementary volume is submitted to an electric body force G_j and a body couple C given by

$$G_j = P_i \frac{\partial E_j}{\partial x_i} \tag{9}$$

$$C = E_j \wedge P_j \tag{10}$$

where E_j is the quasistatic Maxwellian electric field, and P_i is the dipole density. However the Maxwell electrostatic stress tensor can be introduced

$$T^E_{ij} = \varepsilon_o E_i E_j + P_i E_j - \frac{1}{2} \varepsilon_o E_k E_k \delta_{ij} \tag{11}$$

with

$$\frac{\partial T^E_{ij}}{\partial x_i} = P_i \frac{\partial E_j}{\partial x_i} \tag{12}$$

and the electric displacement D_i

$$D_i = \varepsilon_o E_i + P_i \tag{13}$$

ε_o is the permittivity of vacuum, δ_{ij} is the Kronecker's symbol.

2.5. Conservation laws

2.5.1. Conservation of the mass. Let ρ_0 and ρ be respectively the specific mass in the initial and final states. The mass dM of a volume element dV_0 in the initial state, and dV in the final state is conserved, thus

$$\rho_0 \, dV_0 = \rho \, dV \tag{14}$$

which can be written

$$\rho J = \rho_0 \tag{15}$$

where $J = \lim\limits_{dV_0 \to 0} \frac{dV}{dV_0} = \left|\frac{\partial x_j}{\partial a_i}\right|$ is the Jacobian matrix of the deformation.

Conservation of the total mass M of the solid corresponds to

$$\frac{dM}{dt} = 0 \quad \text{with} \quad \dot{M} = \int_V \rho dV \tag{16}$$

Eq. (16) is the integral form of the mass conservation. The local form is easily obtained from Appendix A

$$\frac{\partial \rho}{\partial t} + \frac{\partial}{\partial x_k} (\rho v_k) = 0 \quad \text{in the volume} \tag{17}$$

On the surface limiting the volume

$$[\![\rho v_k]\!] n_k - [\![\rho]\!] \, u_{[n]} = 0 \tag{18}$$

2.5.2. Conservation of the linear momentum. The time derivative of the linear momentum equals the resultant of the surface and body forces

$$\frac{d}{dt} \int_V \rho \, v_j \, dV = \int_S Q_j \, dS + \int_V (F_j + P_i \frac{\partial E_i}{\partial x_i}) \, dV \tag{19}$$

F_j are body forces due to any external field (acceleration field for instance).

Using Eq. (8) (12) and Green theorem

$$\int_V \left\{\frac{\partial(\rho v_j)}{\partial t} + \frac{\partial}{\partial x_k}(\rho v_j v_k) - \frac{\partial}{\partial x_k}(T_{kj}+T^E_{kj}) - F_j\right\} dV$$
$$+ \int_s \{[\![T_{kj}+T^E_{kj}]\!]\, n_k + [\![\rho v_j]\!]\, u_{[n]} - [\![\rho v_j v_k]\!]\, n_k\}\, ds = 0$$

This relation must be satisfied whatever the volume V and the discontinuity surface s are. Therefore

$$\boxed{\rho \dot{v}_j = \frac{\partial}{\partial x_k}(T_{kj} + T^E_{kj}) + F_j} \quad \text{in the volume} \tag{20}$$

$$\boxed{[\![T_{kj}+T^E_{kj}]\!]\, n_k + [\![\rho v_j]\!]\, u_{[n]} - [\![\rho v_j v_k]\!]\, n_k = 0} \quad \text{on the surface} \tag{21}$$

2.5.3. Conservation of the angular momentum. The time derivative of the angular momentum is equal to the sum of the force couples in the solid

$$\frac{d}{dt}\int_V (x_j \wedge \rho v_j)\, dt = \int_S (x_j \wedge Q_j)\, dS$$
$$+ \int_V (x_j \wedge F_j + x_j \wedge P_i \frac{\partial E_j}{\partial x_i} + E_j \wedge P_j) dV \tag{22}$$

Similar transformations give the local forms of this relation

$$\rho\,(x_k \dot{v}_i - x_i \dot{v}_k) = T_{ki} - T_{ik} + x_k \frac{\partial}{\partial x_\ell}(T_{\ell i}+T^E_{\ell i}) - x_i \frac{\partial}{\partial x_\ell}(T_{\ell k}+ T^E_{\ell k})$$
$$+ x_k F_i - x_i F_k + E_k P_i - E_i P_k \tag{23}$$

$$[\![\rho(x_k v_i - x_i v_k) v_m]\!] n_m - [\![\rho(x_k v_i - x_i v_k)]\!] u_{[n]}$$
$$- [\![x_k (T_{\ell i}+T^E_{\ell i}) - x_i (T_{\ell k}+T^E_{\ell k})]\!] n_\ell = 0 \tag{24}$$

From (21), Eq. (24) is always satisfied. Using eq. (20), eq. (23) is simplified, yielding

$$\boxed{T_{ki} - T_{ik} = E_i P_k - E_k P_i} \tag{25}$$

This indicates that in general the stress tensor T_{ki} is not symmetrical. This tensor will be symmetrical only in the absence of electric field, or if the electric field and the polarization are colinear.

2.5.4. Conservation of the energy. The rate of energy supplied to the solid is equal to the variation of its total energy.

The total energy is the sum of the internal energy per unit volume ρU and of the kinetic energy $(1/2)\rho v_k v_k$.

The energy added comes from the rate of working per unit area $Q_k v_k$ of the surface forces, the rate of working $F_k v_k$ of the body forces, the rate of working of the electric force and couple, and the thermal energy flow $n_k q_k$. Thus

$$\frac{d}{dt}\int_V \rho(\frac{1}{2} v_k v_k + U)dV = \int_S (Q_k v_k - n_k q_k)dS + \int_V \sigma\, dV \tag{26}$$

using

$$\sigma = P_i \frac{\partial E_j}{\partial x_j} v_i + E_i \frac{dP_i}{dt} \tag{27}$$

Taking the material time derivative of the left hand side of eq. (26), using the mass conservation law, applying Green theorem and eq. (20), the local form of eq. (26) is obtained

$$\rho \frac{dU}{dt} = T_{ij} \frac{\partial v_j}{\partial x_i} - \frac{\partial q_k}{\partial x_k} + E_i \frac{dP_i}{dt} \tag{28}$$

Introducing the entropy η per unit mass, given by the relation

$$\rho \frac{d\eta}{dt} = -\frac{1}{\theta}\frac{\partial q_i}{\partial x_i} \tag{29}$$

where θ is the temperature, equation (28) becomes

$$\rho \frac{dU}{dt} = T_{ij} \frac{\partial v_j}{\partial x_i} + \rho\, \theta \frac{d\eta}{dt} + E_i \frac{dP_i}{dt} \tag{30}$$

2.6. Constitutive equations

It will be more practical to use the electric field (which can be expressed as a function of the electric potential) and the temperature, as variables rather than the polarization and the entropy in eq. (30). This can be done by means of the following transformation, which introduces the enthalpy (or Gibbs function)

$$\chi = U - \eta\, \theta - \frac{E_i P_i}{\rho} \tag{31}$$

Using also the relation $\frac{\partial v_j}{\partial x_i} = \frac{\partial a_m}{\partial x_i}\frac{d}{dt}(\frac{\partial x_j}{\partial a_m})$, this gives

$$\rho \frac{d\chi}{dt} = T_{ij} \frac{\partial a_m}{\partial x_i}\frac{d}{dt}(\frac{\partial x_j}{\partial a_m}) - P_i \frac{dE_i}{dt} - \rho\, \eta \frac{d\theta}{dt} \tag{32}$$

Taking $\frac{\partial x_j}{\partial a_m}$, E_i, and θ as independent variables, and assuming

$$\chi = \chi\, (\frac{\partial x_j}{\partial a_m}, E_i, \theta),$$

one has

$$\frac{d\chi}{dt} = \frac{\partial \chi}{\partial(\partial x_j/\partial a_m)}\frac{d}{dt}(\frac{\partial x_j}{\partial a_m}) + \frac{\partial \chi}{\partial E_i}\frac{dE_i}{dt} + \frac{\partial \chi}{\partial \theta}\frac{d\theta}{dt} \tag{33}$$

Combining (33) into (32), and considering that the material time derivatives are independent, one obtains the constitutive equations

$$T_{ij} = \rho \frac{\partial \chi}{\partial(\partial x_j/\partial a_m)} \frac{\partial x_i}{\partial a_m} \tag{34}$$

$$P_i = -\rho \frac{\partial \chi}{\partial E_i} \tag{35}$$

$$\eta = -\frac{\partial \chi}{\partial \theta} \tag{36}$$

These expressions can be written with respect to the material coordinate system, as a function of the strains η_{ij}, the electric field W_k of the material state, and the temperature : $\chi = \chi(\eta_{ij}, W_k, \theta)$.

Since

$$W_k = -\frac{\partial \phi}{\partial a_k} = -\frac{\partial \phi}{\partial x_m}\frac{\partial x_m}{\partial a_k} = -E_m \frac{\partial x_m}{\partial a_k} \quad \text{where } \phi \text{ is the potential} \tag{37}$$

the constitutive equations take the form

$$T_{ij} = \rho \frac{\partial x_i}{\partial a_m}\frac{\partial x_j}{\partial a_k}\frac{\partial \chi}{\partial \eta_{mk}} + \rho \frac{\partial x_i}{\partial a_m} E_j \frac{\partial \chi}{\partial W_m} \tag{38}$$

$$P_i = -\rho \frac{\partial x_i}{\partial a_m}\frac{\partial \chi}{\partial W_m} \tag{39}$$

$$\eta = -\frac{\partial \chi}{\partial \theta} \tag{40}$$

2.7. Material constants

The material constants are introduced in a polynomial development of the energy with respect to the independent variables of the material state.

$$\begin{aligned}\rho_o \chi = {} & \frac{1}{2} C_{ijk\ell}\eta_{ij}\eta_{k\ell} + \frac{1}{6} C_{ijk\ell mn}\eta_{ij}\eta_{k\ell}\eta_{mn} - e_{m.ij} W_m \eta_{ij} \\ & - \frac{1}{2} e_{m.ijk\ell} W_m \eta_{ij} \eta_{k\ell} - \frac{1}{2} \varepsilon_{mn} W_m W_n - \frac{1}{6} \varepsilon_{mnp} W_m W_n W_p \\ & - \frac{1}{2} e_{mn.ij} W_m W_n \eta_{ij} + \text{thermal terms} + \text{and higher order terms..}\end{aligned} \tag{41}$$

This development is limited to the third order. $C_{ijk\ell}$ and $C_{ijk\ell mn}$ denote the 2nd and 3rd order elastic constants. $e_{m.ij}$ denote the piezoelectric constants, $e_{mn.ij}$ are the electrostrictive constants, and $e_{m.ijk\ell}$ the electroelastic constants. ε_{mn} and ε_{mnp} are the 2nd and 3rd order dielectric constants.

2.8. Propagation equations in the material state

Let $P_{\ell j}$ be the total stresses (Piola-Kirchoff tensor) represented per unit surface of the initial solid

$$P_{\ell j} = J \frac{\partial a_\ell}{\partial x_i}(T_{ij} + T^E_{ij}) \tag{42}$$

Similarly the electric field $\mathcal{D}_\ell$ in the initial state is

$$\mathcal{D}_\ell = J \frac{\partial a_\ell}{\partial x_i} D_i \tag{43}$$

The total stress $(T_{ij} + T^E_{ij})$ is composed of a symmetrical part and a non-symmetrical part, which can be distinguished

$$T_{ij} + T^E_{ij} = T^S_{ij} + T^{ES}_{ij} \tag{44}$$

with

$$T^S_{ij} = \rho \frac{\partial x_i}{\partial a_m} \frac{\partial x_j}{\partial a_k} \frac{\partial \chi}{\partial \eta_{mk}} \tag{45}$$

$$T^{ES}_{ij} = \varepsilon_o E_i E_j - \frac{1}{2} \varepsilon_o E_k E_k \delta_{ij} \tag{46}$$

The mechanical equilibrium equation is

$$\rho_o \frac{dv_j}{dt} = \frac{\partial P_{\ell j}}{\partial a_\ell} + J F_j \tag{47}$$

with the boundary condition

$N_\ell [\![P_{\ell j}]\!] = 0$ on a free surface of unit normal N_ℓ.

The electric equilibrium equation with the associated boundary condition are

$$\frac{\partial \mathcal{D}_\ell}{\partial a_\ell} = 0 \tag{48}$$

$$N_L [\![\mathcal{D}_\ell]\!] = 0 \quad , \quad [\![\phi]\!] = 0 \tag{49}$$

in the absence of space and surface charges.

3. HARMONIC GENERATION, AMPLITUDE-FREQUENCY EFFECT, INTERMODULATION

3.1. Purely elastic medium

The case of a purely elastic medium can be easily treated. The detailed expression of the Piola-Kirchoff stress tensor P_{ij} is obtained from eqs. (6), (38), (41) and (42)

$$P_{ij} = C_{ijk\ell} u_{k,\ell} + \gamma_{ijk\ell mn} u_{k,\ell} u_{m,n} + \delta_{ijk\ell mnpq} u_{k,\ell} u_{m,n} u_{p,q} \tag{50}$$

(the notation $u_{k,\ell}$ is used for $\partial u_k/\partial a_\ell$).

The tensors $\gamma_{ijk\ell mn}$ and $\delta_{ijk\ell mnpq}$ are characteristics of 3rd and 4th order nonlinearities. The 4th order elastic constants $C_{ijk\ell mnpq}$ were introduced in addition in eq. (41).

$$\gamma_{ijk\ell mn} = C_{ink\ell} \delta_{jm} + \frac{1}{2} C_{ijn\ell} \delta_{km} + \frac{1}{2} C_{ijk\ell mn} \tag{51}$$

$$\delta_{ijk\ell mnpq} = \frac{1}{2} C_{iqn\ell} \delta_{km} \delta_{jp} + \frac{1}{4} C_{ijk\ell qn} \delta_{pm} + \frac{1}{4} C_{ijq\ell mn} \delta_{kp} + \frac{1}{2} C_{iqk\ell mn} \delta_{pj} + \frac{1}{6} C_{ijk\ell mnpq} \tag{52}$$

3.1.1. Travelling bulk wave in infinite medium. If the propagation medium is infinite there is no boundary condition. The propagation equation is from (42) and (50)

$$\rho_o \ddot{u}_j = C_{ijk\ell}\, u_{k,\ell i} + \gamma_{ijk\ell mn}(u_{k,\ell}\, u_{m,n})_{,i} + \delta_{ijk\ell mnpq}(u_{k,\ell}\, u_{m,n}\, u_{p,q})_{,i} \tag{53}$$

The linear approximation corresponds to

$$\rho_o \ddot{\overset{0}{u}}_j = C_{ijk\ell}\, \overset{0}{u}_{k,\ell i} \tag{54}$$

Considering a solution of the form

$$\overset{0}{u}_j = b_j \cos(\omega t - k_o\, N_\ell\, a_\ell) \tag{55}$$

which corresponds to a wave at a frequency ω propagating in a direction of normal unit N_ℓ with a velocity V and wave number $k_o = \omega/V_o$.

This solution satisfies (54) if

$$(C_{ijk\ell}\, N_i\, N_\ell - \rho_o\, V_o^2\, \delta_{kj}) b_k = 0 \tag{56}$$

This equation has three eigen values $\lambda_r = \rho_o {V_o^{(r)}}^2$ (r = 1,2,3) and three eigen vectors of components $b_j^{(r)}$.

The three corresponding modes have director cosines

$$\ell_j^{(r)} = \frac{b_j^{(r)}}{U_o^{(r)}} \quad \text{with} \quad U_o^{(r)} = \sqrt{b_j^{(r)}\, b_j^{(r)}} \tag{57}$$

Since the three modes are orthogonal, they constitute a reference frame. In this system each mode is characterized by a displacement U_r, which can be given as a function of the curvilinear abscisse s

$$U_r = u_j^{(r)}\, \ell_j^{(r)} \quad , \quad s = N_\ell\, a_\ell \tag{58}$$

The system of the equation (54) reduces

$$\rho_o \ddot{U}_r = \lambda_r\, U_{r,ss} \quad \text{with} \quad \lambda_r = C_{ijk\ell}\, N_i\, N_\ell\, \ell_k^{(r)}\, \ell_j^{(r)} \tag{59}$$

The nonlinear equation (53) can be transformed using the same transformation

$$\rho_o \ddot{U}_r = \lambda_r\, u_{r,ss} + \sum_{r',r''} \Gamma_{rr'r''}(U_{r',s}\, U_{r'',s})_{,s} + \sum_{r',r'',r'''} \Delta_{rr'r''r'''}(U_{r's}\, U_{r'',s}\, U_{r''',s})_{,s} \tag{60}$$

with

$$\Gamma_{rr'r''} = \gamma_{ijk\ell mn}\, \ell_k^{(r')}\, \ell_m^{(r'')}\, \ell_j^{(r)}\, N_i\, N_\ell\, N_n \tag{61}$$

$$\Delta_{rr'r''r'''} = \delta_{ijk\ell mnpq}\, \ell_k^{(r')}\, \ell_m^{(r'')}\, \ell_p^{(r''')}\, \ell_j^{(r)}\, N_i\, N_\ell\, N_n\, N_q \tag{62}$$

It appears that the nonlinearities induce a coupling between modes (r', r''≠r and r', r'', r''' ≠ r), and on the other hand a modification of each mode. In the hypothesis of a single mode propagation r = r' = r'' = r , eqs. (60)-(62) become

$$\rho_o \ddot{U}_r = \lambda_r U_{r,ss} + \Gamma_r (U_{r,s})^2_{,s} + \Delta_r (U_{r,s})^3_{,s} \tag{63}$$

where $\Gamma_r = \Gamma_{rrr}$ and $\Delta_r = \Delta_{rrrr}$.

Different methods can be used for solving this equation : successive approximations, characteristics, multiple scales, coupled amplitude [2].

The successive approximations method consists in using a solution of the form

$$U_r = \overset{0}{U} + \overset{1}{U} + \overset{2}{U} \tag{64}$$

where $\overset{1}{U}$ and $\overset{2}{U}$ are correcting terms of order 1 and 2 in the approximation. Using (64) in (63) and identifying yields the system

$$\rho_o \ddot{\overset{0}{U}} - \lambda \overset{0}{U}_{,ss} = 0 \tag{65}$$

$$\rho_o \ddot{\overset{1}{U}} - \lambda \overset{1}{U}_{,ss} = \Gamma (\overset{0}{U}_{,s})^2_{,s} \tag{66}$$

$$\rho_o \ddot{\overset{2}{U}} - \lambda \overset{2}{U}_{,ss} = 2\Gamma (\overset{0}{U}_{,s} \overset{1}{U}_{,s})_{,s} + \Delta (\overset{0}{U}_{,s})^3_{,s} \tag{67}$$

Solution of (65) is obvious

$$\overset{0}{U} = U_1 \cos \Psi \quad \text{with} \quad \Psi = \omega t - k_0 s \tag{68}$$

Solution of (66) is obtained by using (68) in (66) and a solution

$$\overset{1}{U} = A \cos 2\Psi + B \sin 2\Psi + C \tag{69}$$

After identification

$$A = \frac{1}{4}\frac{\Gamma}{\lambda} U_o^2 k_o^2 s \quad ; \quad B = 0 \quad ; \quad C = -\frac{1}{2}\frac{\Gamma}{\lambda} U_o^2 k_o^2 s \tag{70}$$

Solution of (67) is calculated after replacing (68) and (69) into (67). Thus

$$\overset{2}{U} = D \cos 3\Psi + E \sin 3\Psi + F \cos \Psi + G \sin \Psi \tag{71}$$

$$D = +\frac{1}{8}\frac{k_o^4 U_o^3 \Gamma^2}{\lambda^2} s^2 \quad ; \quad E = \frac{1}{6} k_o^3 U_o^3 s \left(\frac{\Gamma^2}{\lambda^2} - \frac{3}{4}\frac{\Delta}{\lambda}\right)$$

$$F = -D \quad ; \quad G = -\frac{3}{4} k_o^3 U_o^3 s \left[\frac{\Gamma^2}{\lambda^2} - \frac{\Delta}{2\lambda}\right]$$

The solution at the fundamental frequency is therefore perturbed, and the phase perturbation is equivalent to a velocity-amplitude dependence

$$V = V_o \left[1 + \frac{3}{4}\frac{\omega^2}{V_o^2} U_o^2 \left(\frac{\Gamma^2}{\lambda^2} - \frac{\Delta}{2\lambda}\right)\right] \tag{72}$$

3.1.2. Standing waves. The propagation medium is limited by two free surfaces located at $s = \pm h$.

The general solution is a superposition of incident and reflected waves. The boundary condition corresponds to surfaces free of stresses ($T_{ij}n_j = 0$) which is verified by

$$U_{,s} = 0 \quad \text{for} \quad s = \pm h \tag{73}$$

therefore

$$\overset{0}{U}_{,s} = \overset{1}{U}_{,s} = \overset{2}{U}_{,s} = 0 \quad \text{for} \quad s = \pm h \tag{74}$$

The linear solution is

$$\overset{0}{U} = U_0 \sin\omega t \sin k_0 s \tag{75}$$

with the resonance condition

$$k_0 h = (2n+1)\frac{\pi}{2} \tag{76}$$

where n is an integer.

Following the same method, the solutions $\overset{1}{U}$ and $\overset{2}{U}$ are calculated. The resonance frequency is amplitude dependent

$$\frac{\omega_r - \omega_0}{\omega_0} = \frac{U_0^2\,\pi^2}{128h^2}\left\{-\frac{21\Gamma^2}{2\lambda^2} + \frac{9\Delta}{\lambda}\right\} \tag{77}$$

where ω_0 is given by (76).

In the case of a quartz resonator, vibrating on the thickness shear mode, with a propagation direction along Oa_2, as shown in Fig. 1, the previous relation becomes

$$\frac{\omega_r - \omega_0}{\omega_0} = \frac{3U_0^2\pi^2}{256h^2}\left[\frac{3C_{22} + 6C_{266} + C_{6666}}{C_{66}}\right] \tag{78}$$

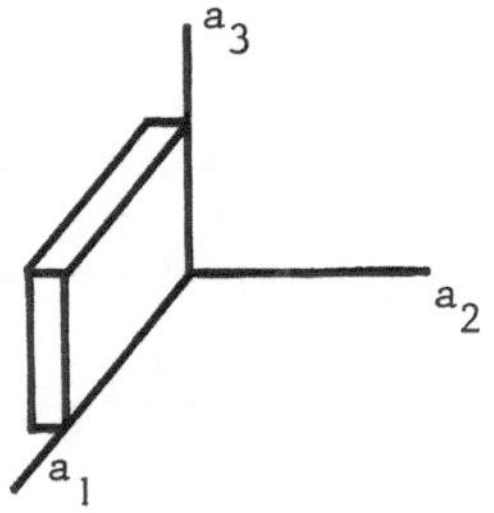

FIGURE 1.

3.2. Piezoelectric medium

The case of piezoelectric medium will be illustrated by the calculation of intermodulation. The driving signal is now composed of two angular frequencies ω_1 and ω_2 [3]. The linear solution is written

$$\overset{0}{U} = U_1 \cos\Psi_1 + U_2 \cos\Psi_2 \tag{79}$$

with

$$\Psi_1 = \omega_1 \left(t - \frac{s}{V_0}\right) \quad \text{and} \quad \Psi_2 = \omega_2 \left(t - \frac{s}{V_0}\right)$$

Using this solution in (66) shows that solution $\overset{1}{U}$ involves harmonic terms at the frequencies $2\omega_1$ and $2\omega_2$, and intermodulation terms at $\omega_1+\omega_2$ and $\omega_1-\omega_2$. Similarly $\overset{2}{U}$ involves fundamental terms at ω_1 and ω_2, harmonics $3\omega_1$ and $3\omega_2$, and intermodulation at $2\omega_1+\omega_2$, $2\omega_2+\omega_1$, $2\omega_1-\omega_2$, and $2\omega_2-\omega_1$.

Let us consider, for example, the case of a resonator driven by the two signals with frequencies ω_1 and ω_2 symmetrically located with respect to the resonance frequency ω_0 and inside the resonator bandwidth. The intermodulation frequencies $2\omega_1-\omega_2$ and $2\omega_2-\omega_1$ will be also symmetrical with respect to ω_0 and inside the resonator bandwidth, as shown in Fig. 2. The other frequencies are largely out of the bandwidth and eliminated.

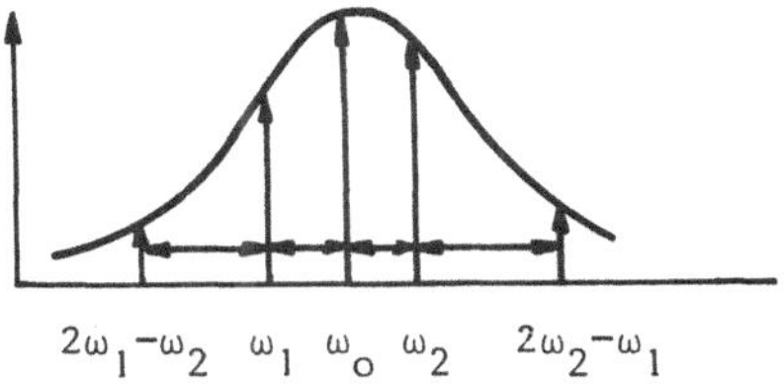

FIGURE 2.

The quartz plate of Fig. 1 again is used. Considering that the lateral dimensions are very large compared to the thickness, the vibration mode is a pure thickness shear propagating in the a_2 direction. In that case the equilibrium equations reduce to

$$\rho_0\ddot{u}_1 - C_{66}\, u_{1,22} - e_{26}\, \phi_{,22} = \Delta(u_{1,2})^3_{,2} \tag{80}$$

$$e_{26}\, u_{1,22} - \varepsilon_{22}\, \phi_{,22} = 0$$

with

$$\Delta = \frac{1}{2} C_{22} + C_{266} + \frac{1}{6} C_{6666}$$

The corresponding boundary conditions are

$$C_{66}\, u_{1,2} + e_{26}\, \phi_{,2} = -\Delta(u_{1,2})^3$$

$$\text{pour } a_2 = \pm h \tag{81}$$

$$\phi = \pm\frac{1}{2}(V_1 \cos\omega_1 t + V_2 \cos\omega_2 t)$$

From the expression of the electric displacement

$$D_2 = e_{26}\, u_{1,2} - \varepsilon_{22}\, \phi_{,2} \tag{82}$$

is obtained the total current through the electrodes of surface S

$$I = \frac{d}{dt}\left(\iint_s D_2\, dS\right) \tag{83}$$

The nonlinear factor Γ is zero in this example and only cubic nonlinearities appear. As a consequence the solutions are written

$$\begin{aligned} u_1 &= \overset{o}{U}_1 + \overset{2}{U}_1 \\ \phi &= \overset{o}{\phi} + \overset{2}{\phi} \\ I &= \overset{o}{I} + \overset{2}{I} \end{aligned} \tag{84}$$

and can be easily calculated.

The resonator is driven, near its resonance frequency, by a source of internal resistance R_G, and loaded by a resistance R_L, as shown in Fig. 3.

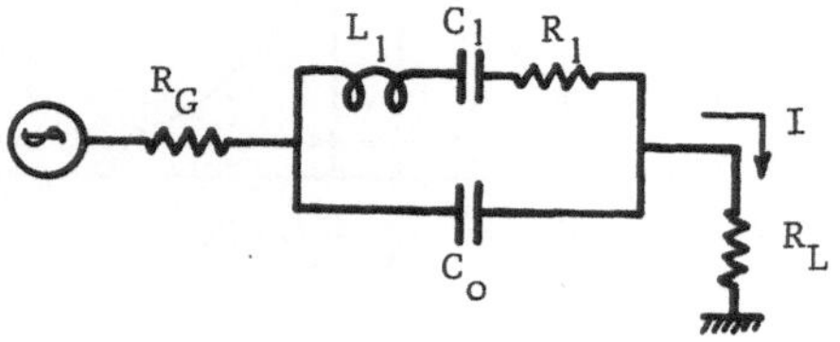

FIGURE 3.

Let P_1, P_2, P_Ω and $P_{\Omega'}$ respectively be the powers on the load R_L at the frequencies ω_1, ω_2, $\Omega = 2\omega_2 - \omega_1$ and $\Omega' = 2\omega_1 - \omega_2$.

The intermodulation ratio is given by

$$\frac{P_1}{P_\Omega} = \frac{R_L^2(R_G + R_L + R_1)^2}{4\Delta^2 G^2} \cdot \frac{1}{P_1 P_2} \tag{85}$$

with

$$G^2 = \frac{9^2 L_1^4 \omega_r^4 \rho_o}{4^3 \pi^2 S^2 N^2 \left(C_{66} + \frac{e_{26}^2}{\varepsilon_{22}}\right)^5}$$

L_1, R_1 are the motional inductance and resistance of the resonator, S the electrode surface, ω the resonance frequency and N the rank of the overtone (N = 1 for the fundamental resonance frequency).

Such a model can be used for characterizing the intermodulation in resonators, but also for determining from intermodulation measurements the value of unknown 4th order elastic constants [3].

4. PROPAGATION IN A MEDIUM WITH A BIAS

If the solid is submitted to static deformations and stresses due to the application of forces, electric fields, ... superimposed on a high frequency vibration, three states are to be considered, as shown in Fig. 4

- the natural state when the solid is at rest
- the initial state after application of the static deformation
- the final state when the vibration is superimposed to the static deformation

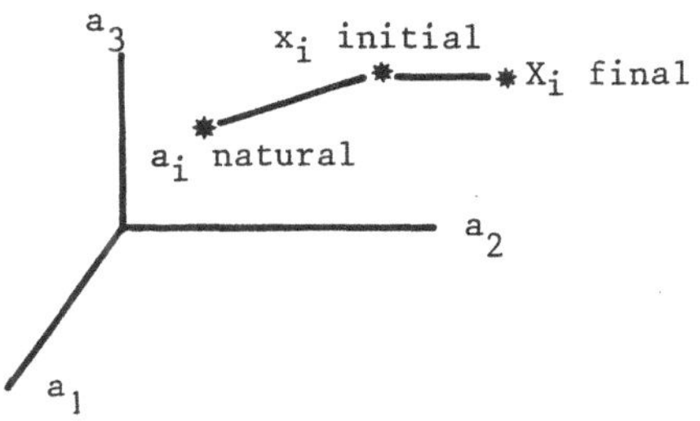

FIGURE 4.

4.1. Definitions

Mechanical displacements

$$\begin{aligned} &\text{static} && \bar{u}_i = x_i - a_i \\ &\text{dynamic} && \tilde{u}_i = X_i - x_i \\ &\text{total} && u_i = X_i - a_i \end{aligned} \tag{86}$$

Specific mass

ρ_0 : natural state

$\bar{\rho}$: initial state

ρ : final state

The same notation $\bar{\eta}_{ij}$, $\tilde{\eta}_{ij}$ and η_{ij} is used for the strains and for the other quantities.

4.2. Mechanical equilibrium equations

- final state

$$\rho \frac{dv_j}{dt} = \frac{\partial}{\partial X_i} (T^S_{ij} + T^{ES}_{ij}) = \frac{\partial \tau_{ij}}{\partial X_i} \tag{87}$$

- initial state

$$\bar{\rho} \frac{dv_j}{dt} = \frac{\partial K_{sj}}{\partial x_s} \tag{88}$$

with

$$K_{sj} = \tilde{J} \frac{\partial x_s}{\partial X_i} \tau_{ij}$$

and

$$\tilde{J} = \bar{\rho}/\rho \ , \quad \frac{\partial}{\partial x_s} (\tilde{J} \frac{\partial x_s}{\partial X_i}) = 0$$

- natural state

$$\rho_o \frac{dv_j}{dt} = \frac{\partial P_{j\ell}}{\partial a_\ell} \tag{89}$$

with

$$P_{j\ell} = \bar{J} \frac{\partial a_\ell}{\partial x_s} K_{sj}$$

and

$$\bar{J} = \rho_o/\bar{\rho}$$

In the expression of $P_{j\ell}$ the static and dynamic parts, respectively $\bar{P}_{j\ell}$ and $\tilde{P}_{j\ell}$, are separated. By difference with the static equilibrium equation $\rho_o \frac{d\bar{v}_j}{dt} = \frac{\partial \bar{P}_{j\ell}}{\partial a_\ell}$ (which supposes that the static deformation is not affected by the vibration and this is true if the wave amplitude is small), the dynamic equation is obtained

$$\rho_o \frac{d\tilde{v}_j}{dt} = \frac{\partial \tilde{P}_{j\ell}}{\partial a_\ell} \tag{90}$$

with

$$\tilde{P}_{j\ell} = \overline{A_{j\ell kr}} \frac{\partial \tilde{u}_k}{\partial a_r} + \overline{A_{rj\ell}} \frac{\partial \tilde{\phi}}{\partial a_r} \tag{91}$$

The $\overline{A_{j\ell kr}}$ and $\overline{A_{rj\ell}}$ coefficients are considered as effective elastic and piezoelectric constants. Separated into symmetrical and nonsymmetrical components. Their expressions are given in Appendix B.

4.3. Electric equilibrium equation

- final state

$$\frac{\partial D_i}{\partial X_i} = 0 \tag{92}$$

(in the absence of space charge)

- natural state

$$\frac{\partial \mathcal{D}_j}{\partial a_j} = 0 \tag{93}$$

with

$$\mathcal{D}_j = J \frac{\partial a_j}{\partial X_i} D_i \quad \text{and} \quad J = \rho_o/\rho$$

The same transformation gives the dynamic equation

$$\frac{\partial \tilde{\mathcal{D}}_j}{\partial a_j} = 0 \tag{94}$$

where

$$\tilde{\mathcal{D}}_j = \bar{A}_{jkr} \frac{\partial \tilde{u}_k}{\partial a_r} - \bar{E}_{jr} \frac{\partial \tilde{\phi}}{\partial a_r} \tag{95}$$

4.4. Boundary conditions

The boundary conditions for a free surface and in the absence of electric charges take the forms

$$N_j \, [\![\tilde{P}_{j\ell}]\!] = 0$$

$$N_j \, [\![\tilde{\mathcal{D}}_j]\!] = 0$$

and $[\![\tilde{\phi}]\!] = 0$

4.5. Perturbation method

The influence of the bias on the wave is determined by applying a perturbation method.

The equilibrium equation without perturbation is written

$$\rho_o \ddot{\tilde{u}}^o_j = \frac{\partial \overset{o}{P}_{ij}}{\partial a_i}$$

$$\mathcal{D}^o_{m,m} = 0 \tag{96}$$

and with perturbation

$$\rho_o \ddot{\tilde{u}}_j = \frac{\partial P_{ij}}{\partial a_i}$$

$$\mathcal{D}_{m,m} = 0 \tag{97}$$

Considering sinusoïdal solutions $\overset{o}{\tilde{u}}_j e^{j\omega_o t}$ and $\tilde{u}_j e^{j\omega t}$ which are used in the mechanical equations (96) and (97), multiplying (96) by $\tilde{u}_j$ and (97) by $\overset{o}{\tilde{u}}_j$, integrating over a volume V and taking the difference between the two equations, gives

$$2\rho_o \omega_o \Delta\omega \int_V \tilde{u}^o_j \, \tilde{u}^o_j \, dV = \int_V (P^o_{ij,i} \, \tilde{u}_j - P_{ij,i} \, \tilde{u}^o_j) dV \tag{98}$$

where ω_0 and ω are the unperturbed and perturbed frequencies and $\Delta\omega = \omega - \omega_0$.

Using in (98) the perturbed and unperturbed constitutive equations (91) and (95) and the electric equilibrium equations (96) and (97), and the effective material constants (B1, B3, B5) gives, after application of the boundary conditions, the perturbation relation

$$2\rho_o\omega_o\Delta\omega \int_V \tilde{u}^o_j \tilde{u}^o_j \, dV = \int_V [\tilde{u}^o_{j,i} \hat{C}_{ijk\ell} \tilde{u}^o_{k,\ell} + 2\tilde{u}^o_{j,i} \hat{e}_{m,ij} \tilde{\phi}^o_{,m} - \hat{\varepsilon}_{mn} \tilde{\phi}^o_{,n} \tilde{\phi}^o_{,m}] \, dV \qquad (99)$$

which is used for calculating either the frequency variation due to the static perturbation, or the velocity or wave number variations from the relation $k = \omega/V$.

5. APPLICATIONS

These methods were applied for studying the influence of temperature, forces, and electric fields on quartz crystal resonators. The problem is decomposed in three steps : the calculation of the static deformation, the determination of the characteristics of the unperturbed wave, then their application in the pertubation relation [4].

The static problem is illustrated by Fig. 5, which represents the calculated temperature distribution in a quartz resonator. The crystal is heated by the dissipated energy due to the vibration at the center, and thermal exchanges take place through the mounting at the two diametric fixation points. The resonator vibration corresponds to a pure thickness shear mode.

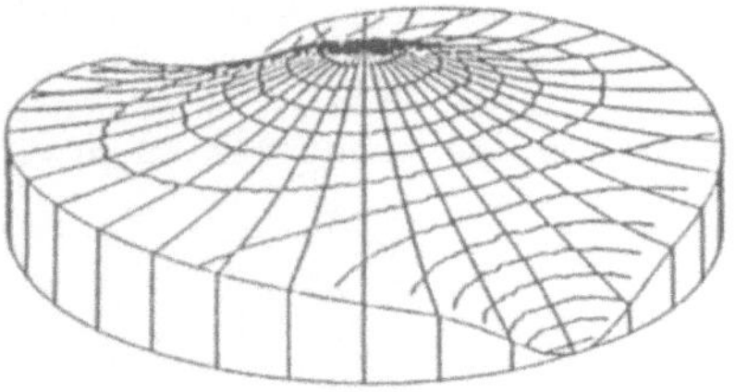

FIGURE 5.

Considering a sinusoïdal variation of the external temperature, the frequency-to-temperature variation corresponds to a cycle, which is called the dynamic thermal behavior, and is characteristic of the temperature gradient distribution. This is shown in Fig. 6.

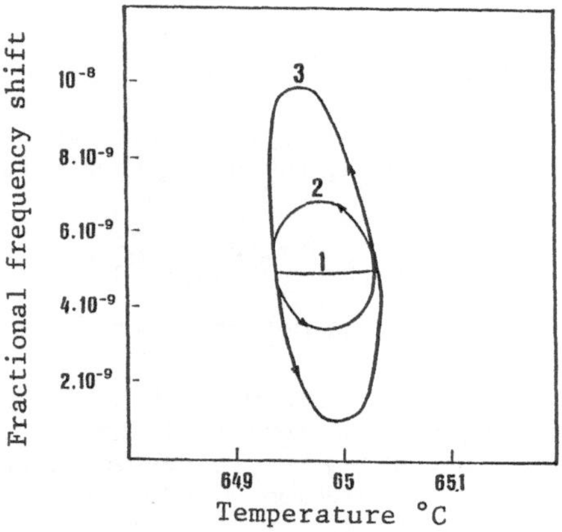

FIGURE 6.

A second example is the influence of external forces. In Fig. 7 is represented the frequency variation induced by two forces, diametrically applied on a 5 MHz quartz oscillator, as a function of the azimuthal angle of the forces. It must be noticed that this effect can be minimized for two particular directions of the application of the forces.

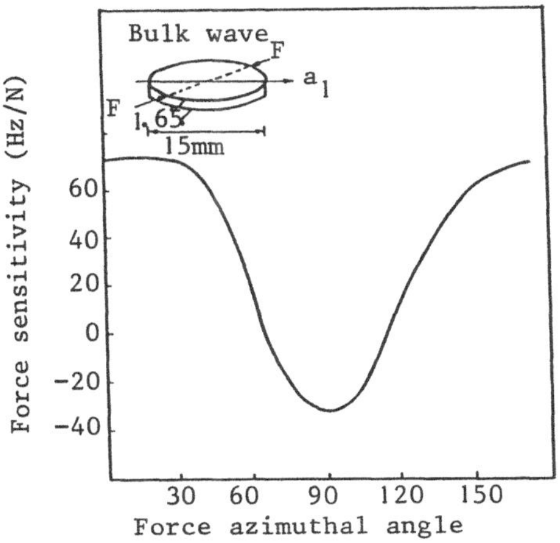

FIGURE 7.

The last example corresponds to the influence of an electric field on the same quartz resonator. The frequency variation, shown in Fig. 8, is mainly due to the electroelastic effect (constants $e_{n.ijk\ell}$) and these models have been used for determining the values of the eight independent electroelastic constants of quartz [5]. Then the initial frequency variation is followed by a relaxation phenomenon. This relaxation is due to the diffusion of ionic impurities in the quartz chanels. This is also a new method, which is actually under development, for characterizing the impurities in the crystal.

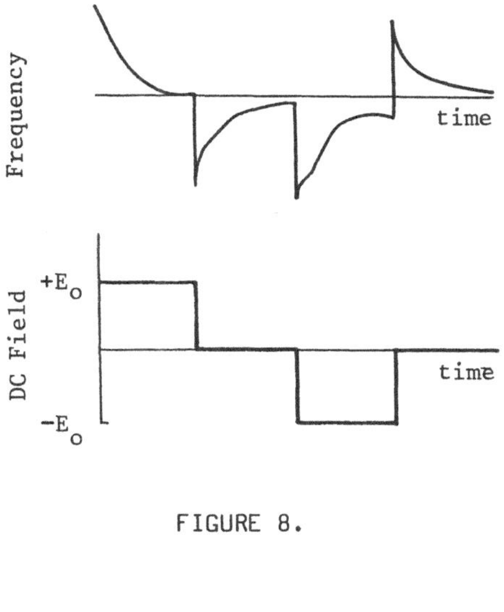

FIGURE 8.

APPENDIX A : GENERALIZED GREEN THEOREM

Let us consider a volume V of surface S, which is separated in two parts V^+ and V^- limited by the surfaces S^+ and S^- and by a discontinuity of surface s, as shown in Fig. I

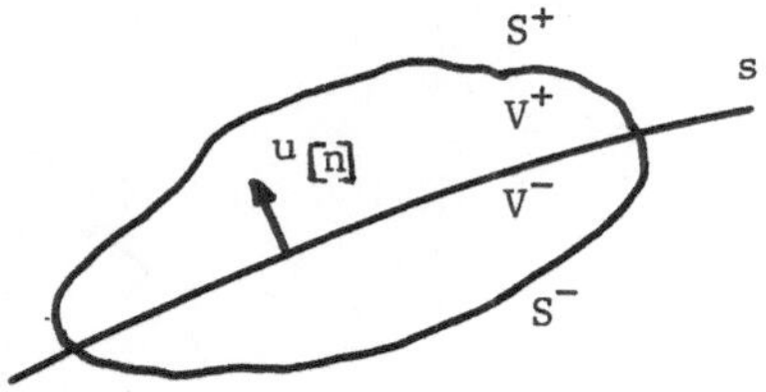

FIGURE I.

The discontinuities of a scalar Ψ across the surface s is noted

$$[\![\Psi]\!] = \Psi^+ - \Psi^- \tag{A1}$$

and of a tensor v_k

$$[\![v_k]\!] = v_k^+ - v_k^- \tag{A2}$$

The generalized Green theorem applied to the tensor v_k is

$$\int_S v_k\, n_k\, dS = \int_V \frac{\partial v_k}{\partial x_k}\, dV + \int_s [\![v_k]\!]\, n_k\, dS \tag{A3}$$

The time derivative of the scalar Ψ is given by

$$\frac{d}{dt}\int_V \Psi\; dt = \int_V \left[\frac{\partial \Psi}{\partial t} + \frac{\partial}{\partial x}(\Psi v_k)\right] dV + \int_s [\![\Psi v_k]\!]\, n_k\, dS - \int_s [\![\Psi]\!]\, u_{[n]}\, dS \tag{A4}$$

$u_{[n]}$ is the normal component velocity of displacement of discontinuity surface.

APPENDIX B : EXPRESSIONS OF THE EFFECTIVE ELASTIC, PIEZOELECTRIC and DIELECTRIC CONSTANTS

Effective elastic constants

$$\overline{A_{j\ell kr}} = C_{j\ell kr} + \hat{C}_{j\ell kr} \tag{B1}$$

where $\hat{C}_{j\ell kr}$ appears as a perturbation of the elastic constant $C_{j\ell kr}$, which depends on the static bias. The expression of $C_{j\ell kr}$ limited to nonlinearities of the 3rd order is

$$\hat{C}_{j\ell kr} = (C_{jrmn}\,\delta_{\ell k} + C_{j\ell nr}\,\delta_{km} + C_{jnkr}\,\delta_{\ell m} + C_{j\ell krmn})\,\frac{\overline{\partial u_m}}{\partial a_n}$$

$$+ (e_{n.jr}\,\delta_{\ell k} + e_{n.j\ell kr})\,\frac{\overline{\partial \Phi}}{\partial a_n} \qquad \text{(B2)}$$

Effective piezoelectric constants

$$\overline{A}_{rj\ell} = e_{r.j\ell} + \hat{e}_{r.j\ell} \qquad \text{(B3)}$$

with

$$\hat{e}_{r.j\ell} = (e_{r.n\ell}\,\delta_{jm} + e_{r.j\ell mn})\,\frac{\overline{\partial u_m}}{\partial a_n}$$

$$+ \left[\varepsilon_0(\delta_{rj}\delta_{\ell n} - \delta_{j\ell}\delta_{rn} + \delta_{r\ell}\delta_{jn}) - e_{rn.j\ell}\right]\frac{\overline{\partial \phi}}{\partial a_n} \qquad \text{(B4)}$$

Effective dielectric constants

$$\overline{E}_{jr} = \varepsilon_{jr} + \hat{\varepsilon}_{jr} \qquad \text{(B5)}$$

with

$$\hat{\varepsilon}_{jr} = -\left[\varepsilon_0\,(\delta_{jr}\delta_{mn} - \delta_{jn}\delta_{rn} - \delta_{rn}\delta_{jm}) + \varepsilon_{jr.mn}\right]\frac{\overline{\partial u_m}}{\partial a_n}$$

$$+ \varepsilon_{jrn}\,\frac{\overline{\partial \phi}}{\partial a_n} \qquad \text{(B6)}$$

REFERENCES

1. H.F. Tiersten : On the nonlinear equations of thermo-electroelasticity. Int. J. Engng. Sci., vol. 9, 1971.
2. M. Planat, E. François : Méthodes d'analyse de la propagation non linéaire d'ondes planes dans un solide. Acustica, vol. 59, n ° 1, 1985.
3. M. Planat, G. Théobald, J.J. Gagnepain : Propagation non linéaire d'ondes élastiques dans un solide anisotrope - Génération d'harmoniques, anisochronisme et intermodulation.
   Onde Electrique, vol. 60, n ° 8-9, 1980.
   Onde Electrique, vol. 60, n ° 11 , 1980.
4. G. Théobald, D. Hauden, J.J. Gagnepain : Propagation des ondes élastiques en milieu perturbé.
   Onde Electrique, vol. 65, n ° 2, 1985.
   Onde Electrique, vol. 65, n ° 3, 1985.
5. R. Brendel : Material nonlinear piezoelectric coefficients for quartz. J.A.P., vol. 54, n ° 9, 1983 + J.A.P., vol. 55, n ° 2, 1984.

DISCUSSION

Comment: Socino
Are the performances of surface acoustic wave oscillators comparable with those of bulk waves?

Reply: Gagnepain
They cannot be compared directly. It is necessary to distinguish first between spectral purity and stability.
Spectral purity of SAW oscillators can be as good as that of BAW oscillators, and sometimes even better. This is because the nonlinearities of SAW are lower; therefore, more power can be applied, and this gives a better signal to noise ratio.
Stability of SAW is generally lower on account of their Q-factor, which can be equal to $50 \cdot 10^3$ or $75 \cdot 10^3$. At the same time, BAW of 5 MHz can reach values of Q equal to $3 \cdot 10^6$. Long term ageing of SAW also is not as good, certainly due to surface poblems, and is of the order of some units times 10^{-7} or 10^{-6} per year.

Comment: Gremaud
What is the dislocation density in synthetic quartz and what could be the effect of the dislocations on the Q factor of the quartz?

Reply: Gagnepain
The dislocation density of synthetic quartz crystals can be very different inside a crystal block. For instance, the ends of the block cannot be used and the central part which contains the seed must be removed. This obviously reduces the useful volume of crystals. Today, it is possible to realize resonators with Q factors close to the one allowed by the material limit. Q factor equal to $3 \cdot 10^6$ at 5 MHz are obtained, which suppose very low defect and dislocation contents.

PHYSICAL AND ELASTIC CHARACTERISTICS OF FIBER REINFORCED COMPOSITES

J.A. GALLEGO-JUAREZ

1. DEFINITIONS AND GENERAL CHARACTERISTICS

In the continuous interest in improving the performance of the currently-used materials, scientists and engineers have produced either advances in the classical materials or new materials. The latter is the case for the modern composite materials.

A material having two or more distinct constituents (or phases) may be considered a composite material. But it is only when significant property changes occur as a result of the combination of constituents that these materials can be recognized as a composite.

The idea of composite in a broad sense can be applied to many different traditional materials which consist of two or more components combined to give a better performance than the individual constituents. This is, among other examples, the case of concrete reinforced with steel bars. For present purposes composites will be considered as materials having two (or more) physically distinct and mechanically separable phases which are mixed in such a way that the dispersion of one in the other can be done in a controlled way to achieve optimum properties. These properties are superior, and possibly unique in some specific respects, to the properties of the individual components. The phase which is harder, stiffer and stronger than the other is referred to as the reinforcement, while the other is referred as the matrix.

Composite materials are usually characterized by the reinforcement. The main types of reinforcement are particles and fibers. Particles are characterized as having all dimensions of the same order of magnitude while fibers are characterized as having a much greater length than cross--sectional dimensions. Table 1 presents a commonly used classification scheme for composite materials. By far the most extensively analyzed composite material is the fiber-reinforced composite of which there are two main types, namely, the continuous fiber and discontinuous (short) fiber composites, (Fig. 1). This work is concerned with continuous fiber-reinforced composites.

Properties of composites are strongly influenced by the properties of the constituent materials, their distribution, and the interaction between them. Thus in describing a composite material one needs to specify the constituent materials and their properties, and the geometry (shape, size and size distribution) of the reinforcement. However, systems

Table 1.- Classification of composite materials

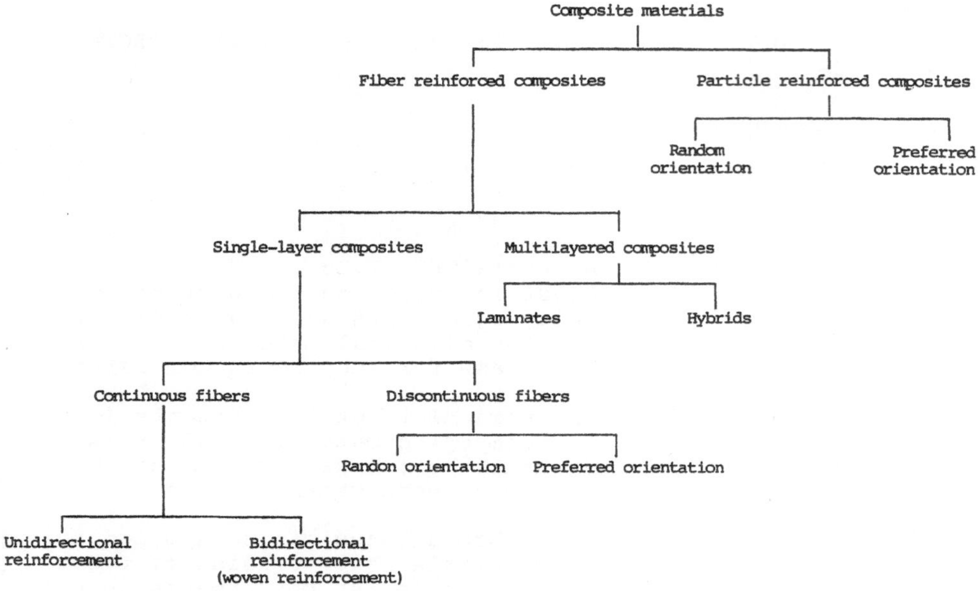

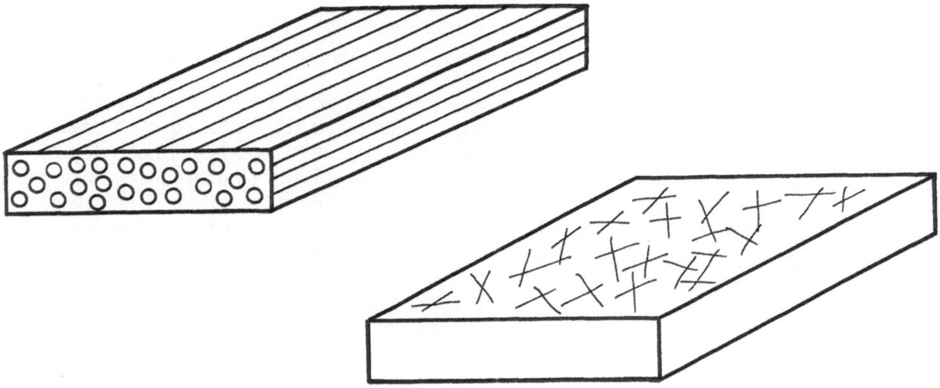

Fig. 1. Continuous and short fiber composites

containing reinforcement with identical geometry may differ in concentration, concentration distribution and orientation. Therefore, all these factors may be important in determining the properties of composites.

The shape of the discrete units of the reinforcement may often be approximated by cylinders or rectangular cross-sectioned prisms. The size and size distribution, together with volume fraction, control the texture of the material and determine the interfacial area which is important in the interaction between the reinforcement and the matrix.

Concentration, usually measured in terms of volume or weight fraction, is generally considered as the single most important parameter which influences the composite properties. The concentration distribution is a measure of homogeneity of the composite. The homogeneity is a characteristic that determines the extent to which a representative volume of material may differ in physical and mechanical properties from the average properties of the material.

The orientation of the reinforcement affects the isotropy of the system. In fact, in continuous fiber reinforced composites the primary advantage is the ability to control anisotropy.

2. FIBERS AND MATRICES

It is well known that the measured elastic modulus and strength of most materials are found to be much smaller than their theoretical values, due to the presence of imperfections or inherent flaws in the material. Flaws lying perpendicular to the direction of applied loads are particularly detrimental to the mechanical characteristics. Thus, man-made filament or fibers, compared with bulk materials, exhibit high modulus and high strength along their lengths, since large flaws are minimized. However, fibers, because of their small cross-sectional dimensions, are not directly usable in engineering structures. They are, therefore, embedded in matrix materials to form composites. The matrix serves to bind the fibers together, transfer loads to the fibers, and protect them against environmental attack. Thus the principal purpose of the matrix is not to be a load-carrying constituent but essentially to bind the fibers together and protect them.

The continuous fibers may all be aligned in one direction to form a unidirectional composite (Fig. 2a). Such composites are very strong in the fiber direction but are generally weak in the direction perpendicular to the fibers. The reinforcement may also be provided in a second direction to give more balanced properties. Such bidirectional reinforcement may be provided by separate unidirectional layers (Fig. 2b) or in a single layer in mutually perpendicular directions as in a woven fabric (Fig. 3).

The basic purpose in the design of fiber composite materials is to find a fiber material of high elastic modulus and strength and preferably low density and then to arrange the fibers in a suitable microstructure giving useful engineering properties to the finished product.

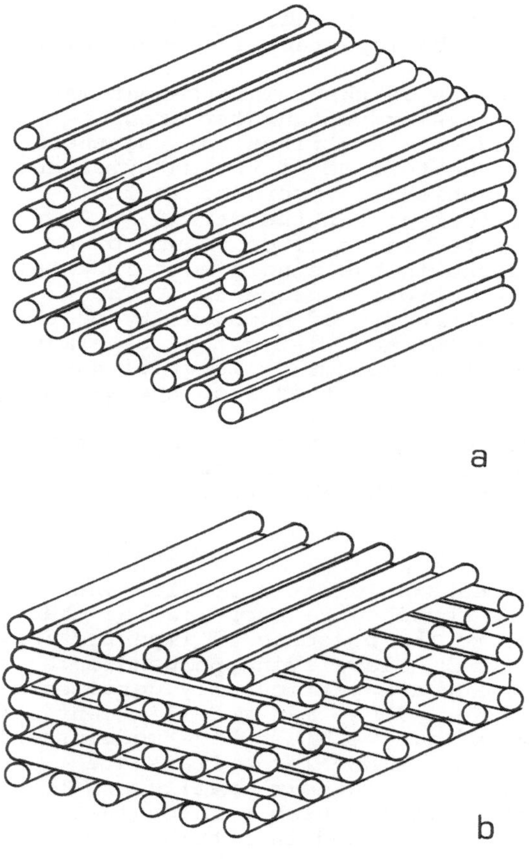

Fig. 2. Unidirectional and 0/90° crossply lay-up of fiber-reinforced material

Many different materials have been used for fiber reinforcement of composites. Examples of the range are given in Table 2, where the fiber materials are arranged in order of increasing specific stiffness (stiffness divided by the density, E/ρ). Properties of some typical bulk materials are included for comparison. It can be seen that while no bulk material has a specific stiffness higher than 27.4, fibers can approach 200.

From the properties of table 2, the importance of carbon fibers in composite materials results evident. These fibers are about 7 to 8 μm in diameter and consist of small crystallites of 'turbostratic' graphite, one of the allotropic forms of carbon. In a graphite single crystal the carbon atoms are arranged in hexagonal arrays. There are two distinct families of carbonaceous fibers which may be distinguished by the highest temperature reached during their manufacture. Fibers heated to less than 2000ºC are designated carbon and those furnaced above 2500ºC are called graphite fibers. The modulus of carbon fibers depends on the degree of perfection of alignment , parallel to the axis of the fiber, of the layer planes of the graphite. That varies considerably with the manufacturing process.

The stress-strain curves of typical fibers are shown in Fig. 4. It may be observed for carbon fibers that the high modulus fibers (Type 1) have a much lower strain to failure compared with the high strength fibers (Type 2).

Other classes of fibers are glass, boron and organic fibers. In particular, the latter constitute a relatively new, and potentially important, class of fibers based on the high strength and stiffness which is possible in fully aligned polymers. The most successful commercial organic fiber to date

Fig. 3. Woven roving composite material

Table 2.- Properties of fibers and bulk materials |4|

	Density ρ (x10^3Kg/m^3)	Young's Modulus E(CN/m^2)	Tensile Stenght σ(CN/m^2)	Specific σ / ρ	Properties E/ρ
FIBERS					
E-glass	2.54	70	1.7	0.67	28
Kevlar 49	1.45	130	2.3	1.59	89.7
Boron	2.65	420	3.5	1.34	158
Carbon, Type 2 (High Stenght)	1.74	230	3.0	1.74	132
Carbon, Type 1 (High Modulus)	2.0	400	2.0	1.00	200
BULK MATERIALS					
High tensile steel	7.8	210	0.75	0.96	26.9
Duraluminium	2.7	74	0.26	0.93	27.4
Magnesium alloy	1.8	46	0.17	0.94	25.5
Tungsten	19.3	350	1.1-4.1	0.06-0.21	18.1

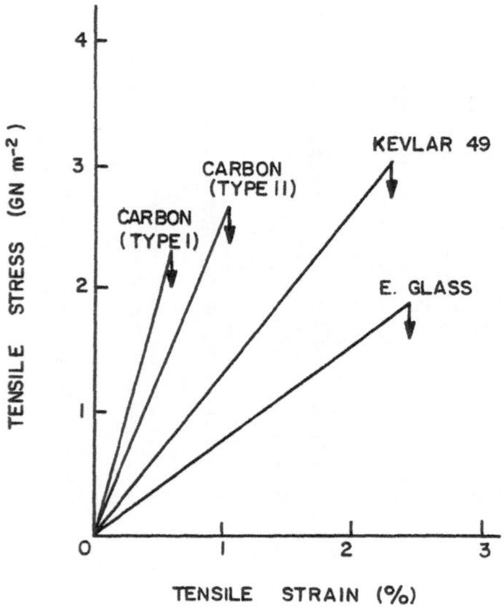

Fig. 4. Stress-strain curves of fibers. The vertical arrows indicate complete failure

has been developed by the Du Pont company with the trade name Kevlar.
To arrange the fibers into an useful structure a matrix is required. This has three main functions: to protect the surface of individual fibres so as to retain their strength, to cement bundles of fibres into a composite solid which retain its shape and to transfer the externally applied loads onto the strong fibers. The most common matrix materials for fiber reinforced composites are polymeric. Thermosetting polymers are resins which cross-link during curing into a glassy brittle solid; they include polyester and epoxy resins. Thermoplastic polymers are high molecular weight long chain molecules which can either become entangled or crystallised at room temperature to give strength and shape; these include nylon, polypropylene and polycarbonate.

Thermosetting resins are usually isotropic. Their most characteristic property is in response to heat since, unlike thermoplastic, they do not melt on heating. However, they lose their stiffness properties at the heat distortion temperature and this defines an effective upper limit for their use in structural components. Epoxy resins are generally superior to polyester resins in this respect but other resins are avalaible which are stable at higher temperatures, such as aromatic polyamides.

Thermoplastics are not cross-linked. They derive their strength and stiffness from the inherent properties of the monomer units and the very high molecular weight. These polymers may have anisotropic properties depending on the conditions during solidification. Thermoplastic matrices are normally used with short fiber reinforcement for applications in products made by injection moulding. Typical mechanical properties of thermosetting resins and thermoplastics are given in Table 3.

Metallic and ceramic matrices are also employed, but for the purposes of this work are not of interest.

Table 3.- Typical mechanical properties of polymeric matrices |2|

	Density (Mg/m^3)	Young's modulus (GN/m^2)	Tensile strength (MN/m^2)
Polyester	1.2-1.5	2-4	40-90
Epoxy	1.1-1.4	3-6	75-100
Nylon	1.1	1.4-2.8	60-75
Polypropylene	0.9	1.0-1.4	25-38
Polycarbonate	1.1-1.2	2.2-2.4	45-70

To conclude this brief information about general characteristics of fiber composite materials, a summary of properties of composites as compared with conventional bulk materials is given in Table 4.

Table 4.- Properties of composite and conventional bulk materials |2|

Material	Density ρ ($\times 10^3 Kg/m^3$)	Young's modulus E (GN/m^2)	Tensile strength σ (MN/m^2)	Elongation to fracture (%)	Coeficient thermal expansion (10^{-6} ºC^{-1})	Specific Young's modulus E/ρ	Specific Tensile strenght σ / ρ	Heat resistance (ºC)
High strength Al-Zn-Mg alloy	2.80	72	503	11	24	25.7	180	350
Quenched and tempered low alloy steel	7.85	207	2050-600	12-28	11	26.4	261-76	800
Nimonic 90 (nickel-based alloy)	8.18	204	1200	26	16	24.9	147	1100
Carbon fiber-epoxy resin unidirectional laminae (V_f = 0.60)								
(i) parallel to fibers	1.62	220	1400	0.8	-0.2	135	865	260
(ii) perpendicular to fibers	1.62	7	38	0.6	30			
Glass fiber-polyester resin unidirectional laminae (V_f = 0.50)								
(i) parallel to fibers	1.93	38	750	1.8	11	19.7	390	250
(ii) perpendicular to fibers	1.93	10	22	0.2				

NOTE: V_f is the volume fraction of fibers

3. UNIDIRECTIONAL LAMINA AND LAMINATES

Many of the properties of fiber composites are strongly dependent on the geometrical structure of the material.

High performance components usually consists of layers of laminae stacked up in a pre-determined arrangement to achieve optimum properties and performance. A stack of laminal is called a laminae.

For the prediction of elastic properties each lamina may be considered as homogeneous in the sense that the fiber arrangement and volume fraction are uniform throughout. The fibres in the laminae may be continuous or in short lengths and can be aligned in one or more directions or randomly distributed. This work is mainly devoted to the analysis of unidirectional laminae because they usually, constitute the basic structural element of continuous fiber-reinforced composites. The laminates are made by stacking together unidirectional layers in predetermined directions and thicknesses to give the desired mechanical properties. Figure 5 shows a unidirectional lamina and a laminate with unidirectional laminae at 90º to each other. In the unidirectional lamina all the fibers are alligned parallel to each other. For theoretical analysis purposes, the fibers can be considered to be arranged on a square or hexagonal lattice with each fiber having a circular cross--section and the same diameter. In practice glass and organic

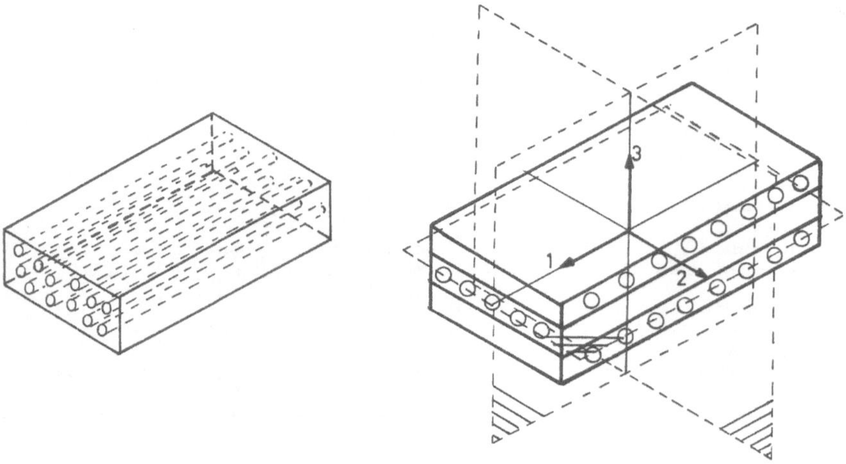

Fig. 5. Unidirectional lamina and cross-ply laminate

fibers closely approximate to a circular cross-section, but carbon fibers, although roughly circular, may have very irregular surfaces. There is some variation in the fiber diameter depending on processing procedure.

In studying the elastic properties of unidirectional laminae hexagonal symmetry is usually assumed, while the laminates are considered as orthotropic bodies with three mutually perpendicular planes of material symmetry.

Several manufacturing techniques may be used. Although the fiber can be supplied dry and subsequently impregnated with the resin matrix material in a mould or in some other wet lay-up process, it is convenient for the manufacture of many structural components to use previously impregnated (pre-preg) sheets. These are sheets of partially cured resin containing a uniform distribution of unidirectionally aligned fibers. A surplus of resin is present and this is bled off when the pre-preg sheets are laid-up at the required orientation and subjected to an appropriate compaction pressure and temperature cycle to form a composite of predetermined volume fraction.

4. ELASTIC PROPERTIES

Directionally reinforced composites are really anisotropic and inhomogeneous materials. However, from the point of view of their overall behaviour, composites can be analyzed as homogeneous, but anisotropic materials, if the distances over which quantities such as stresses are defined are much greater than a characteristic dimension of the reinforcement. In this case, the approach to calculate the elastic moduli of the composite is based upon the assumption of a representative volume element, or an element in which te volume averages of all stress and strain components are identical to the respective quantities in the entire composite. The averaged stresses are related to the averaged strains by means of "effective elastic constants". In this theory, which is known as "effective modulus theory" |6|, the "effective moduli" of the composite are determined in terms of the elastic moduli and the geometric parameters of its constituents.

The advantage of the effective modulus theory for studying the elastic behaviour of a composite is that the actual heterogeneous material is conceptually represented as a homogeneous, anisotropic solid. Thus, instead of having to deal with numerous sets of field equations (one for each inhomogeneity), the approach allows us to work with a single set of equations for the composite medium as a whole. In fact, once the effective elastic moduli are known, the analysis follows the usual path for a classical homogeneous, anisotropic, solid.

The effective modulus theory is rigorous provided that the stress and strain components in regions that are larger than the representative volume are uniform. If these components are variable, the approach is approximate. This theory yields good results for static problems, but for dynamic loading conditions its validity is somewhat restricted.

In the present work, emphasis will be placed on the unidirectional continuous fiber composites because they constitute the basic structural element of the fiber reinforced materials (Fig. 6). Therefore this section is restricted to the linearly elastic behaviour of this kind of composites.

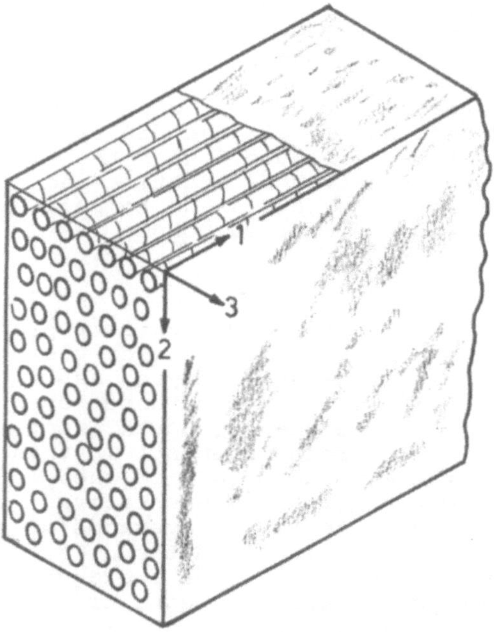

Fig. 6. Unidirectional fiber composite

The elastic properties can be defined by the generalized Hooke's law relating volume averages of the stresses, σ_{ij}, to volume averages of the strains, ε_{kl}, by the efective elastic moduli, C_{ijkl}, in the form

$$\sigma_{ij} = C_{ijkl}\,\varepsilon_{kl} \tag{1}$$

The number of the efective constants C_{ijkl} is determined by elastic symmetry. For a fiber-reinforced composite the significant cases to be considered regarding gross anisotropy are orthotropy, square symmetry and transverse isotropy. An orthotropy elastic body has three mutually perpendicular planes of elastic symmetry. An example is a fiber-reinforced composite with a rectangular array of identical circular fibers (Fig. 7a). A square symmetric material is a particular case of orthotropic material in which two axis are elastically equivalent. A composite with a square array of fibers (Fig. 7b) is an example of this symmetry. For transverse isotropy the material may possess an axis of symmetry in the sense that all rays at right angles to this axis are equivalent. For the hexagonal fiber array of Fig. 7c, the reinforced material has hexagonal symmetry and is thus also transversely isotropic |7|. For random fiber arrangement (Fig. 7d), transverse isotropy may also be assumed.

In dealing with unidirectional fiber-reinforced composite materials, one often assumes transverse isotropy, the axis of isotropy being parallel to the fiber direction. Such an assumption implies that the stress-strain relation may be written in terms of five elastic moduli |8|. If the specimen is referred to a Cartesian-coordinate system $x_1\ x_2\ x_3$ whose

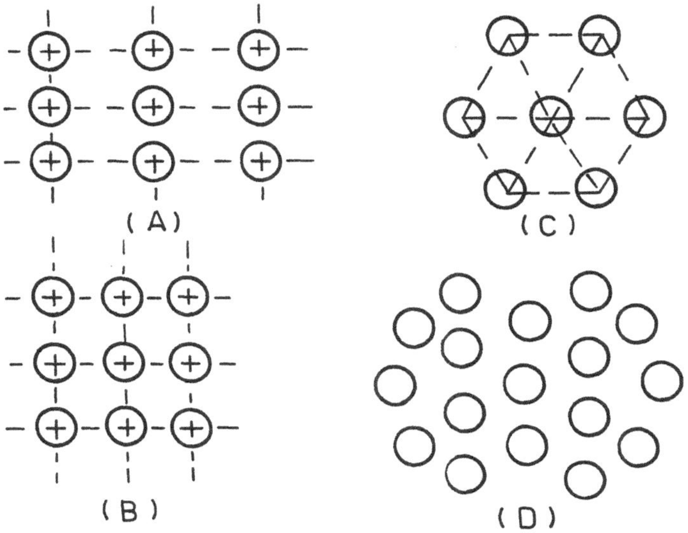

Fig. 7. Directionally reinforced composites:
a) rectangular array, b) square array
c) hexagonal array, d) random array

x_1 - axis is in the fiber direction, then the stiffness matrix has the form |9|:

$$|c_{ij}| = \begin{vmatrix} c_{11} & c_{12} & c_{12} & 0 & 0 & 0 \\ c_{12} & c_{22} & c_{23} & 0 & 0 & 0 \\ c_{12} & c_{23} & c_{22} & 0 & 0 & 0 \\ 0 & 0 & 0 & c_{44} & 0 & 0 \\ 0 & 0 & 0 & 0 & c_{44} & 0 \\ 0 & 0 & 0 & 0 & 0 & c_{66} \end{vmatrix} \qquad (2)$$

where $c_{66} = \frac{1}{2}(c_{22} - c_{23})$, and the five independent stiffness constants are c_{11}, c_{12}, c_{22}, c_{23} and c_{44}. Here, because linearized theory is assumed the stiffness symmetry condition $c_{ijkl} = c_{jikl} = c_{ijlk}$ is satisfied, and we can use the abbreviated subscript notation reducing to two the four subscripts of the general stiffness matrix |8|. For the sake of recalling the physical meaning of the two new subscripts, we remember that the convention is

$$\begin{array}{ll} 1 - & x_1x_1 \\ 2 - & x_2x_2 \\ 3 - & x_3x_3 \\ 4 - & x_2x_3,\ x_3x_2 \\ 5 - & x_1x_3,\ x_3x_1 \\ 6 - & x_1x_2,\ x_2x_1 \end{array}$$

When the x_3 - axis is chosen as the axis of symmetry in the fiber direction, the five independent stiffness constants are c_{11}, c_{12}, c_{13}, c_{33} and c_{44} |10|, and the stiffness matrix has the form:

$$|c_{ij}| = \begin{vmatrix} c_{11} & c_{12} & c_{13} & 0 & 0 & 0 \\ c_{12} & c_{11} & c_{13} & 0 & 0 & 0 \\ c_{13} & c_{13} & c_{33} & 0 & 0 & 0 \\ 0 & 0 & 0 & c_{44} & 0 & 0 \\ 0 & 0 & 0 & 0 & c_{44} & 0 \\ 0 & 0 & 0 & 0 & 0 & c_{66} \end{vmatrix} \qquad (3)$$

where $c_{66} = \frac{1}{2}(c_{11} - c_{12})$ and the five independent stiffness constants are now c_{11}, c_{12}, c_{13}, c_{33} and c_{44}.

The elastic properties can also be defined by the compliance matrix $|S_{ij}|$. The equations which relate stiffness to compliances for the latter group of constants are the following |11|:

$$s_{11} = \frac{c_{11}c_{33} - c_{13}^2}{(c_{11} - c_{12})(c_{11}c_{33} + c_{12}c_{33} - 2c_{13}^2)} \qquad (4)$$

$$s_{33} = \frac{c_{11} + c_{12}}{c_{11}c_{33} + c_{12}c_{33} - 2c_{13}^2} \qquad (5)$$

$$s_{12} = \frac{c_{13}^2 - c_{12}c_{33}}{(c_{11} - c_{12})\ (c_{11}c_{33} + c_{12}c_{33} - 2c_{13}^2)} \quad (6)$$

$$s_{13} = \frac{-c_{13}}{c_{11}c_{33} + c_{12}c_{33} - 2c_{13}^2} \quad (7)$$

$$s_{44} = \frac{1}{c_{44}} \quad (8)$$

$$s_{66} = \frac{1}{c_{66}} \quad (9)$$

These are related to the more commonly used Young's modulus, Poisson's ratio and the shear modulus by the following relationships |12|:

s_{33} is the reciprocal of the longitudinal Young's modulus E_1

$$s_{33} = 1/E_1 \quad (10)$$

s_{11} is the reciprocal of the transverse Young's modulus E_2

$$s_{11} = 1/E_2 \quad (11)$$

s_{13} relates to the axial Poisson's ratio ν_1 such that

$$\nu_1 = -s_{13}/s_{33} \quad (12)$$

s_{12} relates to transverse Poisson's ratio ν_2 such that

$$\nu_2 = -s_{12}/s_{11} \quad (13)$$

s_{44} is the reciprocal of the axial shear modulus G_1

$$s_{44} = 1/G_1 \quad (14)$$

and s_{66} is the reciprocal of the transverse shear modulus G_2

$$s_{66} = 1/G_2 \quad (15)$$

As was previously mentioned, unidirectionally reinforced composites with a rectangular matrix and also cross-ply laminates can be generally considered as orthotropic materials, which have nine independent elastic constants. The stiffness matrix, for this kind of symmetry, has the form |13|:

$$|c_{ij}| = \begin{vmatrix} c_{11} & c_{12} & c_{13} & 0 & 0 & 0 \\ c_{12} & c_{22} & c_{23} & 0 & 0 & 0 \\ c_{13} & c_{23} & c_{33} & 0 & 0 & 0 \\ 0 & 0 & 0 & c_{44} & 0 & 0 \\ 0 & 0 & 0 & 0 & c_{55} & 0 \\ 0 & 0 & 0 & 0 & 0 & c_{66} \end{vmatrix} \quad (16)$$

A major problem in the mechanics of composites consists in deriving relations between the efective elastic moduli and the properties of the constituents. There are several theories that predict with reasonable sucess the elastic properties of composites in terms of the constituent matrix and fiber properties. Expressions for the effective elastic moduli of a unidirectional fiber-reinforced composite have been obtained by Hashin and Rosen |14|, Whitney and Riley |15| and by Greszczuk |16|, among others. Hashin and Rosen, applying variational principles derived bounds and expressions for the moduli of composites with hexagonal and random fiber arrays. They used a model in which the representative volume element was a matrix hexagonal prism or a matrix cylinder surrounding one central fiber. Whitney and Riley employed as representative element a cylindrical fiber embedded in a cylindrical matrix. They developed approximate equations, applying Airy stress functions to the fiber and matrix, with boundary conditions requiring continuity of displacements across the fiber-matrix interface and continuity of the appropriate stresses. Despite the less rigorous mathematics involved in this work, the results are consistent with those of Hashin and Rosen. Finally, Greszczuk, using a mathematical model consisting of a rectangular array of circular filaments embedded in an elastic matrix, developed closed-form-expressions for five elastic constants of a unidirectional composite with transverse isotropy. In all these works it was assumed that the matrix and fiber materials are linearly elastic, isotropic, and homogeneous. The composite is anisotropic because of fiber orientation. Let us consider the fibers in the x_1 direction. In terms of the longitudinal, transverse and shear (in fiber direction) moduli of the composite, E_1, E_2, and G_1, and the Poisson ratios along fiber direction and in transverse plane, ν_1 and ν_2, the effective elastic constants of the composite are given by |17|:

$$c_{11} = \frac{E_1^2\,(1-\nu_2)}{E_1(1-\nu_2) - 2E_2\nu_1^2} \tag{17}$$

$$c_{33} = \frac{E_1E_2}{2E_1(1-\nu_2) - 4E_2\nu_1^2} + \frac{E_2}{2(1+\nu_2)} \tag{18}$$

$$c_{12} = \frac{\nu_1E_1E_2}{E_1(1-\nu_2) - 2E_2\nu_1^2} \tag{19}$$

$$c_{23} = \frac{E_1E_2}{2E_1(1-\nu_2) - 4E_2\nu_1^2} - \frac{E_2}{2(1+\nu_2)} \tag{20}$$

$$c_{44} = G_1 \tag{21}$$

When one assumes the results of Whitney and Riley as representative, the moduli of the composite are related to the moduli of the constituents by the following relatioships (the subscripts f and m refer to fiber and matrix properties, respectively):

$$E_1 = E_f\lambda + E_m(1 - \lambda) + \frac{2(\nu_f - \nu_m)^2 E_m E_f (1 - \lambda)\lambda}{E_m(1 - \lambda)L_f + \left[L_m\lambda + (1 + \nu_m)\right]E_f} \quad (22)$$

$$\nu_1 = \nu_m - \frac{2(\nu_m - \nu_f)(1 - \nu_m^2)E_f\lambda}{E_m(1 - \lambda)L_f + \left[L_m\lambda + (1 + \nu_m)\right]E_f} \quad (23)$$

$$E_2 = \frac{2K(1 - \nu_2)E_1}{E_1 + 4K\nu_1^2} \quad (24)$$

$$\nu_2 = \nu_f\lambda + \nu_m(1 - \lambda) \quad (25)$$

$$G_1 = \frac{\left[(G_f + G_m) + (G_f - G_m)\lambda\right]G_m}{(G_f + G_m) - (G_f - G_m)\lambda} \quad (26)$$

$$K = \frac{(K_f + G_m)K_m - (K_f - K_m)G_m\lambda}{(K_f + G_m) - (K_f - K_m)\lambda} \quad (27)$$

where K is the transverse plane bulk modulus,

$$L_f = 1 - \nu_f - 2\nu_f^2$$

$$L_m = 1 - \nu_m - 2\nu_m^2$$

and λ is the volume fraction of fibers. The third term of Eq. (22) will not be significant for most composites. Thus,

$$E_1 = E_f\lambda + E_m(1 - \lambda) \quad (28)$$

This expresses the well-known "law of mixtures" which arises from the assumption of uniform strain following the Voigt procedure |12|. It seems to be quite adequate for the longitudinal modulus of an unidirectional fiber composite |18|. Note for Eq. (22) that the law of mixtures is an exact solution for the case $\nu_f = \nu_m$.

Similar expressions are obtained by assuming Greszczuk's model, which largely agrees with that of Hashin and Rosen and of Whitney and Riley. In fact, using the expression of the rotated elastic modulus (i.e. the Young's modulus as a function of angle to the composite axis) for hexagonal symmetry

$$\frac{1}{E} = s_{11}\cos^4\Theta + s_{33}\sin^4\Theta + (s_{44} + 2s_{33})\sin^2\Theta\cos^2\Theta \quad (29)$$

where Θ is the angle of fiber orientation with the stress axis, Zimmer and Cost |17| have found a generally good agreement between the Young's modulus values predicted by the three theoretical models and also with experimental results (Fig. 8).

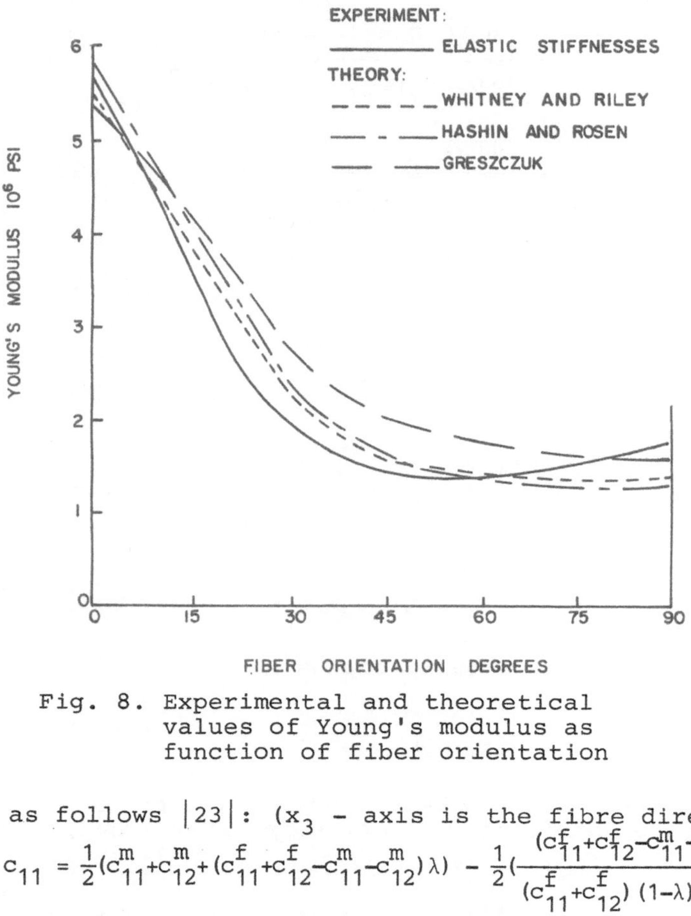

Fig. 8. Experimental and theoretical values of Young's modulus as function of fiber orientation

When we consider unidirectional fiber reinforced materials where the fibers are elastically anisotropic (e.g. carbon fibers), the composite is still transversely isotropic, but the previously mentioned theories for the elastic properties are not appropriate. For this case, using the equations of Halpin-Tsai |19|, |20| Behrens |10|, |21|, Smith |22| proposed a set of expressions for the average elastic constants of the composite. These expressions, that were found by evaluating the dispersion relation of sound waves for long wavelengths, are as follows |23|: (x_3 - axis is the fibre direction).

$$c_{11} = \frac{1}{2}(c^m_{11}+c^m_{12}+(c^f_{11}+c^f_{12}-c^m_{11}-c^m_{12})\lambda) - \frac{1}{2}\left(\frac{(c^f_{11}+c^f_{12}-c^m_{11}-c^m_{12})^2\lambda(1-\lambda)}{(c^f_{11}+c^f_{12})(1-\lambda)+(c^m_{11}+c^m_{12})\lambda+2c^m_{44}}\right) + c^m_{44}\left(\frac{1+\xi\eta\lambda}{1-\eta\lambda}\right) \quad (30)$$

$$c_{12} = \frac{1}{2}(c^m_{11}+c^m_{12}+(c^f_{11}+c^f_{12}-c^m_{12})\lambda) - \frac{1}{2}\left(\frac{(c^f_{11}+c^f_{12}-c^m_{11}-c^m_{12})^2\lambda(1-\lambda)}{(c^f_{11}+c^f_{12})(1-\lambda)+(c^m_{11}+c^m_{12})\lambda+2c^m_{44}}\right) - c^m_{44}\left(\frac{1+\xi\eta\lambda}{1-\eta\lambda}\right) \quad (31)$$

$$c_{13} = c^m_{12} + (c^f_{13}-c^m_{12})\lambda - \left(\frac{(c^f_{13}-c^m_{12})(c^f_{11}+c^f_{12}-c^m_{11}-c^m_{12})\lambda(1-\lambda)}{(c^f_{11}+c^f_{12})(1-\lambda)+(c^m_{11}+c^m_{12})\lambda+2c^m_{44}}\right) \quad (32)$$

$$c_{33} = c_{11}^{m} + (c_{33}^{f} - c_{11}^{m})\lambda - 2\left(\frac{(c_{13}^{f} - c_{12}^{m})^{2}\lambda(1-\lambda)}{(c_{11}^{f} - c_{12}^{f})(1-\lambda) + (c_{11}^{m} + c_{12}^{m})\lambda + 2c_{44}^{m}}\right) \tag{33}$$

$$c_{44} = c_{44}^{m} + (c_{44}^{f} - c_{44}^{m})\lambda - \left(\frac{(c_{44}^{f} - c_{44}^{m})^{2}\lambda(1-\lambda)}{c_{44}^{f}(1-\lambda) + c_{44}^{m}(1+\lambda)}\right) \tag{34}$$

where the parameter ξ is a "reinforcing factor" its value dependent on the properties of the matrix and the elastic constant considered, and

$$\eta = \frac{c_{66}^{f} - c_{44}^{m}}{c_{66}^{f} + \xi c_{44}^{m}} \tag{35}$$

where $c_{66}^{f} = 1/2\,(c_{11}^{f} - c_{12}^{f})$. The superscripts f, m refer to fiber and matrix properties, respectively.

REFERENCES

1. Agarwall BD, Broutman LJ: Analysis and Performance of Fibre Composites, Wiley, 1980.
2. Hull D: An Introduction to Composite Materials. University Press 1981.
3. Jones RN: Mechanics of Composite Materials. Mc Graw Hills 1975
4. Matthews FL(ed): Introduction to Fibre-Reinforced Composites. Imperial College, London 1984.
5. Lubin G(Ed): Handbook of Composites. Van Nostrand 1982.
6. Achenbach JD: In Composite Materials Vo. 2. Academic Press 1974.
7. Love AEH: A Treatise on the Mathematical Theory of Elasticity. Dover 1944.
8. Audl BA: Acoustic Fields and Waves in Solids. Wiley 1973.
9. Dean GD, Lockett FJ: ASTM STP 521, Am. Soc. Test. Mat. 326-346, 1973.
10. Behrens E: J. Acoust. Soc. Amer. 45, 102-108, 1969.
11. Markham MF: Composites 1, 145-149, 1970.
12. Brody H, Ward IM: Polym. Eng. Sc. 11, 139-151, 1971.
13. Tauchert TR, Guzelsu AN: J. Appl. Mech. 39, 98-102, 1972.
14. Hashin Z, Rosen BW: J. Appl. Mech. 31, 223-232, 1964.
15. Whitney JM, Riley MB: AIAAJ. 4, 1537-1542, 1966.
16. Greszczuk LB: AIAAJ. 9, 1274-1280, 1971.
17. Zimmer JE, Cost JR: J. Acoust. Soc. Amer. 47, 795-803, 1970.
18. Sendeckyj GP: In Composite Materials Vo. 2. Academ. Press 1974.
19. Halpin JC, Tsai SW: Report AFML-TR-67-423.
20. Nielsen LE: J. Appl. Phys. 41, 4626-4627, 1970.
21. Behrens E: J. Acoust. Soc. Amer. 45, 1567-1570, 1969.
22. Smith RE: J. Appl. Phys. 43, 2555-2561, 1972.
23. Martin BG: J. Appl. Phys. 48, 3368-3373, 1977.

ULTRASONIC EVALUATION OF ELASTIC PROPERTIES OF DIRECTIONAL FIBER REINFORCED COMPOSITES

J. A. GALLEGO-JUAREZ

1. INTRODUCTION

Fibre reinforced composites have been developed over the last two decades to provide a range of materials which combine high strength with low density. The properties of these materials are strongly influenced by their constituents (fibers and matrix), their distribution and the interaction between them.

Propagation of ultrasonic waves is one of the more useful methods of studying structure properties and elastic behaviour of fiber composites. Consequently ultrasonic methods for evaluating these materials are widely employed. Theoretical and experimental work on composites has shown correlations between ultrasonic velocity and the elastic properties of the material. Analysis of these results has enabled the principal independent elastic constants to be derived and to obtain data on the quantitative behaviour of composites as a function of the volume fraction of its constituents.

Structural designs which incorporate fiber-reinforced composites are based on the elastic response of the simplest structural unit: the unidirectional plate in which all the fibers are parallel. This paper deals with a survey of the basic aspects of acoustic wave propagation in unidirectional fiber-reinforced composites and their relationship with the characterization of the elastic properties of the material. Then, ultrasonic methods for measuring the elastic constants of composite materials are described and some typical results, together with a discussion about the influence of the different parameters, are presented.

It ought to be noted that this paper is not intended as a thorough review of a rather extensive subject but as an introduction to be supplemented with the wide range of current literature.

2. ACOUSTIC WAVE PROPAGATION

As was discussed before, the simplest elastic symmetry which is likely to represent the properties of unidirectional composites of continuous fibers is transversely isotropic. This is the case of a material which has a preferred direction (the fiber direction) but is isotropic in planes normal to that direction. For such a composite, there are five independent elastic constants that comprise the stiffness matrix. These are the same five stiffnesses required to describe fully the elastic properties of a single crystal with hexagonal symmetry. Hence, the procedures developed in studying the propagation of elastic waves in single crystals of hexagonal symmetry |1|, |2|, can be easily adapted to composites.

In this section a brief introduction to the propagation of acoustic waves in anisotropic solids is given, before proceeding to discuss the theory for a unidirectional fiber composite.

2.1. Anisotropic materials

In an anisotropic material there is no simple relation between the direction of propagation of a wave and the direction of the particle displacement. In other words, in general the waves are not pure longitudinal or transverse types.

The equation of motion for an anisotropic body may be obtained by substituting stress σ_{ij} from the anisotropic linear stress-strain relation

$$\sigma_{ij} = C_{ijkl}\,\varepsilon_{kl} \qquad (i,\ j,\ k,\ l = 1,\ 2,\ 3) \tag{1}$$

into Newton's law $\dfrac{\partial \sigma_{ij}}{\partial x_j} = \rho \dfrac{\partial^2 u_i}{\partial t^2}$, (2), to give $\rho \dfrac{\partial^2 u_i}{\partial t^2} = C_{ijkl} \dfrac{\partial \varepsilon_{kl}}{\partial x_j}$ (3)

where ρ is the volume density and u_i the displacement vector. Substitution of the linearized strain-displacement relation

$$\varepsilon_{kl} = \frac{1}{2}\left(\frac{\partial u_k}{\partial x_l} + \frac{\partial u_l}{\partial x_k} \right)$$

in Eq. (3), yields

$$\rho \frac{\partial^2 u_i}{\partial t^2} = \frac{1}{2} C_{ijkl} \left(\frac{\partial^2 u_k}{\partial x_j \partial x_l} + \frac{\partial^2 u_l}{\partial x_j \partial x_k} \right) \tag{4}$$

which, because of the symmetry with respect to k and l, reduces to

$$\rho \frac{\partial^2 u_i}{\partial t^2} = C_{ijkl} \frac{\partial^2 u_l}{\partial x_j \partial x_k} \tag{5}$$

Let us now consider a plane harmonic wave of the form

$$u_i = (A\, d_i)\, \exp\left[i\omega(x_j s_j - t)\right] \tag{6}$$

where A is the amplitude, d_i are the components of a unit vector defining the direction of the particle displacements, ω is the angular frequency and s_j the components of the slowness vector defined as $s_j = n_j/v$ (7)
Here n_j are the components of a unit vector defining the direction of propagation and v is the phase velocity.
Substituting Eq. (6) in Eq. (5) gives $(C_{ijkl}\, s_j s_l - \rho\, \delta_{ik}) u_k = 0$ (8)
where δ_{ik} is the Kronecker delta. Equation (8) is known as Christoffel's equation. It represents a set of three homogeneous equations of the first degree with u_1, u_2 and u_3 as unknowns. For a nontrivial solution the determinant of the coefficients must be zero, that is

$$\left| C_{ijkl}\, s_j s_l - \rho\, \delta_{ik} \right| = 0 \tag{9}$$

Equation (9) can be rewritten in the form $\left|\Gamma_{ik} - \rho v^2 \delta_{ik}\right| = 0$ (10)
where $\Gamma_{ik} = C_{ijkl}\, n_j n_l$ is known as the Christoffel stiffness matrix. Evaluation of Eq. (10) leads to a cubic equation in (v^2):

$$\begin{vmatrix} (\Gamma_{11} - \rho v^2) & \Gamma_{12} & \Gamma_{13} \\ \Gamma_{12} & (\Gamma_{22} - \rho v^2) & \Gamma_{23} \\ \Gamma_{13} & \Gamma_{23} & (\Gamma_{33} - \rho v^2) \end{vmatrix} = 0 \tag{11}$$

where $\Gamma_{11} = c_{11}\, n_1^2 + c_{66}\, n_2^2 + c_{55}\, n_3^2 + 2c_{56}\, n_2 n_3 + 2c_{15}\, n_1 n_3 + 2c_{16}\, n_1 n_2$

$\Gamma_{12} = c_{16}\, n_1^2 + c_{26}\, n_2^2 + c_{45}\, n_3^2 + (c_{46} + c_{25}) n_2 n_3 + (c_{14} + c_{56}) n_1 n_3 + (c_{12} + c_{66}) n_1 n_2$

....

Here for the stiffnesses, we use the abbreviated subscript notation discussed in section 4 of the previous Chapter.

The three roots of this equation are in general different, giving rise to three different velocities of propagation.

From the properties of C_{ijkl} it follows that Γ_{ik} is a symmetric and positive definite matrix. That is $\Gamma_{ik} = \Gamma_{ki}$, $\Gamma_{ik}d_i d_k \geq 0$ for all d_i. It follows that all of the eigenvalues of Γ_{ik} are real and positive and their corresponding eigenvectors are orthogonal. Therefore, for a given direction of propagation, defined by the vector n_j, there will be three waves with mutually perpendicular displacement vectors but with different phase velocities (Fig. 1). In general these waves will not be pure longitudinal or pure transverse, but one will be quasilongitudinal (QL) and the other two will be quasi transverse (QT) (Fig. 2). However, for certain special directions of propagation, which are usually directions of symmetry, one wave is pure longitudinal and the other two are pure transverse.

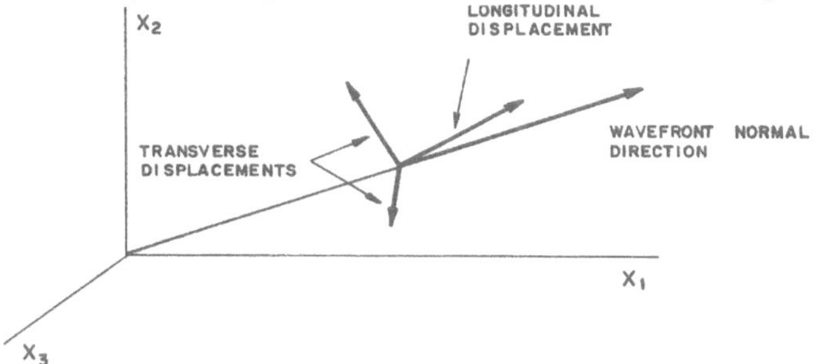

Fig.1. Relationships between the wavefront normal direction and the displacement vectors

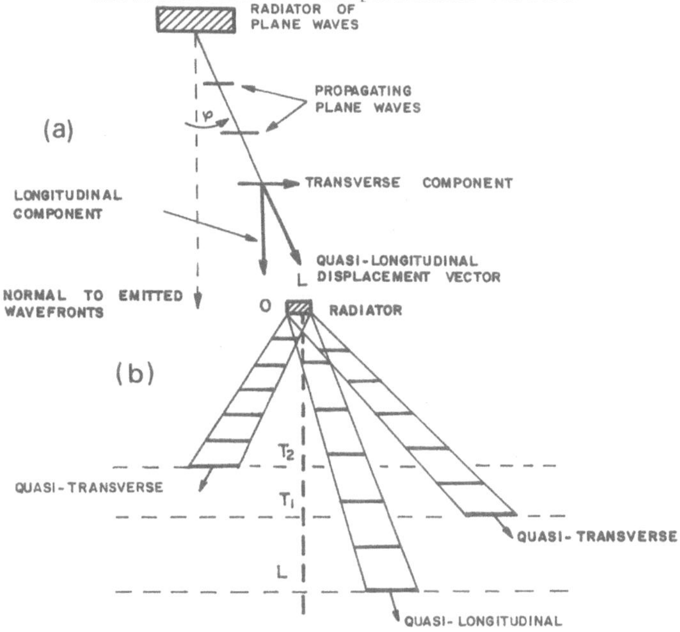

Fig.2. (a) The propagation of a QL wave in an anisotropic material. (b) The three possible energy beams in an anisotropic solid.

Another aspect of wave propagation which is of great importance in anisotropic materials is the propagation of energy. The energy flow vector is defined by the product of the stress tensor and the particle velocity |3|:

$$E_i = \sigma_{ij}\, \dot{u}_j \quad (12)$$

From a point source in an anisotropic solid, three wavefronts (also known as wave surfaces) spread out: one corresponds to the longitudinal motion and two correspond to the transverse waves (These surfaces are generally non--spherical). The three phase velocities are related to the velocities which the wavefront normals propagate. The relationships between the wave normal and the displacement vectors are shown in Fig. 1. The deviation ψ of any of the displacement vectors from the wave normal (Fig. 2) is given by an expression of the form $\cos\psi = n_i d_i$ for each wave type. We conclude from these properties that energy may travel across a specimen with parallel sides along a path making an oblique angle with the normals to the specimen surfaces, |4|, (Fig. 3). This phenomenon of the deviation of the energy from the direction of wave propagation in anisotropic solids has many practical consequences to be taken in consideration when experimental studies to characterize these materials are to be performed.

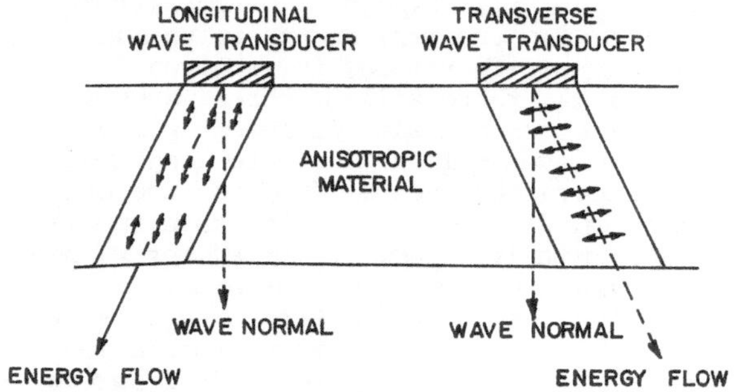

Fig. 3. Energy flow deviation in an anisotropic material

2.2. Transversely isotropic composite materials

It is of interest now to adapt the general analysis to the study of the propagation of plane waves in unidirectional fiber-reinforced composites.

For a unidirectional composite with transverse isotropy, the preferred direction is the fiber direction. A rectangular cartesian coordinate system (x_1, x_2, x_3) is defined with the x_1 - axis in the preferred direction. As we discussed in Section 4 of the precedent Chapter the stiffness matrix for uniaxial materials has five independent constants c_{11}, c_{12}, c_{22}, c_{23} and c_{44}. Bearing in mind the symmetry it can be assumed, without loss in generality, that the (x_1, x_2)-plane is defined by the preferred direction and the wave direction |5|. Thus, the plane harmonic wave is written as

$$u = (u_1,u_2,u_3)\ \exp\left[i\omega(s_1x_1 + s_2x_2 - t)\right] \tag{13}$$

where u_i are the amplitudes of the displacement components. The Christoffel's equation for these amplitudes may lead to the following set of equations

$$(c_{11}s_1^2 + c_{44}s_2^2 - \rho)u_1 + (c_{44} + c_{12})s_1s_2\ u_2 = 0 \tag{14}$$

$$(c_{44} + c_{12})s_1s_2\ u_1 + (c_{44}s_1^2 + c_{22}s_2^2 - \rho)u_2 = 0 \tag{15}$$

$$(c_{44}s_1^2 + c_{66}s_2^2 - \rho)u_3 = 0 \tag{16}$$

Here ρ and c_{ij} are the effective mass density and the effective elastic moduli, respectively. Equation (16) represents a pure transverse wave with displacement vector perpendicular to the plane containing the preferred direction and the direction of propagation. Since $s_1 = \cos\phi/v$, $s_2 = \sin\phi/v$, (17), where ϕ denotes the angle between the preferred direction and the direction of propagation, the velocity of propagation will be

$$v = \left(\frac{c_{44}\ \cos^2\phi + c_{66}\ \sin^2\phi}{\rho}\right)^{1/2} \tag{18}$$

with $c_{66} = (c_{22} - c_{23})/2$. Waves in the preferred direction and in any perpendicular direction have velocities $(c_{44}/\rho)^{\frac{1}{2}}$ and $(c_{66}/\rho)^{\frac{1}{2}}$ respectively.

In general, the pair of equations (14) and (15) do not have solutions for which $u_1/u_2 = s_1/s_2$ or $u_1/u_2 = - s_2/s_1$. Thus, the waves governed by these equations, except for some very special cases, are not purely longitudinal, nor purely transverse, because the displacement vector is neither in the direction of nor perpendicular to the direction of propagation. The velocities of propagation are obtained by equating to zero the determinant of the coefficients:

$$(c_{11}\cos^2\phi + c_{44}\sin^2\phi - \rho v^2)\,(c_{44}\cos^2\phi + c_{22}\sin^2\phi - \rho v^2) = (c_{44} + c_{12})^2 \sin^2\phi\cos^2\phi \qquad (19)$$

The aforementioned exceptions occur when the wave propagates along the preferred direction or normal to it, that is, along a symmetry axis of the composite. Equations for both cases are obtained from Eqs. (14), (15) and (17) by setting $\phi = 0$ and $\phi = 90^\circ$, respectively. For propagation along the fiber direction ($\phi = 0$), $(c_{11}/v^2 - \rho)u_1 = 0$ (20); $(c_{44}/v^2 - \rho)u_2 = 0$ (21), there results a purely longitudinal wave travelling with velocity $(c_{11}/\rho)^{\frac{1}{2}}$ and a purely transverse wave with velocity $(c_{44}/\rho)^{\frac{1}{2}}$. For propagation normal to fiber axis ($\phi=90^\circ$) $(c_{44}/v^2 - \rho)u_1 = 0$ (22) ; $(c_{22}/v^2 - \rho)u_2 = 0$ (23), there results a purely transverse wave with velocity $(c_{44}/\rho)^{\frac{1}{2}}$ and a purely longitudinal wave with velocity $(c_{22}/\rho)^{\frac{1}{2}}$. The results for propagation of pure waves are summarized in Table 1.

Table 1.- Propagation of pure waves in an unidirectional fiber composite with transverse isotropy (fiber direction: x_1-axis)

Direction of Propagation	Direction of Displacement Vector	Type of wave	Velocity of Propagation
x_1	x_1	L	$(c_{11}/\rho)^{1/2}$
x_1	x_2	T	$(c_{44}/\rho)^{1/2}$
x_1	x_3	T	$(c_{44}/\rho)^{1/2}$
x_2	x_1	T	$(c_{44}/\rho)^{1/2}$
x_2	x_2	L	$(c_{22}/\rho)^{1/2}$
x_2	x_3	T	$(c_{66}/\rho)^{1/2}$

2.3. Orthotropic composite materials

As mentioned in Section 4 of the previous Chapter, some unidirectional fiber composites, such as a composite with a rectangular array of identical circular fibers, can be studied more accurately as orthotropic materials. In addition, cross-ply laminates have orthotropic elastic properties since there are three mutually perpendicular planes of material symmetry (see Fig. 5, previous Chapter). Thus, it seems advisable to have a brief look at the propagation of plane waves in an orthotropic elastic solid.

For an orthotropic elastic material the stiffness matrix has nine independent elastic constants (see Section 4, previous Chapter). The Christoffel stiffnesses Γ_{ij} are given by

$$\begin{aligned}
\Gamma_{11} &= n_1^2\, c_{11} + n_2^2\, c_{66} + n_3^2\, c_{55} \\
\Gamma_{22} &= n_1^2\, c_{66} + n_2^2\, c_{22} + n_3^2\, c_{44} \\
\Gamma_{33} &= n_1^2\, c_{55} + n_2^2\, c_{44} + n_3^2\, c_{33} \\
\Gamma_{23} &= n_2\, n_3\, (c_{23} + c_{44}) \\
\Gamma_{13} &= n_1\, n_3\, (c_{13} + c_{55}) \\
\Gamma_{12} &= n_1\, n_2\, (c_{12} + c_{66})
\end{aligned} \qquad (24)$$

Consider, for example, a plane wave travelling in the x_1 - direction ($n_1 = 1$, $n_2 = n_3 = 0$). From (24): $\Gamma_{11} = c_{11}$, $\Gamma_{22} = c_{66}$, $\Gamma_{33} = c_{55}$, $\Gamma_{23} = \Gamma_{13} = \Gamma_{12} = 0$ (25)
so that, from equation (10)

$$\begin{vmatrix} c_{11} - \rho v^2 & 0 & 0 \\ 0 & c_{66} - \rho v^2 & 0 \\ 0 & 0 & c_{55} - \rho v^2 \end{vmatrix} = 0 \qquad (26)$$

Hence there exists the possibility of three different waves which have velocities $(c_{11}/\rho)^{1/2}$, $(c_{66}/\rho)^{1/2}$ and $(c_{55}/\rho)^{1/2}$ By substituting each of these velocities into Christoffel's equation, Eq. (8), it is found that the first wave is purely longitudinal, while the other two are purely transverse having displacement vectors in the x_2 adn x_3-directions, respectively. Similar results are obtained for waves propagating in the directions of the other coordinate axes.

Consider, now, a wave transmitted along an axis at 45 deg to the x_2 and x_3-axes ($n_2 = n_3 = \frac{1}{\sqrt{2}}$, $n_1 = 0$). In this case Eq. (24) gives

$$\Gamma_{11} = (c_{55}+c_{66})/2;\ \Gamma_{22} = (c_{22}+c_{44})/2;\ \Gamma_{33} = (c_{33}+c_{44})/2;$$
$$\Gamma_{23} = (c_{23}+c_{44})/2;\ \Gamma_{13} = \Gamma_{12} = 0 \qquad (27)$$

and again from Eq. (10)

$$\begin{vmatrix} (c_{55}+c_{66})/2 - \rho v^2 & 0 & 0 \\ 0 & (c_{22}+c_{44})/2 - \rho v^2 & (c_{23}+c_{44})/2 \\ 0 & (c_{23}+c_{44})/2 & (c_{33}+c_{44})/2 - \rho v^2 \end{vmatrix} = 0 \qquad (28)$$

One solution to this equation represents a purely transverse wave having a displacement vector in the x_1 - direction and a phase velocity given by $v = ((c_{55} + c_{66})/2\rho)^{\frac{1}{2}}$ (29). The displacement vectors for the other two waves occur in the $x_2 - x_3$ plane; neither of these waves are purely longitudinal or purely transverse and propagate with velocities given by the roots of the equation

$$((c_{22}+c_{44})/2 - \rho v^2)((c_{33}+c_{44})/2 - \rho v^2) - ((c_{23}+c_{44})/2)^2 = 0 \qquad (30)$$

The faster wave is quasilongitudinal (QL) and the slower one is quasitransverse (QT). Similar equations are obtained when one considers transmission along lines bisecting any other two coordinate axes.

The results for the propagation in an orthotropic solid are summarized in Table 2. From these results it can easily be deduced the relations for transverse isotropy, as discussed earlier, by simply considering the restrictions to be satisfied for the stiffnesses in this particular case.

2.4. Dispersion of waves

We recall that the approach we have employed here in describing the elastic behaviour of the fiber composites has been based on the "effective modulus theory", where the composite is considered as a homogeneous but anisotropic medium in which the volume averages of all stress and strain components in a representative volume element are identical in all the material. This theory yields good results in the study of many static and dynamic problems, but in considering wave propagation it is not able to

Table 2. Propagation of waves in an orthotropic fiber composite

Direction of Propagation	Direction of Displacement vectors	Type of wave	Velocity of Propagation
x_1-axis	x_1	L	$v=(c_{11}/\rho)^{1/2}$
$(n_1=1$	x_2	T	$v=(c_{66}/\rho)^{1/2}$
$n_2=n_3=0)$	x_3	T	$v=(c_{55}/\rho)^{1/2}$
x_2-axis	x_2	L	$v=(c_{22}/\rho)^{1/2}$
$(n_2=1$	x_1	T	$v=(c_{66}/\rho)^{1/2}$
$n_1=n_3=0)$	x_3	T	$v=(c_{44}/\rho)^{1/2}$
x_3-axis	x_3	L	$v=(c_{33}/\rho)^{1/2}$
$(n_3=1$	x_1	T	$v=(c_{55}/\rho)^{1/2}$
$n_1=n_2=0)$	x_2	T	$v=(c_{44}/\rho)^{1/2}$
45° to x_2,x_3-axes	x_2x_3-plane	QL,QT	(*)
$(n_1=0$			
$n_2=n_3=1/\sqrt{2})$	x_1	T	$v=((c_{55}+c_{66})/2\rho)^{1/2}$
45° to x_1,x_3-axes	x_1x_3-plane	QL,QT	(**)
$(n_2=0$			
$n_1=n_3=1/\sqrt{2})$	x_2	T	$v=((c_{44}+c_{66})/2\rho)^{1/2}$
45° to x_1,x_2-axes	x_1x_2-plane	QL,QT	(***)
$(n_1=n_2=1/\sqrt{2}$			
$n_3=0)$	x_3	T	$v=((c_{44}+c_{55})/2\rho)^{1/2}$

(*) $c_{23} = ((c_{22}+c_{44}-2\rho v^2)(c_{44}+c_{33}-2\rho v^2))^{1/2}-c_{44}$

(**) $c_{13} = ((c_{11}+c_{55}-2\rho v^2)(c_{55}+c_{33}-2\rho v^2))^{1/2}-c_{55}$

(***) $c_{12} = ((c_{11}+c_{66}-2\rho v^2)(c_{66}+c_{22}-2\rho v^2))^{1/2}-c_{66}$

describe the effect of dispersion because a classical anisotropic continuum model cannot account for the dispersion which should occur in a composite medium if the wavelength is of the same order of magnitude as a characteristic length of the structuring. This is an inconvenience, since for a fiber reinforced composite, one would expect for a large frequency range that the medium should include wave dispersion.

Herrmann and Achenbach |6| have proposed a conceptually different approach for the dynamic analysis of fiber-reinforced composites. Instead of introducing a representative homogeneous medium, representative elastic moduli are used for the matrix, and the elastic and geometrical properties of the fiber are combined into "effective stiffnesses". The so-called "effective stiffness" theory for unidirectionally reinforced composites rests on the assumption that the characteristic dimensions of the reinforcing elements and the distances between them are small when compared to the dimensions of the body and the deformations.

In the "effective stiffness" theory for fiber-reinforced composites, the fibers are considered as long and slender structural elements of circular cross section which are endowed with stiffnesses relative to flexure, torsion and extension. Thus, bending, shear and extensional stiffnesses of the reinforcing elements enter the strain energy of the composite. It is considered that the interaction of the fiber and matrix deformations takes place only through the displacement of the center line of the fiber and the displacement of the matrix. (In the real composite it takes place through continuity of displacements and tractions at the fiber-matrix interface). Thus, the fiber does not affect deformations of the composite in planes normal to the fiber direction. However, the contribution of the fibers to the stiffnesses of the composite in planes normal to the fiber direction can be included by redefining the matrix as a transversely isotropic material. So, this "effective matrix" includes the part of the effect of the fibers, which is not coupled to deformation of the fibers in flexure, torsion and extension.

Achenbach and Herrmann, |7|, derived representative kinetic and strain energy densities and obtained, by the subsequent use of Hamilton's principle, the displacement equations of motion. According to their results for plane wave propagation, longitudinal waves propagating along the fibers (x_1-axis) and waves propagating along x_2 or x_3 are not dispersive. Instead, transverse waves propagating in the fibre direction are dispersive, the dispersion equation being,

$$\left[(\gamma(1+\eta)+(1-\eta))/(\gamma(1-\eta)+(1+\eta)) + \eta\chi\gamma - \beta^2(\eta\Theta+1-\eta)\right] \cdot \left[1/2(1+\nu_f)\,\gamma\xi^2 + \chi\gamma - 1/4\Theta\,\beta^2\xi^2\right] - \eta\chi^2\gamma^2 = 0 \tag{31}$$

where, $\Theta=\rho_f/\rho_m$, $\beta=v/(\mu_m/\rho_m)^{\frac{1}{2}}$, $\xi=k\,a$, $\gamma=\mu_f/\mu_m$, $\eta=n\,A/S$ the fiber density, χ= the Timoshenko shear coefficient, ν_f=Poisson's ratio of the fibers. Here f and m refer to fiber and matrix, respectively; ρ is the mass density, v is the phase velocity, μ is the shear modulus, k is the wave number, $A=\Pi a^2$ is the cross-sectional area of a fiber and S is the cross-sectional area of a cylindrical element of the fiber-reinforced composite containing n fibers. The independent parameters in Eq. (31) are γ, Θ, η and ν_f.

An examination of the propagation of dispersive transverse motion according to the theory just described, can be made through Fig. 4 where dispersion curves for various values of the fiber density are shown. It is noted that the dispersion curves depart at small values of the dimensionless wave number ξ from the horizontal lines representing the phase velocities,

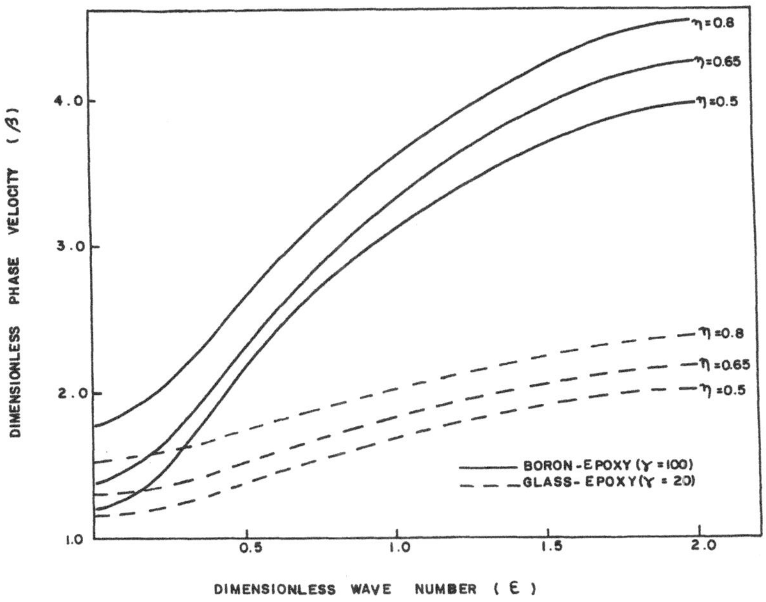

Fig. 4. Dispersion curves for transverse waves according to the effective modulus theory.

Most of the analysis about dispersion has been directed towards harmonic wave propagation. However, from a practical point of view the specific study of dispersive pulse propagation should be of interest, since ultrasonic pulse techniques are often used in characterizing composite materials. For linear theories the propagation of a stress pulse can be studied by using a Fourier decomposition of the pulse. Equations governing the phase velocities of the individual Fourier components provide the basic information on geometrical dispersion. A detailed analysis of the dispersion of transient waves in a laminated composite was developed by Peck and Gurtman |8| applying Fourier transforms to time and the coordinate in the propagation direction. Inversion of the spatial transform by residues yields a formal solution in the form of an infinite series of integrals, each of which represents the contribution to the transient response from a mode of sinusoidal wave propagation. Long-time approximations indicate that the low-frequency portion of the integral for the first mode gives the dominant contribution (head-of-the-pulse approximation). The form of the expression for the head-of-the-pulse approximation leads to the definition of a characteristic dispersion time of the composite which describes the dispersion of the wave. The characteristic dispersion time depends algebraically on the material properties and geometry of the composite and the propagation distance.

3. EXPERIMENTAL METHODS

The experimental methods, in which properties are deduced from velocity measurements for ultrasonic waves propagation, present several advantages over classical methods. In particular, ultrasonic thechniques can be used with small specimens or with large specimen analyzing small regions, and can be applied to determine material properties at a large range of frequencies, and are also relatively fast, easy to apply and don't require special specimen preparation. However, it must be realized that ultrasonic wave propagation in materials even if easily achieved, is not always easily and unambiguously understood.

As early described a unidirectional fiber-reinforced composite can be characterized by five elastic constants. In particular, if the x_1-axis is chosen in the fiber direction these constants are c_{11}, c_{12}, c_{22}, c_{23} and c_{44}. From Table 1 (Section 2.2) it is clear that the constants c_{11}, c_{22}, c_{44} and c_{66} (and hence c_{23}) can be determined by measuring the velocities of appropriate waves propagating along symmetry directions. Thus, if the material specimen is a rectangular parallelepiped, with the fiber direction parallel to one set of edges (Fig. 5), this requires normal propagation across the specimen faces. To determine the constant c_{12} is necessary to use Eq. (19) and then measure the velocity for a wave propagation at an intermediate angle to the preferred direction.

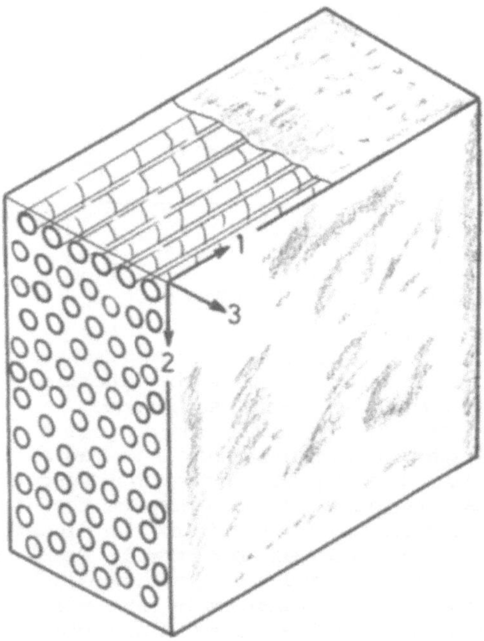

Fig.5. Parallelepipedic specimen of unidirectional fiber composite

Different experimental methods have been developed to achieve the situations just discussed. These methods can be classified in the following general groups: ultrasonic pulse velocity, resonance, and ultrasonic goniometry.

The first method is by far the most frequently used and to apply it several different techniques have been implemented. The ultrasonic pulsed through--transmission technique (see, for example |9|) involves the measurement of the transit time along specimens. The transmitter and receiver transducers are bonded directly to the faces of the specimen. A schematic diagram of the apparatus is shown in Fig. 6. The time between an initial or undelayed pulse and a delayed pulse is measured with the time-delay multiplier on the oscilloscope and this time is used to calculate velocity. For the determination of c_{12} the required direction of propagation can be achieved by cutting a specimen with faces at intermediate angle to the preferred direction. This technique suffers from the need to bond transducers to the specimen, and, for accurate measurements, the presence of the bond must be taken into account. In addition, to produce a polar diagram of velocity it is necessary to cut specimens which have several orientations. These deficiencies can be avoided by immersing the specimen in a liquid (usually water). The specimen is mounted on a turntable inside a water tank and two ultrasonic transducers are placed on either side, one for transmission T_E and the other for reception T_R, |10|

(Fig.7). By rotation of the turntable, the angle of incidence may be set at any desired value. At normal incidence, the wave on entering the specimen will continue to travel as a longitudinal wave. By measuring the difference in arrival time with and without the specimen, the velocity at normal incidence may be easily calculated. By rotation of the turntable and thus increasing the angle of incidence, the pulse on entering the solid will split by mode conversion into two components, one being quasi-longitudinal and the other quasi-transverse (Fig. 8). After leaving the solid, they will each arrive at the receiver and will be seen separated in time on the oscilloscope. If the specimen has been rotated from normal incidence about an axis perpendicular to the line joining the transducers, then the direction of the polarization of the transverse wave lies in the plane defined by this line and the specimen normal. Beyond the critical angle for the longitudinal mode, the transverse wave alone will be propagated. In all cases Snell's law is obeyed and the direction of propagation will depend on the angle of incidence and the ratio of the wave velocities in the composite and in water. By measuring

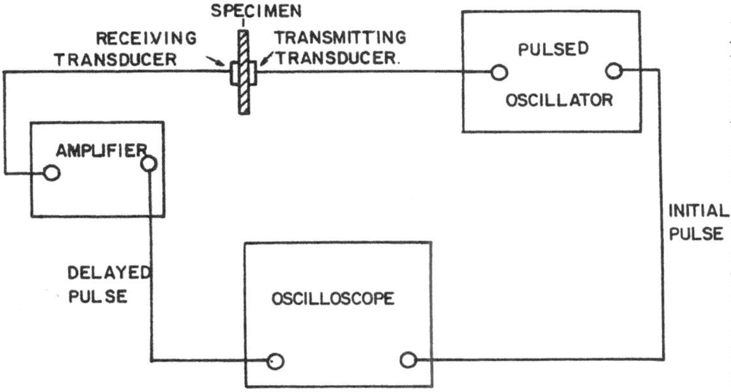

Fig.6. Apparatus for ultrasonic pulsed through-transmission technique

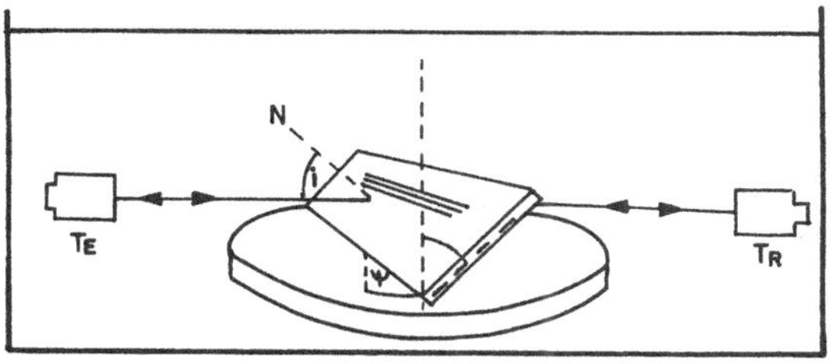

Fig.7. Arrangement for ultrasonic immersion technique

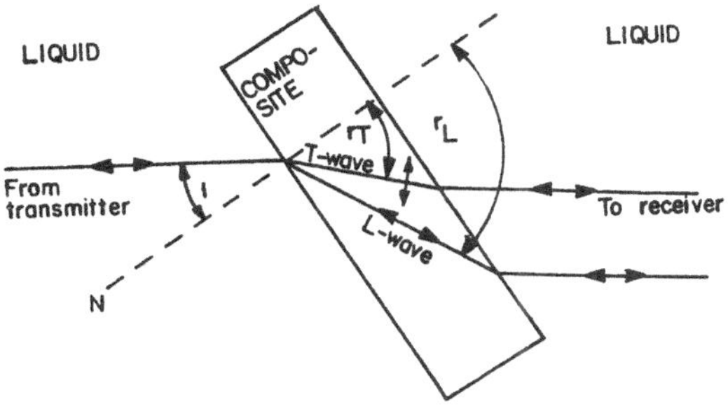

Fig.8. Propagation of longitudinal and transverse waves at a liquid-solid interface

the difference in time τ for the waves to travel from the transmitter to the receiver with and without the composite sample, and also the angle of incidence i, the thickness of the specimen d and the velocity in the liquid v_w, the angle of refraction may be found from $\tan r = \sin i/(\tau v_w/d+\cos i)$, and from it the wave velocity may be obtained. Waves with polarization perpendicular to the plane containing fiber direction can be generated by rotating the specimen about two axes as shown in Fig. 7. One limitation of this method is caused by the longitudinal character of the incident wave. In fact, if specimen faces are parallel and normal to the fiber axis, it is not possible to excite transverse waves at normal incidence in the specimen. This may be achieved by passing the incident wave at an oblique angle through a prism bonded to the specimen (Fig. 9), but this solution again suffers from the problems associated with the bond.

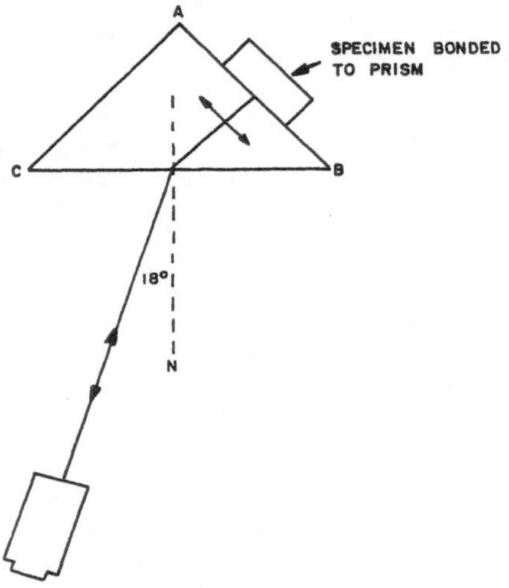

Fig.9. Generation of transverse waves normal to the surface of the specimen

A procedure for determining the five independent elastic constants using only the immersion technique was proposed by Dean and Lockett |15|. In this procedure the velocities of the longitudinal and transverse wave $(c_{22}/\rho)^{1/2}$ and $(c_{66}/\rho)^{1/2}$ are obtained from oblique incidence of the incident longitudinal wave on a transverse x_2x_3-plane. Consider now a plane containing x_1-axis (fiber direction) (Fig. 10). When the incident beam is directed in the x_1-direction, a longitudinal wave is transmitted which travels with velocity $(c_{11}/\rho)^{1/2}$. When the x_1-axis lies parallel to the incident face, the transmitted waves for oblique incidence are governed by Eq. (19) and vary with the transmission angle as shown in Fig. 11. Following Table 1 (Section 2.2) the velocities of the points A, B and C are $(c_{44}/\rho)^{1/2}$, $(c_{44}/\rho)^{1/2}$ and $(c_{11}/\rho)^{1/2}$ respectively. Here the value of c_{44} cannot be determined directly from experimental data, but it can be obtained by a process of extrapolation in the neighborhood of point B. Since ϕ is small, from eq. (19) it is possible to express $\rho v2$ in the form :

$$\rho v_T^2 = c_{44} + (c_{22} - c_{44} - \gamma)\sin^2\phi \qquad (32)$$

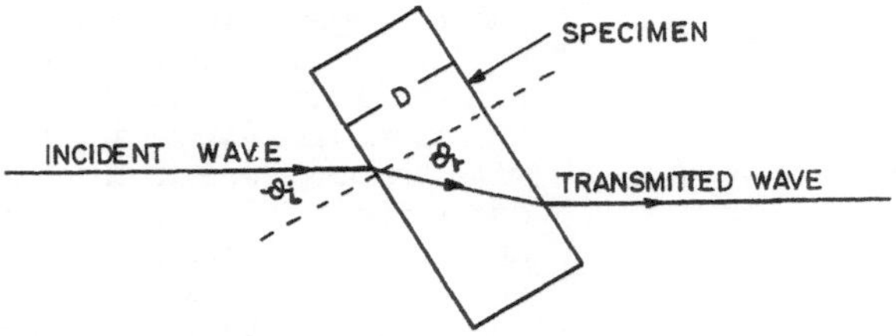

Fig.10. Wave path through the specimen

where v_T is the velocity of the quasi-transverse wave and $\gamma = (c_{44}+c_{12})^2/(c_{11}-c_{44})$. When $\phi = \pi/4$ the approximation of Eq. (32) and the exact solution, for a typical directionally reinforced carbon--fiber composite, give values which differ by

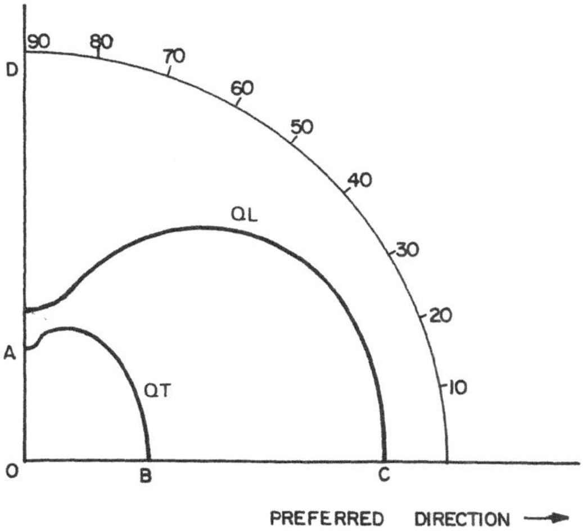

Fig. 11. Variation of the velocities of quasi-longitudinal (QL) and quasi-transverse (QT) waves with direction of propagation in a plane containing the preferred direction

less than one percent. From Eq. (32) it follows that the plot of ρv_T^2 againts $\sin^2\phi$ is a straight line. Thus, the intercept of the experimental line determines c_{44} and the slope determines γ and hence c_{12}.

One of the key factors involved in these methods is that the angle of refraction of the longitudinal waves for practical composites becomes large for even small angles of incidence. Therefore, in order to transmit a longitudinal wave it is necessary to use small angles of incidence and in this case its measurement cannot be very accurate. For this reason is better to use transverse wave measurements.

When the pulse through-transmission methods are applied to thin specimens, there is difficulty in separating the signals associated with the front and back surfaces of the specimen. In this case detection and measurement of the resonances of the specimen can give a more useful information. A resonance technique for compressional and shear waves velocity measurements has been described by Chang et al. |11|. This technique utilizes the principle of ultrasonic interference spectroscopy. A pulsed wave is transmitted through a coupling medium to the composite plate and is received by the same transducer. The signal reflects from the front and back surfaces and produces interference phenomena. When the output signal of the transducer contains a wide band of continuous frequency components, the interference effects in multiple reflected sound waves give rise to anti-resonance dips in the frequency spectrum of the received signal. In the case of normal incidence the period in frequency Δf of these anti-resonance dips is related to the plate thickness d and the velocity of sound v_L by $\Delta f = v_L/2d$. If a shear wave is generated in the composite specimen by the oblique incidence at an angle Θ of a longitudinal wave, then the velocity of that wave can be computed from |11|

$$v_T = 2dv_o\Delta f/(4d^2\Delta f^2\sin^2\Theta + v_o^2)^{1/2} \tag{33}$$

where v_o is the longitudinal wave velocity of the coupling medium.

To specify the modulus of the material and assess its degree of elastic anisotropy is not a quick process by pulse transmission and resonance methods, and it often requires the cutting of the specimen in special shapes. Ultrasonic goniometry |12| offers a more sensitive method in some situations. If an ultrasonic beam is incident on a liquid-solid interface the refracted wave decomposes into a longitudinal wave and a transverse wave. For angles of incidence slightly greater than the critical angles for longitudinal and transverse waves, surface waves on the solid are generated. These surface waves extract energy from the reflected wave and cause spectacular drops in the reflected amplitude. The critical conditions stand out well and, using Snell's law, the velocities, and hence the elastic moduli can be determined. The accuracy of the method is very high because the angular measurements may be made with a precision of ±1°. The method proves to be very useful for the study of the degree of anisotropy of composites. In fact, by the goniometer method the material surface is analysed to a depth of approximately a wavelength. By varying the wave--length it will be possible to detect elasticity changes as a function of depth.

The choice of one of the described methods depends largely upon the geometry of the specimen, its elastic complexity and the precision required. However at present, with the improvements reached in the sensitivity, response and power capacity of the pulse transducers and the availability of higher frequencies and a better understanding of wave propagation, the pulse velocity methods are clearly the most widely used.

In the measurement procedures it is usually assumed that the energy of the plane wave propagates in the same direction as the wave normal. But, as we discussed in Section 2.1, this is not exactly true and in some cases the deviation between the direction of the energy propagation and the wave normal may be very large (as large as 40°). If thick specimens are used to determine wave velocity, the direction of the propagating energy should be calculated in order to assure that the propagating wave is not reflecting off unexpected boundaries, as shown in Fig. 12. Equations to calculate these angles were derived by Kriz and Stinchcomb |13|, |14|. Using these equations, deviation of the energy-flow vector from the wave normal were calculated for a unidirectional graphite/epoxy composite. The results are listed in Table 3.

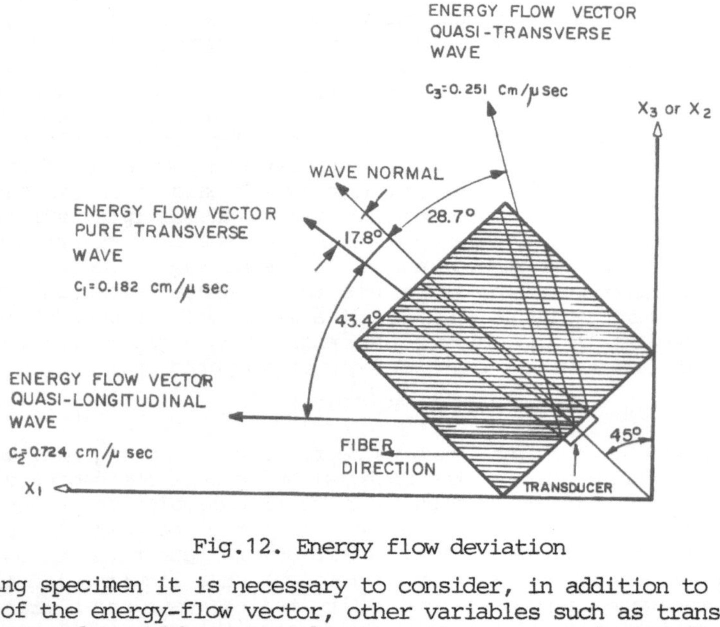

Fig.12. Energy flow deviation

In sizing specimen it is necessary to consider, in addition to the deviation of the energy-flow vector, other variables such as transducer size, maximum pulse width, material attenuation, separation of QL and QT waves, and effect of boundaries on the character of the propagating wave. For specimens in which no deviation in energy flow is expected, the only

Table 3.- Deviation of energy flow direction from wave normal in a unidirectional graphite/epoxy composite (fiber direction x_1).

Direction of wave normal	Direction of Displacement vectors	Type of wave	Deviation Energy-flow vector $\Theta°$
x_1-axis	x_1	L	0
($n_1=1$	x_2	T	0
$n_2=n_3=0$)	x_3	T	0
x_2-axis	x_2	L	0
($n_2=1$	x_1	T	0
$n_1=n_3=0$)	x_3	T	0
x_3-axis	x_3	L	0
($n_3=1$	x_1	T	0
$n_1=n_2=0$)	x_2	T	0
45° to x_2,x_3-axes	x_1	T	0
($n_1=0$	x_2x_3-plane	L	0
$n_2=n_3=1/\sqrt{2}$)	x_2x_3-plane	T	0
45° to x_1,x_3-axes	x_1x_3-plane	QL	43.4
($n_2=0$	x_2	T	17.8
$n_1=n_3=1/\sqrt{2}$)	x_1x_3-plane	QT	28.7
45° to x_1,x_2-axes	x_1x_2 plane	QL	43.4
($n_3=0$	x_1x_2 plane	QT	28.7
$n_1=n_2=1/\sqrt{2}$)	x_3	T	17.8

restriction on thickness is that it has to be large enough for accurate wave-velocity measurement but small enough so that the received pulse is not very attenuated. Instead, for specimens in which deviation in energy flow is present, the restrictions are more severe. In fact, the specimens must be sufficiently thin so that the energy flow propagated from one face is not received as a reflection from some other surface, and sufficiently thick to allow separation of the QL and QT waves. Kriz and Stichcomb |13| made simple calculations for graphite/epoxy parallelepipedic samples cut at 45° to the fiber direction, and found that for a sample length of 2.54 cm, a transducer diameter of 0.63 cm and a pulse width of 2.25μs, the minimum and maximum allowed specimen thicknesses were of 0.98 cm and 1.02 cm, respectively. A lower limit to the size of a specimen is also imposed by the requirement that waves should have the character of waves in an infinite medium unaffected by the presence of boundaries. Since no rigorous theoretical criterion exists for predicting the minimum size, it must be established experimentally.

4. RESULTS AND DISCUSSION

The first study that made use of velocity measurements of ultrasonic waves to completely describe the elastic behaviour of a fiber composite was developed by Zimmer and Cost |9|. Using the pulsed through-transmission technique, they measured the velocity of longitudinal and shear waves propagated and polarized in specified directions to give values corresponding to each of the elastic stiffness constants. In so doing, Zimmer and Cost measured the five stiffnesses required to describe fully the elastic properties of a unidirectional glass fiber reinforced composite. One of the main purposes of this work was to test the various theories for the elastic properties of fiber composites, and in particular those of Whitney and Riley |15|, Hashin and Rosen |16| and Greszczuk |17|. Table 4 shows the comparison between experimental and theoretical results. It can be

Table 4.- Comparison between experimental and theoretical elastic constants for a glass-epoxy composite (units N/m^2)

Constants	Experimental (Zimmer-Cost)	Theoretical (Whitney-Riley)	(Hashin-Rosen)	(Greszczuk)
c_{11}	$4.15\pm0.25\times10^{10}$	4.00×10^{10}	4.99×10^{10}	4.23×10^{10}
$c_{22}=c_{23}$	1.78 ± 0.1	1.17	1.24	1.78
$c_{12}=c_{13}$	0.48 ± 0.48	0.36	0.92	0.79
c_{23}	0.98 ± 0.15	0.43	0.60	1.06
$c_{44}=c_{55}$	0.34 ± 0.02	0.41	0.41	0.54

seen that the values computed from the three different theories show general agreement with experimental values. However, it may be still noted that, in particular, the predictions of Greszczuk's theory show an agreement that in most cases is within the experimental error of the measurement.

Points which arise in obtaining experimental values are specially related with the constants c_{44} and c_{13}. The two velocities which determine the stiffness c_{44} correspond to transverse waves propagating either along fiber direction or normal to it (see Table 1). In the experimental works referred in the literature |15|, |18|, a discrepancy between these two velocities by nearly a factor of 2 can be seen. This suggests an effect of dispersion in some of the transmitted waves. The approximate theory of Achenbach and Herrmann about dispersion, that we discussed in Section 2.4, predicts that the only dispersive wave in an unidirectional fiber-reinforced composite is a plane transverse wave propagating in the fiber direction. Also, an exact treatment for laminate materials, developed by Sun, Achenbach and Herrmann |19|, shows that dispersion is most pronounced for a transverse wave propagating parallel to the interface and with displacement vector normal to it. These two predictions, and the fact that the wave along the fiber direction shows a higher velocity, indicate that c_{44} should be obtained from the velocity of the waves normal to the preferred axis.

The elastic constant $c_{13}=c_{12}$ is determined from waves at an angle to the fiber axis propagating with velocities governed by Eq. (19). These waves are, in general, quasilongitudinal (QL) or quasitransverse (QT). Values calculated by several authors for c_{13} show that they are extremely sensitive to the accuracy with which the phase velocity is measured. In fact in the experimental values given in Table 4 the uncertainty in $c_{13}=c_{12}$ is estimated to be 100%. The problem is that the expression for wave velocities at an angle to the fiber axis contains c_{13} together with other stiffness constants and is very sensitive to the magnitude of c_{13}. Thus, any reasonable accuracy in c_{13} can only be obtained if measurement of velocity and the other stiffness components are made very precisely. However, variations in the specimen constituents, along with errors introduced by the measuring procedure, make it unrealistic to pretend a good accuracy in the experimental values of c_{13}. Fortunately, some procedures have been successfully developed to obtain the five characteristic properties of a transversely isotropic composite. We will mention here the technique developed by Dean and Turner |20| and improved by Kriz and Stinchcomb |13|. These investigators demonstrate that it is possible to obtain the complete set of transversely isotropic properties even though the ultrasonic data contain inherent scatter. This can be done by curve fitting ultrasonic data using the Hashin-Rosen equations. In fact, transverse isotropy is demonstrated by taking any of the four known properties and calculating the fifth. Therefore, accurate measurement of $c_{12}=c_{13}$ is not necessary.

In addition to the determination of the elastic constants of composite materials, the ultrasonic techniques described earlier in this report are very useful in assessing some special properties of the material such as homogeneity, symmetry, fiber misorientation, fiber volume fractions, etc. An indication of specimen homogeneity may be obtained from a series of velocity measurements along the same direction but at different positions within the specimen. For the determination of the axes of symmetry, measurements of wave velocities in three orthogonal planes allow the calculation of the elastic moduli and from examination of the obtained values the class of symmetry can be deduced. To estimate the fiber misorientation it is necessary to bear in mind that a departure from uniaxial fiber alignment in a composite will result in a reduction of modulus along the preferred direction together with an increase in the transverse direction. A simple model has been proposed which relates elastic properties with the degree of orientation |21|.

The dependence of the elastic stiffness components of fibre composites upon fibre fraction and, also, upon void content has been reported in several papers |22|, |5|, |13|. Isotropic and anisotropic fibers were considered in relation to the typical fiberglass and carbon-fiber composites. Results show that increasing the fiber content increases the ultrasonic longitudinal and transverse velocities. This can be attributed to the fact that the stiffnesses of the fibers are larger than those of the matrix, so that increasing in fiber content results in larger stiffnesses of the composite and hence in larger velocities. Considering the presence of voids in the matrix, increasing the void content decreases the longitudinal and shear velocities. This is due to a decrease of the elastic constants of the composite with an increase of void content. In addition, the presence of fiber reinforcement makes that the velocities decrease more rapidly with increasing void content than in the case of no fibers. This behaviour can be imputed to the assumption that voids were taken to be only in the matrix. Thus for higher fiber content a given void content will occupy less volume of the matrix altering its moduli. Finally, it may be pointed out that the behaviour of the velocities as a function of fiber and void content is very similar in isotropic and anisotropic fiber composites.

We would not like to end these comments without saying some words about experimental results on the dispersion of ultrasonic waves. Earlier in this section we discussed the effect of dispersion on the determination of the stiffness constant c_{44}. According to approximate and exact theories transverse waves propagating in the fiber direction present the most pronounced dispersion. Experimental work |23| confirms this behaviour. The velocity of transverse waves in the fiber direction experiences a considerable increase with frequency from even rather low frequencies. This increase indicates that as the wavelength shortens, the energy is transmitted to a larger extent through the fibers and to a lesser extent through the matrix. Transverse waves propagating in a direction perpendicular to the fibers and having particle motion in the fiber direction suffer much less dispersion. No significant dispersion was observed for transverse waves travelling perpendicular to the fiber direction and having the displacement vector in the plane of transverse isotropy. Longitudinal waves propagating either parallel or perpendicular to the fiber direction are dispersive for frequencies higher than a certain value which seems to be related to the magnitude of the fiber diameter or the distance between fibers. In fact, dispersion of longitudinal waves, measured in boron-epoxy specimens,happens when the wavelength in the matrix is of the same order of magnitude as the diameter of the fibers. It may be noted that for longitudinal waves the velocity decreases with increasing frequency.

REFERENCES

1. Musgrave MJP: Proc. Roy. Soc. A 226, 339-355, 1954.
2. Musgrave MJP: Proc. Roy. Soc. A 226, 356-366, 1954.
3. Pollard HF: Sound Waves in Solids. London: Pion Ltd., 1977.
4. Henneke EG, Duke JC: Materials Evaluation 43, 740-745, 1985.
5. Dean GD, Lockett FJ: ASTM STP521, 326-346, 1973.
6. Herrmann G, Achenbach JD: Proc. AIAA/ASME 8th Structures, Structural Dynamics and Materials Conference, 112-118, 1967.
7. Achenbach JD, Herrmann G: AIAA J. 6, 1832-1836, 1968.
8. Peck JC, Gurtman GA: J. Appl. Mech. 36, 479-484, 1969.
9. Zimmer JE, Cost JR: J. Acoust. Soc. Am. 47, 795-803, 1970.
10. Markham MF: Composites 1, 145-149, 1970.
11. Chang FH et al.: J. Comp. Mat. 8, 356-363, 1974.
12. Curtis G: Ultrasonics for Industry 1969, 4-7, 1969.
13. Kriz RD, Stinchcomb WW: Experim. Mech. 19, 41-49, 1979.
14. Kriz RD, Stinchcomb WW. VPI-E-77-13, May 1977.

15. Whitney JM, Riley MB: AIAA J. 4, 1537-1542, 1966.
16. Hashin Z, Rosen BW: J. Appl. Mech. 31, 223-232, 1964.
17. Greszczuk LB: AIAA J. 9, 1274-1280, 1971.
18. Smith RE: J. Appl. Phys. 43, 2555-2561, 1972.
19. Sun CT et al.: J. Appl. Mech. 35, 467-475, 1968.
20. Dean GD, Turner P: Composites 174-180, July 1973.
21. Brody H, Ward IM: Polym. Eng. Sc. 139-151, 1971.
22. Martin BG: J. Appl. Phys. 48, 3368-3373, 1977.
23. Tauchert TR, Guzelsu AN: J. Appl. Mech. 39, 98-102, 1972

CHARACTERIZATION OF POROUS MEDIA WITH ELASTIC WAVES

B.R. Tittmann

INTRODUCTION

Elastic wave propagation in porous media will be introduced by a discussion of scattering to show the use of broad frequency band pulses in relating the attenuation, velocity and backscattering to information on porosity. Examples will be given for microporosity in castings and powder metal components. This will be followed by a discussion of the influence of volatiles within the pores on the absorption of elastic waves in porous ceramics and rocks. Implications of these findings will be given for ultrasonic device applications and for the interpretation of the lunar seismic experiments carried out as part of the Apollo missions to the moon.

HOMOGENEOUS ELASTIC SOLID

The wave equation for a homogeneous isotropic linear elastic solid may be written in vector form for the displacements

$$\mu\nabla^2 s + (\lambda + \mu)\nabla\nabla\cdot s = \rho\ddot{s} \tag{1}$$

where λ and μ are Lame's constants and ρ is the density. The displacement may be conveniently expressed as

$$s = \nabla\phi + \nabla\Lambda\psi \tag{2}$$

where ϕ and ψ are the scalar and vector potentials, respectively, leading to

$$\nabla^2\phi = (1/C_L^2)\ddot{\psi} \tag{3}$$

for longitudinal waves with velocity $C_L^2 = (\lambda + \mu)/\rho$ and

$$\nabla^2 \psi = (1/C_T^2)\ddot{\psi} \tag{4}$$

for transverse waves with velocity $C_T^2 = \mu/\rho$

SCATTERING FROM ISOLATED SPHERICAL PORE

The general solution to the wave equations has been found to be of the form (1)

$$\sum_{m=0}^{\infty} c_m z_m(\ell r) P_m(\cos\theta) \tag{5}$$

where(r, θ, ϕ) is the system of coordinates with origin at the center of the sphere, the c_m's are constants, z_m is a linear combination of spherical Bessel and Neumann functions of order m, ℓ is a wave number (k or K), and P_m is the Legendre polynomial of degree m. The normalized, dimensionless scattering cross section is found to be[1]

$$\gamma_N = \Omega(a,f)/\pi a^2 \tag{6}$$

$$\gamma_N = 4\sum_{m=0}^{\infty} (2m + 1)[|A_m|^2 + m(m + 1)(k_1/\kappa_1)|B_m|^2] \tag{7}$$

where $k_1 = \omega[\rho(\lambda + 2\mu)]^{1/2}$ and $\kappa_1 = \omega(\rho/\mu)^{1/2}$ for the host material, the series coefficients A_m and B_m are evaluated by matching the boundary conditions between the sphere and the matrix, $\Omega(a,f)$ is the total scattering cross section and a is the radius of the sphere. For the case of the spherical cavity, no wave is inside the sphere and only two boundary conditions are necessasry, namely, that the components of stress vary continuously across the boundary.

For small ka (Rayleigh limit):

$$\Omega(a,f) \simeq g_0 f^4 a^6 \tag{8}$$

where g_0 depends on materials properties.

For large ka (geometric optics limit):

This regime is computationally difficult for ka > 10 and it is sufficient to assume γ_N is constant

$$\Omega(a,f) \simeq [\gamma_N(ka = 10)][\pi a^2] = g_, a^2 \tag{9}$$

where $g_1 \approx \gamma_N(10)\pi$.

These features are shown in Figure 1, which displays the normalized backscattered amplitudes for a single spherical void in Ti-alloy as a function of ka, where k is the longitudinal wave number.[2] Here both incident and backscattered waves are longitudinal. In the range $ka < 1$ the amplitude rises steeply in accordance with Eq. (1); at $ka = 1$ the response goes through a knee and after some oscillations levels off at a constant value in accordance with Eq. (9). The oscillations near $ka = 1$ are believed to arise as a result of interference between the primary backscattered waves and socalled creep waves that travel around the perimeter of the sphere.[3]

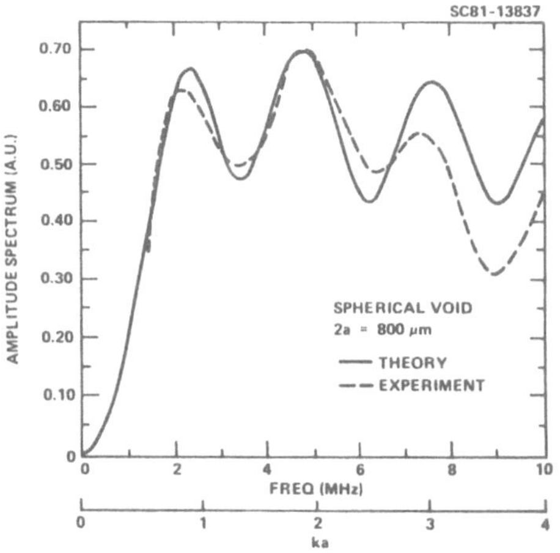

Figure 1. Theoretical and experimental backscattered amplitude for spherical void in Ti-alloy, with radius a = 400 μm.

SPARSE DISTRIBUTION OF PORES

Consider a volume element δV_j such as shown in Figure 2. Let $N_{3j}(a)$ be a random variable representing the number of pores per unit volume with radii $\leq$ a and centers near r_j. The observed attenuation due to scattering from a sparse distribution of pores, allowing no multiple scattering, is given by[4]

$$\hat{\alpha}_j = \frac{1}{2} \sum_{k=1}^{\delta V_j N_{3j}(\infty)} \Omega(a_k, f)/\delta V_j$$

$$\equiv \frac{1}{2} \int_0^\infty \Omega(a,f) dN_{3j}(a) \tag{10}$$

Here $\hat{\alpha}_j$ has units of nepers/cm and may be converted to dB/cm by multiplication with 20 $\log_{10}e$; the pore radius a is in cm, the volume element δV_j in cm^3 and the scattering cross section in cm^2. For a large volume V and a certain range of pore sizes (a_1,a_2)

$$V = \sum_{j=1}^{J} \delta V_j \tag{11}$$

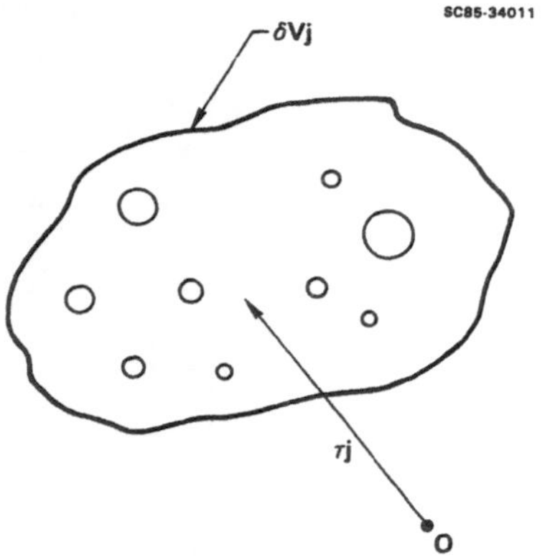

Figure 2. Schematic diagram showing volume element δV_j a distance r_j from the origin.

$$\hat{a}(a_1,a_2) = \frac{1}{2} \int_{a_1}^{a_2} \Omega(a,f) dN_3(a) \tag{12}$$

where

$$N_3(a) = \frac{1}{V} \sum_{j=1}^{J} \delta V_j N_{3j}(a) \tag{13}$$

ATTENUATION MEASUREMENTS

For many porous materials the attenuation is reasonably high, so that the conventional technique of measuring the decay of multiple echoes becomes infeasible because often too few echoes may be observed. However, an alternative method may yield good estimates of the attenuation, even if only one echo is observed.[5]

In addition to the sample and a single transducer, a buffer made from material with acoustic impedance similar to that of the sample but with much lower attenuation (Fig. 3). In step one just the echo from the buffer-air interface is recorded for calibration purposes, giving the echo amplitude A. In step two the sample is bonded to the buffer with a bond that is much thinner than the ultrasonic wavelength and therefore assumed to make a negligible contribution. The echo from the buffer-sample interface is recorded and is the product of A and the absolute value of the reflection coefficient $|R|$. The use of the echoes in steps one and two allows solving for $|R|$. Finally, in step three, one (C) or more echos (C', C" ...) are recorded after they have travelled one or more corresponding round trips through the sample under the assumption of negligible interface losses

$$1 = \vec{T} + \vec{R} \tag{14}$$

$$|T| = A^{1/2}(1 - |R|^2)^{1/2} \tag{15}$$

$$C = |T|^{-1/2}(1 - |R|^2)^{1/2} \exp(-2\alpha\ell) \tag{16}$$

$$\alpha = 1/2\ell[\ln(1 - |R|^2) + \ln A/C] \tag{17}$$

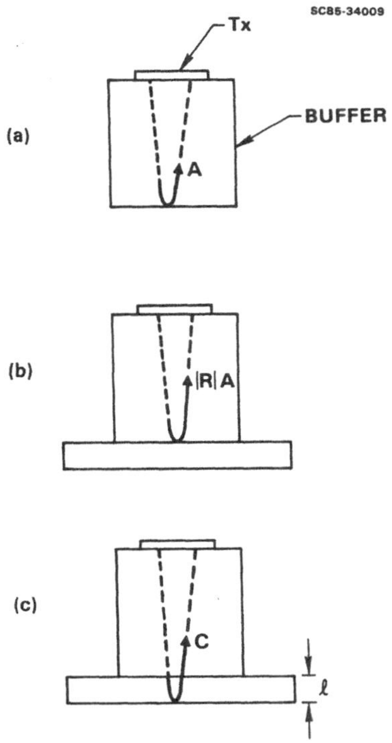

Figure 3.
Schematic diagram for attenuation measurements.

where α is the attenuation in nepers/cm and ℓ(cm) is the thickness of the sample.

This measurement is subject to two corrections: diffraction losses due to the frequency dependent spreading of the ultrasonic beam and apparent losses due to lack of parallelism and flatness of the sample surfaces. How these losses are minimized and taken into account in a quantitative way is discussed in some length in Ref. (6).

POROSITY IN CASTINGS

The characterization of porosity in castings is a technologically important example which has received some recent attention by Adler et al.[7] Using the results of Gubernatis and Domany[8] he shows the feasibility of separately determining pore size and pore density. Here are shown some results[9] on aluminum-silicon castings with intentional porosity. In Figure 4 a representative metallograph shows grains and pores on a polished surface of the sample. Based on a statistical evaluation of many such graphs, the void volume was estimated at 4% and the mean pore diameter at 0.25 μm assuming an equivalent circle for the outline of the voids. Ultrasonic measurements were carried out with a broad-band transducer, such that the return echoes could be digitized and deconvolved with the calibration echo to determine the longitudinal wave attenuation for the frequencies contained in the waveform as shown in Figure 5. Under the assumptions of spherical pores and no multiple scattering, Gubernatis and Domany[8] predict that the knee, which marks the transition from the Rayleigh regime to the geometric optics regime, gives the mean pore radius from the consideration that ka = 1, according to

$$a = \left(\frac{C_L}{2\pi}\right) \frac{1}{f_p} \tag{18}$$

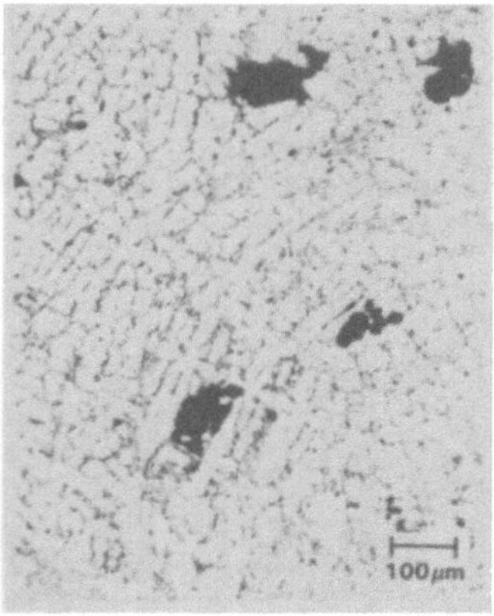

Figure 4.
Representative metallographic picture of casting. From statistics; estimated void volume is 4% and mean pore diameter is 0.25 mm.

where f_p is the frequency corresponding to the knee. They predict that the attenuation at frequencies above f_p provides the mean volume fraction of pores according to

$$VF = N_3 \frac{4}{3} \pi a^3 \tag{19}$$

where N_3 is obtained from Eq. (12) in terms of attenuation α_p, as indicated in Figure 5. Using these data, one obtains for the porosity 3% and for the mean pore diameter 0.40 mm in good agreement with the metallographic estimates. In field-type applications a preferred technique is to calculate $d\alpha/df$, since this gives a cleaner interpretation of the value of f_p (7).

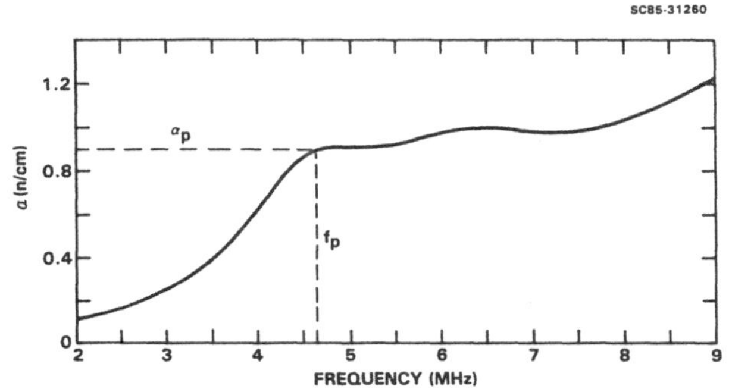

Figure 5. Attenuation (α nepers/cm) as a function of frequency for Al-Si castings with porosity.

MICROPOROSITY

In many applications, the mean pore radius is much smaller than the wavelength of the ultrasonic waves used in the inspection. One example is the microporosity found in powder metal alloys. Figure 6 shows micropore size distributions for two samples of a Ni-base powder alloy. The grain size distribution is typically constrained to lie in the 5-10 μm range. For sample A the pore density is ≃ 330 pores/mm^3 and the pore radii are as great as 40-50 μm, an unacceptable condition requiring detection before the cyclic fatigue associated with service produces void coalescence and crack initiation. Ultrasonic shear wave measurements were carried out[10] to obtain the backscattering noise shown in Figure 7. Each time trace represents the reflected amplitude for a different direction of the ultrasonic beam. The more porous sample shows consistantly higher microstructural noise. To obtain a good statistical average, the noise taken from 20 different sections of the time traces corresponding to 20 different regions of each sample. The corresponding power spectra including the electronic noise are shown in Figure 8. After deconvolution with reference signals obtained from corner reflections in the same samples the data appeared as in Figure 9. Also shown are theoretical results (dashed lines) based on Rayleigh scattering, i.e., Eq. (8). the full expression used for the deconvolved power spectrum is

$$\left|\frac{Y(\omega)}{R(\omega)}\right|^2 = \frac{2C_T^2}{ka^2D_o} \quad \frac{1}{(R_1R_2)^2} \quad Vol(\omega) \quad \omega^4 \quad \eta\rho_v E(r^6) \qquad (20)$$

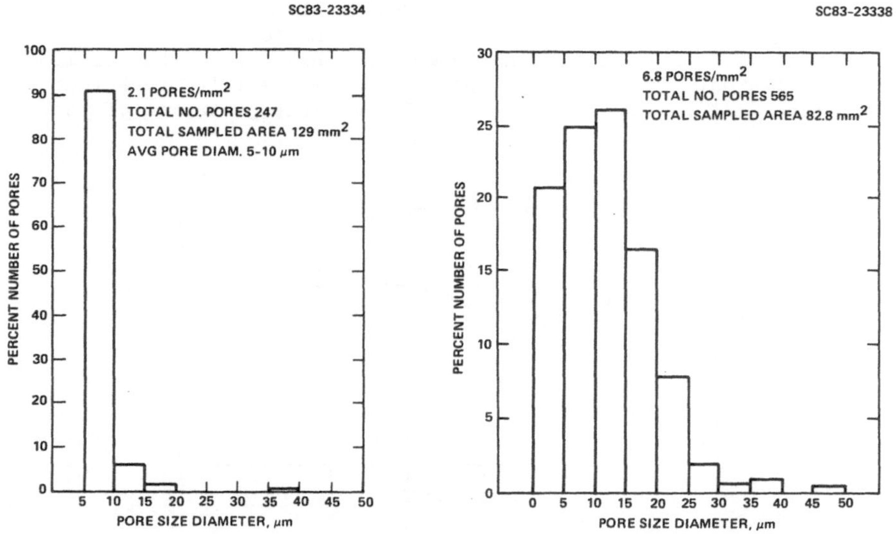

Figure 6. Micropore size distributions for two samples of IN-100. A nickel-base powder alloy used in aircraft engines.

where the first term on the right describes the scattering from a distribution of scatterers, the second term the Rayleigh term, the third term the frequency dependent volume of Gaussian profile ultrasonic beam, the fourth term the corner reflection coefficients and the last the on-axis diffraction. In the first term $E(r^6)$ is the expectation value of the sixth power of the pore radius, ρ_v the pore density and η a proportionality coefficient. Note that the third term (which is not reproduced because of its complexity) introduces an additional frequency dependence that gives the total expression with $\omega^{3.25}$ variation in good agreement with the data. The theory was fitted to the data curve of sample A at one point which provided a calibration for η. Using this value and the metallographic data an absolute determination was made is shown in Figure 9 for sample C, which is in good agreement with the experimental data.

Table 1

Summary of Results for IN-100

	Pore Density (mm^{-3})	Pores in 5 μs Beam Vol - 0.23 cm^3	$\left\lvert\frac{Y}{R}\right\rvert^2$ at 10 MHz	Scattering $\langle r^6\rangle^{1/6}$ (μm)	Micrograph $\langle r^6\rangle^{1/6}$ (μm)
Research Sample A	329	80,000	1×10^{-5} (-50 dB)	5.34	4.1-19.5
Spin Pit Sample C	89	20,000	5.01×10^{-8} (-73 dB)	2.74	2.4-10.0

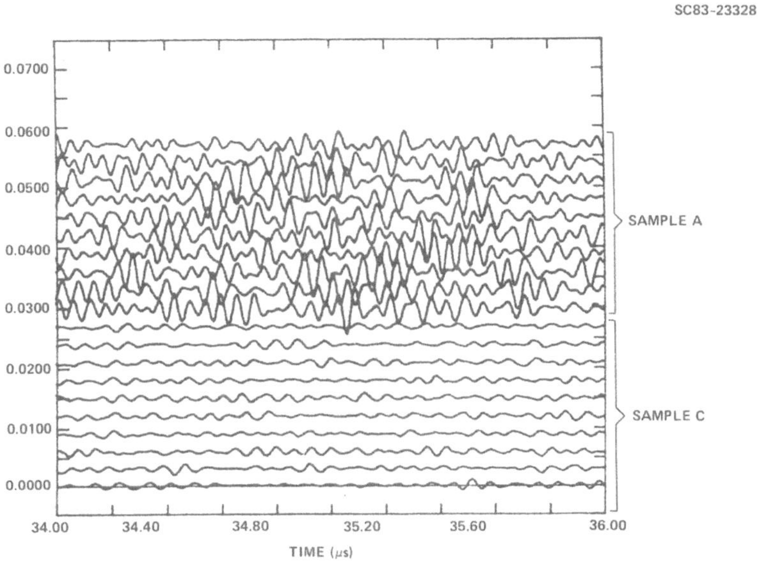

Figure 7. Ultrasonic backscattering noise in two samples of IN-100.

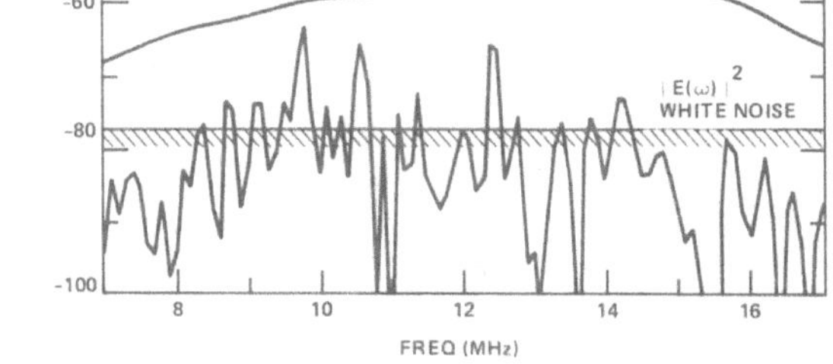

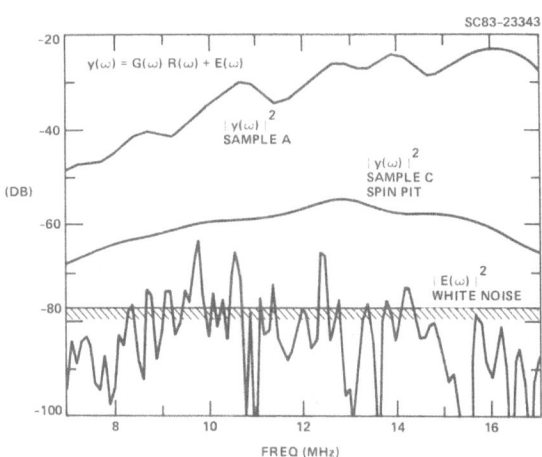

Figure 8. Ultrasonic backscatter power spectra for two samples of IN-100 before deconvolution.

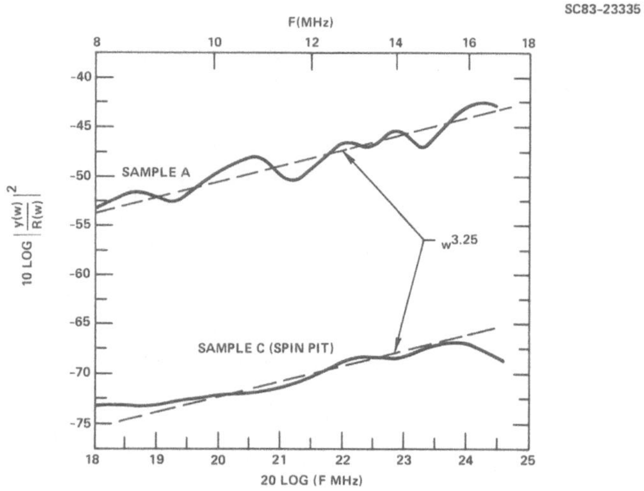

Figure 9. Ultrasonic backscattering spectra for two different samples of IN-100. Comparison between theoretical (dashed line) and experimental (solid line) results.

Table 1 summarizes the results of the ultrasonic and micrographical examination.

EFFECT OF FRACTURES ON RAYLEIGH WAVE VELOCITY

Ever since the lunar seismic experiments of the Apollo missions to the moon showed very low elastic wave velocities in the lunar crust, scientists have become keenly aware of the role of fractures in dramatically reducing the elastic stiffness. The same trends were demonstrated in laboratory experiments on lunar return samples.[11] Figure 10 shows some representative data on lunar return sample 15450.65. This rock is a coarse breccia from the rim of Spur Crater. When one face of the sample was polished it showed pronounced fractures running more or less parallel to one another at a spacing of about 1 to 2 μm. The Rayleigh wave velocity was measured with the impulse technique[12] in directions parallel and perpendicular to the fractures. In the impulse technique small piezoelectric chips are pressed against the surface, one as a transmitter, another as a receiver. A voltage spike is applied to the transmitter chip and responds by giving the surface an elastic stress impulse generating bulk wave and Rayleigh wave pulses. For suitably chosen travel distances, the Rayleigh wave pulses arrive as clean, well-defined signals before the bulk wave pulses arrive, reflected from the opposite end of the sample. The data are plotted as graphs of Rayleigh wave signal arrival time as a function of transducer separation, which in general gives a straight line, the reciprocal of whose slope is the Rayleigh wave velocity v_R. The most important feature to notice in Figure 10 is that the velocity parallel to the fractures is very high, i.e., $v_R \simeq 1.95$ km/s. This value is higher than that for most of the

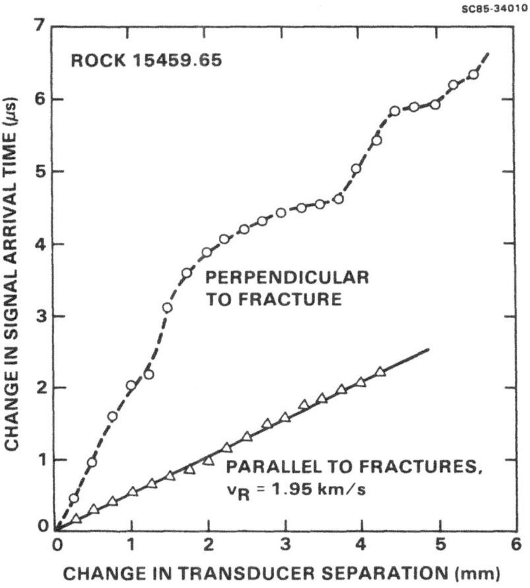

Figure 10. Sample data of change in signal arrival time vs change in transducer separation for rock 15459. The data were obtained by the impulse technique. The reciprocal of the slope gives the Rayleigh wave group velocity.

lunar rocks measured in the laboratory and approaches the value found in synthetic analogs that are of similar composition but typically are very compact. The data in the direction perpendicular to the fractures do not give a straight line plot but show several steps in which the slope changes its value abruptly, reflecting the crossing of the receiving transducer over a fracture. A line drawn through the data points with the steps viewed as deviations from a mean gives a velocity of $v_R \simeq 1.1$ km/s. In the vicinity of the fractures, however, the velocity dips to values as low as $v_R \simeq 0.5$ km/s. The high value of $v_R \simeq 1.95$ km/s is, therefore, most likely closest to the intrinsic velocity value for this lunar breccia.

These results and others are summarized in Table 2. The Rayleigh wave velocity values cover a range of about one order of magnitude from $v_R \simeq 0.3$ km/s to $v_R \simeq 2.0$ km/s. These velocities have all been obtained at zero confining pressure and are therefore representative of the solid material in the upper few kilometers of the lunar surface. In view of this spread in velocities it is not surprising that seismic experiments show not only low velocities but also a high degree of scattering in this layer. SEM studies on these rocks reveal an apparent correlation between degree of fracture and compliance. While any quantitative relationship is difficult to determine at this time, there seems little doubt that the presence of microfractures created in an environment devoid of liquids and gases contributes greatly to increasing the compliance of the lunar rock concomitant with a decrease in velocity. Furthermore, the existence of a very low elastic wave velocity in these generally competent, ingeneous rocks indicates that the existence of a thick, low seismic velocity zone near the lunar surface does not necessarily require postulating the existence of a thick layer of dust accumulations in the lunar mare regions, as has been proposed by some.

Table 2

Laboratory Rayleigh Wave Velocity Data

Sample	Velocity (km/s)	Comments
Apollo 12038	0.97-1.45	Granular basalt[11]
Apollo 12063	0.94-1.59	Diabase[11]
Apollo 14310	1.20	Basalt[11]
Apollo 14321	0.90 ± 0.15	Complex breccia[11]
Apollo 15459	0.50 fractures 1.95 fractures	Breccia[11]
Apollo 15555	0.28-0.34	Highly fractured basalt[11]
Apollo 10046	0.7 (calc)	Micro breccia[13]
Synthetic rock	2.21-2.26	Analog of Apollo 10017 basalt[11]
Terrestrial rock	2.97-3.15	Olivine basalt with structure and minerology similar to lunar basalts[11]

EFFECT OF PORE VOLATILES ON ATTENUATION AND INTERNAL FRICTION IN COMMERCIAL CERAMICS

The attenuation of surface acoustic waves SAW by fluids, volatiles, or gas surrounding the surface of propagation is well known and has its source in frictional losses determined by the viscosity of the fluid or gas and in the emission of compressional waves into the environment.[14-17] These effects are especially pronounced at high frequencies, i.e., 1-3 GHz, and vacuum-tight encapsulation of the SAW devices is necessary to remove the extra attenuation. Studies[18] of the effects of environment on low frequency measurements and the operation of devices have revealed surprisingly substantial effects. Representative data for 2 MHz, 55 μs delay line constructed from high density lead-zirconate-titanate (PZT-8) are shown in Table 3. The total straight-through transmission loss was 25 dB in air so that the total changes of about 18 dB are a substantial contribution that cannot be accounted for by the above mentioned causes, which dimenish rapidly with decreasing frequency ($\sim \omega^x$, $1 < x < 3/2$). The presence of an additional loss mechanism is confirmed by internal friction measurements on vibrating bar samples resonating at about 10 kHz. These results are summarized in Table 4. Optical micrographs of polished surfaces of the high density PZT-8 revealed some microporosity (< 1%) with pore diameters in the 1 μm to 30 μm range. Apparently, volatiles trapped in the interconnected pore volume are responsible for the increases in attenuation and internal

Table 3

Relative Transmission Loss in 2 MHz PZT-8 Delay Line (55 μs delay)

Change in Transmission Loss (dB)	Outgassing Conditions
0 (reference)	After washing in ethyl alcohol and allowing to dry in atmosphere for 0.5 h at 25% humidity
-6	After gentle heating (≃ 40°C) for 12 h at 10^{-3} Torr
+10	After 3 h exposure to high-humidity atmosphere (mixture of steam and air) and some condensation on surface and then allowed to dry
-8	After 12 h evacuation at 10^{-2} Torr at 25°C

Table 4

Internal Friction Quality Factor Q and Resonance Frequency γ_{Res} for PZT-8

Q and γ_{Res} (Hz)	Outgassing Conditions
1070 (12795 Hz)	At 25°C in air (1 atm) after cleaning and drying
1425 (12890 Hz)	At 25°C in 10^{-4} after drying below 40°C
1730 (12804 Hz)	At 25°C in 10^{-8} Torr after long term (6 days) outgassing with one thermal treatment to 50°C

friction. Measurements on fully dense materials, both metals and ceramics, showed no such behavior.

INTERNAL FRICTION IN LUNAR RETURN SAMPLES

The measurements on the commercial ceramics and the observations on the effects of volatiles were triggered by earlier measurements on lunar return samples. Ever since the lunar seismic experiments[19] revealed Q values for the crust in the range of 3000-5000 in contrast to similar terrestrial experiments giving Q values in the 30-60 range, seismologists have been wondering about the cause for this contrast. It was suggested that laboratory-measured Q of rocks completely free of volatiles should be similar.[20] In vibrating bar resonance measurements[11] the quality factor Q (the reciprocal of the interval friction Q^{-1}) in a lunar basalt (14310) increased from about 70 at 1 atm (in laboratory air) top 800 at 10^{-8} Torr. This result was confirmed by several laboratories on different rock samples. The highest Q achieved in the laboratory was reported for a sample of lunar basalt[21] that had an initial Q of 60 and that was raised to a value of 4800 by repeating heating cycles under vacuum with pressures ranging up to 10^{-11} Torr. Data showing the systematic increase in Q with outgassing at intermediate states of the heat treatments and vacuum are shown in Table 5.

Table 5

Effects of Ultrahigh Vacuum on Q and Resonance Frequency of Lunar Rock 70215,85[22]

Pressure (Torr)	Exposure Time (h)	Q	Increase in Frequency (%)
1×10^{-7}	12	1851	0
1×10^{-8}	12	2381	0.06
1×10^{-10}	12	3330	0.21

This result and similar observations on terrestrial analogs led to the conclusion that only small amounts of volatiles are sufficient to lower Q dramatically. Measurements carried out at 50 Hz showed qualitatively similar increases in Q over the range of outgassing feasible with the low frequency apparatus.[23] Figure 11 shows an example of the signal decay at 56 Hz for an analog of lunar basalt after heating to 200°C for 48 h in a vacuum of 10^{-6} Torr. As shown in Figure 11, at room temperature, Q of about 110 was obtained as shown in Figure 11 in contrast to Q ≈ 50-60 when the sample was tested before outgassing. These results suggest that high Q values for volatile-free rock may be expected down to the seismic frequencies in which the lunar seismic experiments were carried out, supporting the postulation that the high lunar crustal Q values are consistent with rock devoid of volatiles, such as water, typically found on earth.

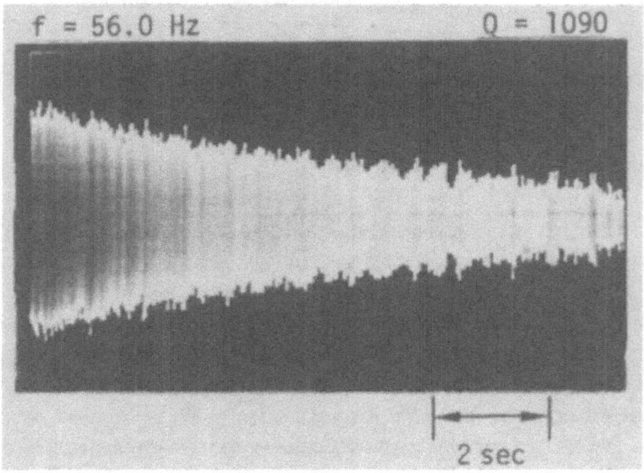

Figure 11 Oscilloscope trace of signal decay pattern with an outgassed olivine basalt samples.

Investigations into the mechanism for the effect of volatiles on the internal friction have led to some interesting results,[25] but a completely satisfactory explanation is not available at this time. Q_s^{-1} was measured for shear waves as a function of relative partial pressure P/P_o for benzine, hexane, ethanol, methanol and water. The measurements were carried out at about 10 kHz with the vibrating bar technique in a chamber in which P/P_o was varied between almost zero and about 0.9. As shown in Figure 12, the results revealed that in the regime of one- or two-monolayer coverage of adsorbed volatiles Q_s^{-1} increased dramatically with

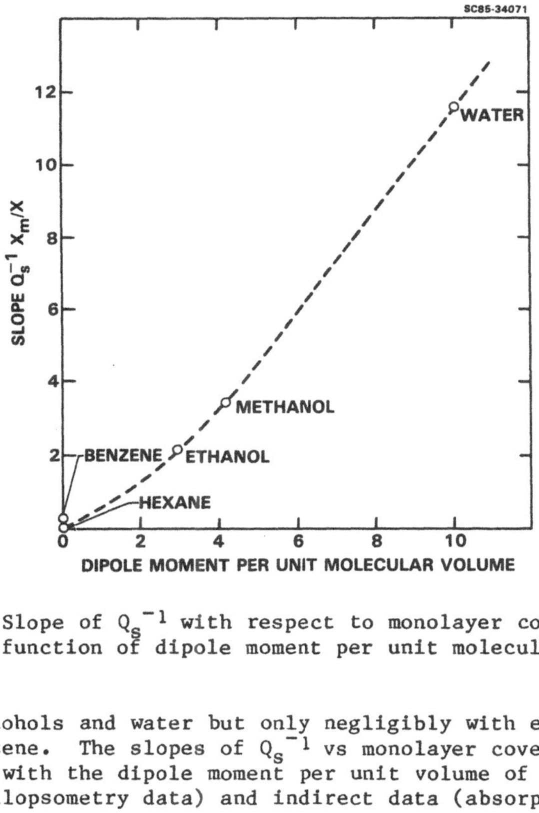

Figure 12 Slope of Q_s^{-1} with respect to monolayer coverage as a function of dipole moment per unit molecular volume.

exposure to alcohols and water but only negligibly with exposure to the hexane and benzene. The slopes of Q_s^{-1} vs monolayer coverage appeared to correlate well with the dipole moment per unit volume of the volatiles. Both direct (ellopsometry data) and indirect data (absorption isotherm data) point to the presence of thin films of adsorbed volatiles. On the basis of these observations it was postulated that the stresses associated with the passage of the elastic wave induce deformations and displacements involving the adsorbed layers of volatiles. The associated internal friction losses are significantly influenced by the surface chemistry and the nature of the volatiles.[24]

REFERENCES

1. C.F. Ying and R. Truell, "Scattering of a Plane Longitudinal Wave by a Spherical Obstacle in an Isotropically Elastic Solid," J. Appl. Phys. 27, 1086 (1956).
2. B.R. Tittmann, "Scattering of Elastic Waves From Simple Defects in Solids, A Review," Wave Motion 5, 299 (1983).
3. F. Cohen-Tenoudji, L. Ahlberg, B.R. Tittmann, J.L. Opsal, V.V. Varadan and G. Quentin, "The Role of Creep Rays in the Scattering from Spheroids Cavities and Inclusions in Solids," 1981 Ultrasonics Symposium Proceedings, IEEE, New York, 236 (1981).
4. K. Fertig, private communication.
5. A.G. Evans, B.R. Tittmann, L. Ahlberg, B.T. Khuri-Yacub, G.S. Kino, J. Appl. Phys. 49, 1669 (1978).
6. R. Truell, C. Elbaum, B. Chick, "Ultrasonic Methods in Solid State Physics," (Academic Press, New York, 1969), 53-121.
7. L. Adler, Proceedings of the Review of Progress in Quantitative Nondestructive Evaluation, 3A, edited by D.O. Thompson and D.E. Chimenti (Plenum Publishing) (1984).
8. J. Gubernatis and E. Domany
9. L. Ahlberg and B. Tittmann, private communication.
10. B.R. Tittmann, L.A. Ahlberg and K.W. Fertig, "Ultrasonic Microsructural Noise Parameters in a Powder Metal Alloy," Review of Progress in Quantitative Nondestructive Evaluation, 3A Edited by D.O. Thompson and D.E. Chimenti (Plenum Publishing), 57-63 (1984).
11. B.R. Tittmann, M. Abdel-Gawad and R.M. Housley, "Elastic Velocity and Q Factor Measurements on Apollo 12, 14 and 15 Rocks," Proc. 3rd Lunar Sci. Conf. Geochim. Cosmochim. Acta, Suppl. 2 MIT Press, 3, 2337-2343 (1972).
12. B.R. Tittmann, "A Technique for Precision Measurements of Elastic Surface Wave Properties on Arbitrary Materials," Rev. of Sci. Instr. 42, 1136-1142 (1971).
13. O.L. Anderson, C. Scholz, N. Soga, N. Warren and E. Schreiber, "Elastic Properties of a Micro-Breccia, Igneous Rock and Lunar Pines from Apollo 11 Mission," Prof. Apollo 11 Lunar Sci. Conf. Geochim. Cosmochim. Acta. Suppl. 1, Pergamon Press, 3, 1959-1973.
14. R.M. Arzt, E. Salzmann and K. Dransfeld, "Elastic Surface Waves in Quartz at 316 MHz," Appl. Phys. Lett. 10, 165-169 (1967).
15. A.J. Slobodnik, Jr., "Attenuation of Microwave Acoustic Surface Waves Due to Gas Loading," J. Appl. Phys. 43, 2565-2568 (1972).
16. K. Dransfeld and E. Salzmann, "High-Frequency Elastic Surface Waves," Physical Acoustics (W.P. Mason and R.N. Thurston, eds), Academic Press 8, 219-272 (1970).
17. B.A. Auld, "Acoustic Fields and Waves in Solids," 2, 297-300, Wiley (1973).
18. B.R. Tittmann, "Reduction of Losses in Low Frequency (2 MHz) Surface Wave Delay Line by Outgassing," Electron. Lett. 12, No. 9 (1976).
19. G.V. Latham, M. Ewing, J. Dorman, F. Press, N. Toksaz, G. Sutton, R. Meissner and M. Yates, "Seismic Data from Man-Made Impacts on the Moon," Science 170, 620-626 (1970).
20. B.I. Pandit and D.C. Tozer, "Anomalous Propagation of Elastic Energy Within the Moon," Nature 226, 335 (1970).
21. B.R. Tittmann, "Lunar Rock Q in 3000-5000 Range Achieved in Laboratory, " Philos. Trans. R. Soc. London, Ser. A. 285, 475-479 (1977).

22. B.R. Tittmann, L. Ahlberg, H. Nadler, J. Curnow, T. Smith and E.R. Cohen, "Internal Friction Quality Factor Q Under Confining Pressure," Proc. Lunar Sci. Conf. 8th, 1209-1224 (1977).
23. B.R. Tittmann, "Internal Friction Measurements and Their Implications in Seismic Q. Structure Models of the Crust," in the Structure and Physical Properties of the Earth, Geophys. Monogr. Ser. edited by J.G. Heaclck, AGU, Washington, D.C. 20, 197, 1978.
24. B.R. Tittmann, V.A. Clark, J.M. Richardson and T.W. Spencer, "Possible Mechanism for Seismic Attenuation in Rocks Containing Small Amounts of Volatiles," J. Geoph. Res. 85, 5199-5208 (1980).

DISCUSSION

Comment: Schmitt

I would like to add the following comments.

1. Effects of absorption on the acoustic response of rocks.

a) The activation energy for the process is 4 kcal/mole, which is equivalent to the breaking of two hydrogen bonds of the water molecule (Sponcer, 1982).

b) The peak of attenuation at one statistical monolayer is tied to the comment of Prof. Zarembowitch regarding a commensurate-incommensurate transition in absorbed layers.

c) The characteristic relaxation frequency for the attenuation is in the kHz range, which corresponds to a millisecond time scale. This would correspond to length scales of micrometers for a diffusion activated process:

$$\ell = \sqrt{6Dt},$$

where ℓ is the diffusion length and D the diffusion constant. These are the pore sizes of rocks so that we can see a strong suggestion of a diffusion limited process.

2. Utilization of NMR

From proton NMR of rocks, three important constants can be obtained.

a) Volume percent of surface adsorbed water and from this the surface area of the rock as well as surface to volume distribution of the pore space.

b) The rate of the exchange reactions between the surface adsorbed water and the bulk water, by looking at the temperature variation of the NMR.

c) Since proton NMR is only responsive to the hydrogen nuclei in the sample, it is possible to obtain sample porosity from the NMR signal intensity.

TISSUE CHARACTERIZATION BY LINEAR AND NON-LINEAR ULTRASOUND

R. SCHMITT
Fraunhofer Inst. für zerstörungsfreie Prüfverfahren
Saarbrücken, FRG

Tissue characterization currently pursued in ultrasounds in medicine is based upon attenuation (i.e. scattering and absorption) and velocity property of the tissue. Whereas absorption can be related to the molecural weight of macromolecules and to multiple relaxation processes, the scattering property is related to larger structures at the subcellular and cellular level.

Speed of sound in tissues depends strongly on the content of the main tissue component: water, collagen and fat. Speed of sound increases with increasing collagen tissue content and decreases with increasing water and fat content.

To map ultrasonic parameters quantitatively in cross section areas of the human body, ultrasonic computed tomography (UCT) is the most accurate and the most spatial resolving method. Unfurtunately, UCT is restricted to parts of the body having diameters less than 15 to 20 cm, because of the unsufficient dynamic range of the equipment currently available.

Current research focusses on the precise determination of attenuation from backscattered ultrasounds with conventional B-Mode Scanners. Different methods have been developed estimating the attenuation coefficient and attenuation as functions of frequency employing either the backscattered amplitude or the backscattered power spectrum. One approach developed for grain size estimates in NDT models the slope of attenuation versus tissue depth using backscattered amplitudes /1/. Another approach /2/ uses the echo amplitude from two C-planes at a certain depth separated by distance d.

From the log-difference of echo amplitude obtained and averaged over the two C-planes having a distance of at least a few centimeters, the attenuation can be calculated.

In a third approach /3/, cross section regions of interest are divided in segments. By taking the differences of the log of the power spectrum averaged over all possible Hamming weighted segments, the attenuation can be obtained as a function of the frequency.

The combination of both methods takes the power spectrum of segments located in two C-planes. By averaging power spectra in each C-plane separated by a distance d and by taking the log difference, the attenuation versus frequency will also be obtained.

In another procedure, the mean or the centeroid of the power spectrum is used as a spectral estimator.

Since higher frequency components of an ultrasound beam are usually more attenuated then lower ones, a shift of the mean or the centeroid of the power spectrum will be observed when an ultrasonic pulse is attenuated by tissue. The determination of this spectral down shift in small segments of an A-mode line, has also been used for estimating tissue attenuation /8/.

Since it is difficult to obtain reliable results from one A-Mode line, estimates over several adjacent scan lines have to be performed.

Homomorphic filtering /5/ seems to be the most sophisticated attenuation estimation developed so far. In this approach a single A-mode line is considered as the convolution of the incident waveform with reflectors spatially distributed along this particular A-mode line. Attenuation estimates are then obtained by employing several signal processing steps.

Although this method models more detailinly the physical situation of a pulse travelling through tissue, it has not reached its full potentiality.

Miller et al. /6/ have pushed successfully tissue characterization of the heart muscle by devoloping methods for monitoring the backscatter coefficient and the integrated backscatter by means of a substitution technique. They also found that the actual values of these parameters are depending on the contractile states of the muscle fiber. Thus the dynamic variation can be used to monitor the functional muscle performance. This dynamic variation resulted in a clear distinction between an ischemic and a normal heart tissue.

All the ultrasonic procedures mentioned above are relying upon a linear model: speed of sound is independent of pressure amplitude, which is an assumption we can consider true in medical ultrasound. Recently, it has been shown by several groups that this assumption does no longer hold if pressure amplitudes are raised to several atmospheres. It has been proved that the extent of non-linearity, summarized in the B/A parameters, is a useful parameter for tissue characterization.

1. Goebbels, K., Structure analysis by scattered ultrasound, Res. Techn. in NDT. Academic Press, Sharpe, 1980.
2. Ophir, I., Maklad, N.F., A new stochastic C-scan technique for attenuation coefficient measurements in tissue equivalent material. Proc. 23 rd. Ann. Mat. AIUM 1978.
3. Kuc, R., Schwartz, M., Estimating the acoustic attenuation coefficient slope for liver from reflected ultrasound signals. IEEE Trans. On Sonics And Ultrasonics SU-26 1979, S. 353-362.
4. Hutchins, D.A., Leemann, R., Pulse-estimation from medical ultrasound signals. Ultrasonics International 1981. Z. Novak. Ed. IPC-Press (1981), S. 427-433.
5. Jones, J.P., Gontaks, S., Kovack, A., Differentation of abdominal organs and pathologies by the analysis of A-mode ultrasound waveforms. IEEE Ultrasonic Symp. Proc. 1980.
6. Miller, J.G., Myocardial tissue characterization. IEEE Ultrasonic Symp. Proc. 1983 S. 782-793.
7. Zhu, Z., Roos, M., Cobb, W., Jensen, K., Determination of the acoustic nonlinearity parameter B/A from phase measurements. Journal of Acoust. Soc. of Ann. 74, S. 1518-1521.
8. Fink, M., Hottier, F., Ultrasonic signal processing for in vivo attenuation measurements short time Fourier analysis. Ultrasonic Imaging, 5, 1983, p. 117-135.

DISCUSSION

Comment: Adler
Is there any concern in the medical ultrasonic community about the fact that microcavitation bubbles have been found even when only very low intensity short pulses were used, typical for medical diagnostic application?

Reply: Schmitt
When the pulse length is reduced in order to increase axial resolution, the frequency content of that pulse is increased, too. Since power level for creating microbubbles decreases with increasing frequency, creating microbubbles is not only a question of power levels, but also of spectral content. The medical community, and the medical physicists as well, should be aware of that effect. Current dosimetric considerations, however, are related to the AIUM statement of 1978 on ultrasound safety.

Comment: Sliwinski
What is the relation between standard doses of ultrasound used in medicine and the non-linear effects that you mentioned? Are non-linear effects taken into account in establishing the standard, and, if so, how are they taken into consideration, if non-linear mechanisms of interaction in tissues are not yet clear enough?

Reply: Schmitt
The occurrence of non-linear effects depends on several factors, such as the intensity, the frequency, the pathlength in the medium, the beam geometry, the acoustic properties of the medium, etc. If the ultrasonic device produces levels below 100 mW/cm^2, one generally would not expect non-linear effects; however, it also depends upon individual circumstances.
Dose levels were established and set below intensities which have produced damages in a few experiments. Consequently, if non-linear effects are responsible for the damage observed, then they are included implicitly in dose considerations.

Comment: Chivers
I would like to make four points.
1. The concept of dose in connection with the potential hazardous effects of ultrasound may be a misleading one. It is derived from the analogy with ionising radiation where a distinction is made between "exposure" and "dose", the latter being the measure of energy absorbed to which harmful effects may be directly related. For ultrasound the distinction between exposure and dose is more critical since, while some of the acoustic energy is obsorbed and converted to heat, there are mechanical mechanisms for damage which are not primarily related to absorption - these used to be classified as cavitational and "non-thermal, non-cavitational".
2. There are two types of cavitation: "collapse" cavitation and the excitation of pre-existence bubbles. It is believed that we do not cause the former in tissue with the pulses

used in diagnosis and until recently it was believed that the latter would only be significant if the pulses were long enough for the microbubbles to grow by rectified diffusion and resonate. The recent work of Conn and others has shown that significant effects can be produced by pulses as short as those used in diagnostic ultrasound. The main scientific question that confronts us is whether or not suitable microbubbles exist in intact living tissues.
3. Regarding the question of non-linearity and potential damage mentioned by Prof. Sliwinski, it is probable that more energy will be absorbed in the non-linear region due to the energy transfer to higher harmonics and the increase of absorption with frequency. In addition, cavitation is itself a non-linear phenomenon and some coupling between the two processes may be expected.
4. Finally, it is not clear to me that we should be looking for an absolute threshold, below which exposure is "safe" and above which it is forbidden. Clinical practice with almost all investigative techniques is based on a balance between potential benefit to the patient and the potential hazard that may result. It is an assessment made by the clinician for each individual patient. Thus, many clinicians would, under certain circumstances, be prepared to increase the potential hazard if there would be a significant benefit for the patient in terms of more or better diagnostic information. The need is for an understanding of the mechanisms of damage and their relation to exposure criteria that are relatively easy to measure.

Comment: Cheeke
Regarding the threshold intensity for damage in tissue, could one link this to non-linear effects by postulating that it is due to cavitation caused by extreme non-linearity?

Reply: Schmitt
Cavitation is affected by different parameters such as frequency, temperature, stress amplitude, etc. Thus non-linearity certainly plays a role in the generation of cavitation. Whether non-linearity is the threshold for damage in tissues is still an unsolved problem, as fas as I know.

WOOD CHARACTERISATION THROUGH ULTRASONIC WAVES

V. BUCUR

1. INTRODUCTION

The Chartesian orthotropic model is used as a highly orthotropic model for wood structure to study the plane wave propagation and to characterize its elastic behaviour by nine constants, three Young's moduli, three shear moduli and three Poisson's ratios. The stiffness characteristics of wood are required to enable the calculation of these technical constants. The most rapid way to accede to stiffnesses is offered by the ultrasonic velocity method.

2. NOTATION

The following notations will be used throughout the present paper.

1,2,3,	principal anisotropic directions of wood L, R, T
Γ_{ij},	Christoffel's stiffness, a function of terms of matrix [C] and of components of unit wave normal $\vec{n}$
i,j,k,l,	used as indices: integer taking the values 1,2, or 3
C_{ijkl},	elastic stiffnesses
S_{ijkl},	elastic compliances
E,	Young's modulus
G,	shear modulus
ν,	Poisson's ratio
V,	phase velocity of plane elastic wave. Suffixed used to distinguish different velocities associated with direction cosines of normal vector
ρ,	density of medium
V^*_{QL}, V^*_{QT},	measured values of respectively quasi-longitudinal and quasi-transversal velocities
n_i,	components of unit displacement vector $\vec{n}$ (n_1, n_2, n_3)
p_k,	components of displacement vector $\vec{p}$ (p_1, p_2, p_3)
θ,	angle of wave vector orientation
δ,	displacement from wave normal

3. THEORETICAL CONSIDERATIONS

The relationships between the terms of the stiffness matrix and the ultrasonic velocities are given by Christoffel equations (1,2,3):

$$[\Gamma_{ij} - \rho V^2 \delta_{ik}]\ p_k = 0 \tag{1}$$

Equations (1) form a set of simultaneous equations in p_k and for a unique solution to those we have the condition:

$$[\Gamma_{ij} - \rho V^2 \delta_{ik}] = 0 \tag{2}$$

The components of the Christoffel tensor in terms of stiffness matrix for an orthotropic solid are given in (3).

3.1. Eigen values

If (1) is solved for wave propagation along the symmetry axes of an orthotropic solid, we get three solutions, the eigen values, which show that along every axis it is possible to have three types of waves: i.e. one longitudinal and two transversal. The difference between each type of wave is determined by the displacement vector components p_k.

Such solutions enable us to calculate the 6 diagonal terms of the stiffness matrix [C] by a relation presenting the general form:

$$C_{ii} = V^2 . \rho \tag{3}$$

The 3-off-diagonal stiffness components can be calculated using the general equation:

$$C_{ij} = \Gamma_{ij}/n_k.n_1 - C_{ii} \tag{4}$$

From Eq. (4) we calculate Γ_{12} or C_{12} such:

$$(C_{12}+C_{66}).n_1 n_2 = \pm[(C_{11} n_1^2 + C_{66} n_2^2 - \rho V^2)(C_{66} n_1^2 + C_{22} n_2^2 - \rho V^2)]^{0.5} \tag{5}$$

By permutations of indices we obtain the corresponding expression for C_{13} and C_{23}. If we admit that the matrix [C] > 0 and consequently $C_{ij} > 0$, then for:

$\vartheta \epsilon\ (0; \pi/2)$ or $\vartheta \epsilon\ (\pi; 3\pi/2)$	the expression under square root must be considered with the sign (+)
$\vartheta \epsilon\ (\pi/2; \pi)$ or $\vartheta \epsilon\ (3\pi/2; 2\pi)$	the same expression must be considered with the sign (-)

In the interest of clarity, we insist on the fact that for the calculation of the non-diagonal terms of the stiffness matrix we need the value of a quasi-longitudinal or a quasi-transverse wave, or both of them. It must also be noted that those values are dependent on the propagation vector and consequently on the orientation of the specimen (angle $\theta°$).

With this in mind and having all nine terms of the stiffness matrix, the calculation of elastic constants can easily be done by inverting the matrix to get compliance terms and then by calculating Young's moduli and Poisson's ratios with simple relations (4).

It is well know that the physical properties of wood are strongly dependent on orientation of reference coordinates or in other words dependent on the angle α mentioned before. This directional dependence of wood constants makes conventional averaging techniques inapplicable to redundant measurements. For this reason we provide herein a rational procedure for data averaging of directionally dependent measurements. Our optimisation procedure is based on the assumption that both quasi-longitudinal and quasi-transversal velocity values of redundant measurements could be used for non-diagonal stiffness term calculation (5). The relationships between those velocities and the diagonal terms of the stiffness matrix can be deduced from Eq. (2).

$$(\rho V^2)^2 - \rho V^2 (C_{11} n_1^2 + C_{22} n_2^2 + C_{66}) + (C_{11} n_1^2 + C_{66} n_2^2)(C_{66} n_1^2 + C_{22} n_2^2) - (C_{12} + C_{66})^2 n_1^2 n_2^2 = 0 \qquad (6)$$

There are several points of interest relating the roots V_{QL} and V_{QT} and the coefficients of equation (6) such that:

$$\rho V_{QL}^2 + \rho V_{QT}^2 = C_{11} n_1^2 + C_{22} n_2^2 + C_{66} \qquad (7)$$

or

$$V_{QL}^2 + V_{QT}^2 = V_{11}^2 n_1^2 + V_{22}^2 n_2^2 + V_{66}^2 \qquad (7')$$

and

$$\rho V_{QL}^2 \cdot \rho V_{QT}^2 = (C_{11} n_1^2 + C_{66} n_2^2) \cdot (C_{66} n_1^2 + C_{22} n_2^2) - (C_{12} + C_{66})^2 n_1^2 n_2^2 \qquad (8)$$

which gives:

$$C_{12}/\rho = n_1^{-1} n_2^{-1} [(V_{11}^2 n_1^2 + V_{66}^2 n_2^2)(V_{66}^2 n_1^2 + V_{22}^2 n_2^2) - V_{QL}^2 \cdot V_{QT}^2]^{0.5} - V_{66}^2 \qquad (9)$$

Experimental measurements of V_{QL}^* and V_{QT}^* does not satisfy Eq. (7). From this complementary condition we can deduce the correction factor k which is:

$$k = (V_{11}^2 n_1^2 + V_{22}^2 n_2^2 + V_{66}^2) / (V_{QL}^{*2} + V^{*2}_{QT}) \qquad (10)$$

This correction must be then introduced into Eq. (9) for the term:

$$V_{QL}^2 \cdot V_{QT}^2 = k^2 \, V_{QL}^{*2} \cdot V_{QT}^{*2} \qquad (11)$$

corresponding to the smallest value deduced for all experimentally redundant measurements determined on separate specimens cut at convenient angle to the principal directions. This optimisation leads to a new matrix [C]. These results are of considerable interest; first of all for the calculation of the characteristic surfaces (velocity, slowness) and secondly for the calculation of technical constants of wood. The velocity surface is known to be a surface of three sheets, deduced from the eigen values of Eq. (2) as follow:

$$2\rho V_{QL}^2 = (\Gamma_{11} + \Gamma_{22}) + [(\Gamma_{11} - \Gamma_{22})^2 + 4\Gamma_{12}^2]^{0.5} \quad (12)$$

$$2\rho V_{QT}^2 = (\Gamma_{11} + \Gamma_{22}) - [(\Gamma_{11} - \Gamma_{22})^2 + 4\Gamma_{12}^2]^{0.5} \quad (13)$$

$$\rho V_T^2 = C_{55} \cdot n_1^2 + C_{44} \cdot n_2^2 \quad (14)$$

The slownesses are inversely related to the velocity of propagation.

For the calculation of technical constants, the stiffness [C] measured by ultrasonic methods must be converted to compliance [S] matrix. This conversion is simple because $[C]^{-1} = [S]$ and the elastic moduli are the reciprocal of the compliance coefficients of the principal diagonal of [S], such as $E_1 = 1/S_{11}$. etc...

Wood material exhibits six Poisson's coefficients defined in the form $-\nu_{ij} = S_{ij}/S_{ii}$ or in the form $-\nu_{ij} = S_{ij}/S_{jj}$. In this report we choose the first definition because of the necessity of comparing the data with the elastical data of Hearmon (6).

3.2. Eigen vectors

From Eq. (1) we can get the eigen vectors in a very simple way if two off-diagonal components of the tensor are zero. The linear equations for the displacements p_k (p_1, p_2, 0) associated with the quadratic factor of Eq. (2) for $\vec{n}(n_1, n_2, 0)$ are:

$$\begin{aligned}(\Gamma_{11} - \rho V^2)p_1 + \Gamma_{12}p_2 &= 0\\ \Gamma_{12}p_1 + (\Gamma_{22} - \rho V^2)p_2 &= 0\end{aligned} \quad (15)$$

whence $p_1/p_2 = \Gamma_{12} / (\rho V^2 - \Gamma_{11}) = (\rho V^2 - \Gamma_{22}) / \Gamma_{12}$ (16)

If we let the polarization correspond to the same sign i.e. : $p_1 = \sin\delta$ and $p_2 = \cos\delta$, we shall obtain:

$$\mathrm{tg}\delta = \Gamma_{12}/(\rho V^2 - \Gamma_{11}) \quad (17)$$

and $p_2 = \cos\delta = (\rho V^2 - \Gamma_{11}) / [(\rho V^2 - \Gamma_{11})^2 + \Gamma_{12}^2]^{0.5}$ (18)

From Eq. (18) the particle velocity polarisation can be calculated in terms of phase velocity V, propagation direction and stiffness constants of the solid for each plane of symmetry.

Now let $n_1 : n_2 = \cos\theta : \sin\theta$ and let the polarisation correspond to the alternative sign in equation (16):

$$p_1 : p_2 = -\sin\delta : \cos\delta \quad (19)$$

It is readily shown that:

$$\mathrm{tg}\delta = \Gamma_{12} \ / (\Gamma_{11} - \rho V^2) = (\Gamma_{22} - \rho V^2) / \Gamma_{12} \tag{20}$$

We deduce from Eq. (17) and for θ (0... $\pi/2$) that:

1) When $\Gamma_{12} / (\rho V^2 - \Gamma_{11}) > 0$ and $\Gamma_{12} > 0$ and $\rho V^2 - \Gamma_{11} > 0$, tg$\delta$ may take values ranging from 0 ... $+\infty$

2) When $\Gamma_{12} / (\rho V^2 - \Gamma_{11}) < 0$ and $\Gamma_{12} > 0$ and $\rho V^2 - \Gamma_{11} < 0$, or $\Gamma_{12} < 0$ and $\rho V^2 - \Gamma_{11} < 0$, tg$\delta$ may take walues ranging from $-\infty$... 0.

3) When $\Gamma_{12} / (\rho V^2 - \Gamma_{11}) = 1$ or $\Gamma_{12} + \Gamma_{11} = \rho V^2$, tg$\delta = 1$ and $\delta = \pi/4$ or $3\pi/4$.

4) When $\theta = 0$, $\Gamma_{12} = 0$, tg$\delta =$ 0 and the polarisation is longitudinal.

5) When $\theta = \pi/2$; $\Gamma_{12} = 0$, tg$\delta =$ 0 and the polarisation is transversal.

6) When $\theta = \pi/4$, then $\Gamma_{12} + \Gamma_{11} > 2\rho V^2$.

The analysis or waves of $\vec{n}(0, n_2, n_3)$ or $\vec{n}(n_1, 0, n_3)$ can be carried out in a manner strictly analogous to the foregoing.

As far as we know data on eigen vectors related to wood were reported only by Hearmon (7) in his pioneering paper on ultrasonic methods used for the characterization of this natural fibrous material.

4. MATERIAL AND METHOD

Measurements of ultrasonic velocities were carried out on cubic specimens of some species (spruce - Picea rubens Sarg., curly maple - Acer rubrum L., and poplar wood - Populus sp.) using a classical transmitting pulse technique using a buffer block. The ultrasonic pulse was generated by an ultrasonic analyser 5052 U.A. Panametrics. The receiver and transmitter piezo-electric ceramic transducers were identical (1 MHz). Panametrics transducers V 103 and V 156 were used to propagate longitudinal and shear waves. The accuracy of time measurement was 0.01 μs. Measurements were performed on 16 mm cubes. This size gives a good approximation of cartesian orthotropie, so that the curvature of annual rings may be neglected. For the three species mentioned before, the cubes were cut at 0°, 15°, 30°, 45°, with respect to the main orthotropic axis in 3 planes. The specimens were conditioned at a 12% moisture content.

In the second part of this article some practical applications of ultrasonic velocity method on standing trees are presented.

The velocity of propagation of surface waves on douglas tree circumferences in the longitudinal anisotropic direction was measured using an AU 80 ultrasonic device from Sattec (France).

5. RESULTS

5.1. Elastic constants on cubic specimens

Tables 1 and 2 summarize some values of ultrasonic velocities on spruce, maple and poplar wood. It is to be noted that the maximum velocity was measured on spruce along its longitudinal axis (V_{LL} = 5588 m/s). The minimum velocity was measured on poplar in the transversal plane (RT) using shear waves (V_{TR} = 642 m/s).

Furthermore, the terms of the stiffness matrix were calculated using the experimental values of velocity. Table 3 gives the diagonal stiffness. As it was shown in a previous paper (8) the ordering of principal stiffness of extension C_{11}, C_{22}, C_{33} and of shear C_{44}, C_{55}, C_{66} for maple and poplar satisfies the inequality $C_{11} > C_{22} > C_{33} > C_{66} > C_{55} > C_{44} > 0$, but we must set aside the special case of spruce (in this table resonance spruce) for which the extensional term C_{33} is to be inserted between shear terms.

TABLE 1 : Velocities (m/s) on principal axes of wood

	Density	Velocities (m/s)					
		longitudinal waves			transversal waves		
	kg/m3	V_{LL}	V_{RR}	V_{TT}	V_{TR}	V_{LT}	V_{RL}
Spruce	500	5588	2192	1557	1978	1295	1301
Poplar	326	5074	2201	1211	642	1250	1536
Maple	561	3995	2517	1900	727	1466	1792

TABLE 2 : Velocities (m/s) out of principal axes of wood

	Quasi-longitudinal waves						Quasi-transversal waves					
	SPRUCE			MAPLE			SPRUCE			MAPLE		
Angle	LR	RT	LT	LR	RT	LT	LR	RT	LT	LR	RT	LT
15°	5588	1890	5158	3785	2348	3745	1252	1236	1242	1397	1703	1520
30°	4524	1691	3004	3620	2289	3303	1124	1197	1160	1297	1621	1516
45°	3104	1403	2374	3075	2104	2419	1473	1108	1326	1064	1577	1027
60°	2477	1579	1968	2815	1933	1820	1532	1157	1383	1536	1482	818
75°	2199	1469	2004	2658	1790	2282	1156	1139	1517	1679	1443	941

TABLE 3 : Diagonal stiffness terms (x 10^8 N/m^2)

	Stiffness terms						Inequality
	C_{11}	C_{22}	C_{33}	C_{44}	C_{55}	C_{66}	
Spruce	179.76	24.02	12.12	19.56	8.39	8,46	$C_{11}>C_{22}>C_{44}>C_{33}>C_{66}>C_{55}$
Poplar	83.93	15.79	4.79	1.34	5.09	7.68	$C_{11}>C_{22}>C_{33}>C_{66}>C_{55}>C_{44}$
Maple	89.53	35.54	20.25	2.97	12.06	18.02	$C_{11}>C_{22}>C_{33}>C_{66}>C_{55}>C_{44}$

This is perhaps due to the particular anatomical structure, having very fine annual rings of 1.5 mm width and an important proportion of late wood.

Table 4 gives the off-diagonal term C_{12} calculated for several angles on poplar wood for quasi-longitudinal velocity, quasi-transversal velocity and for both of them using the procedure mentioned in chapter 2. For each angle the average value was calculated. The minimum average value was considered the optimum value of C_{12} for the new optimized matrix stiffness, which allows us to deduce technical constants. Data about the optimum values of C_{13} and C_{23} appear in table 5. With optimized values of the off-diagonal terms of the stiffness matrix the slowness curve may be calculated (figure 1). A good agreement between ultrasonic experimental values and theoretical optimized slowness is to be noted. For more than 60 years a considerable number of tables on wood properties have been published. These deal with the determination of mechanical properties of this material using static methods. However, very few data concern ultrasonic measurements.

For the species analyzed here, no data are available in the literature. Faced with this situation, we accepted to check the validity of the procedure to calculate velocity or slowness curves by fitting optimized values for [C] in Eqs. (12,13) and to compare those graphs with experimental values. A good agreement between theoretical and experimental values could be considered a test of validity of proposal optimization procedure.

Table 6 gives the technical constants of wood. Our results are compared with statical results of some similar, but not identical species, reported by Hearmon (6) and Bodig, Goodman (9). The Young's moduli and the shear moduli are in the same range as indicated in the literature, but there are important differences in Poisson's ratios.

We have seen that an important aspect of wood characterization from a mechanical point of view is the analysis of experimental measurement. Variations in those data are due to two principal sources; the instrumentation accuracy and resolution as well as the inherent sample to material variability.

We note on one hand that error measurements estimated by the procedure of the most conservative error prediction, using the formulae of standard error calculation with partial derivatives, are less then 1 p.c. on ultrasonic velocity and less then 4 p.c. on diago-

a) MAPLE. Rotation T in LR plane

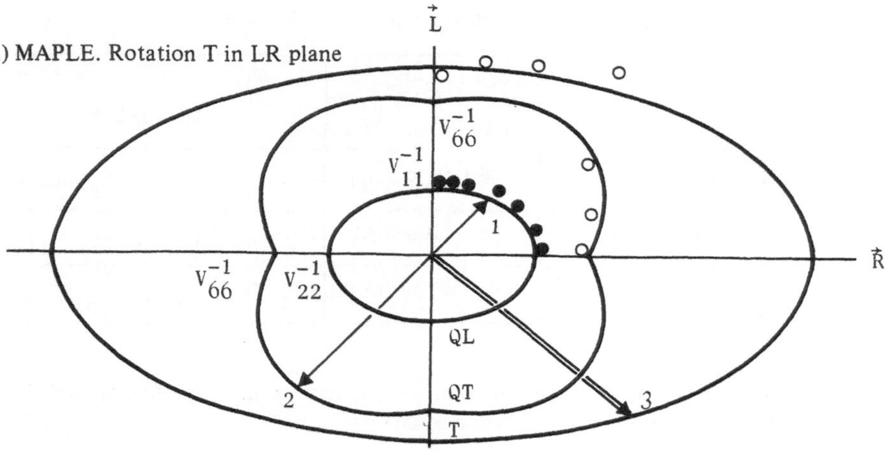

1. $V^{-1}_{QL} = [4\rho / (C_{11} + C_{22} + 2C_{66} - \sqrt{(C_{11} + C_{22})^2 + 4C_{12}^2})]^{0.5}$

2. $V^{-1}_{QT} = [4\rho / (C_{11} + C_{22} + 2C_{66} - \sqrt{(C_{11} + C_{22})^2 + 4C_{12}^2})]^{0.5}$

3. $V^{-1}_{T} = [2\rho / (C_{44} + C_{55})]^{0.5}$

FIGURE 1: Slowness surface in maple and spruce.

b) SPRUCE. Rotation R in LT plane

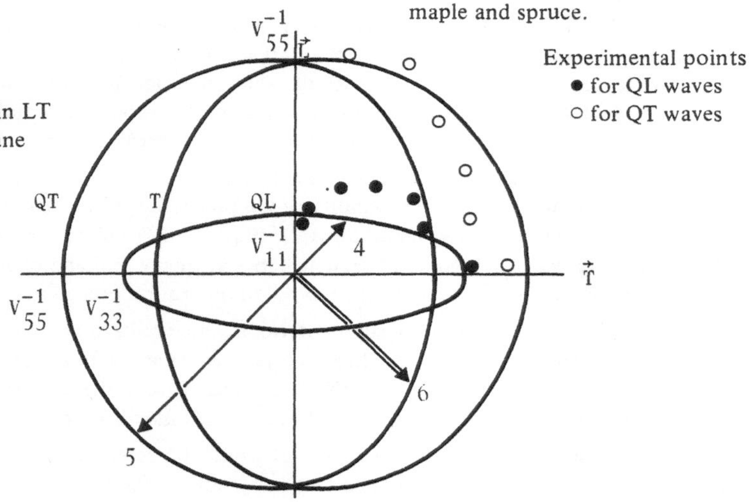

4. $V^{-1}_{QL} = [4\rho / (C_{11} + C_{33} + 2C_{55} + \sqrt{(C_{11} + C_{33})^2 + 4C_{13}^2})]^{0.5}$

5. $V^{-1}_{QT} = [4\rho / (C_{11} + C_{33} + 2C_{55} - \sqrt{(C_{11} + C_{33})^2 + 4C_{13}^2})]^{0.5}$

6. $V^{-1}_{T} = [2\rho / (C_{44} + C_{66})]^{0.5}$

nal terms of the stiffness matrix. For off-diagonal terms the error is very important. The amount of variation can range from 20 p.c. to 100 p.c..

On the other hand the experimental error measurement on velocity and diagonal terms of the stiffness matrix is inferior to the inherent sample-to-sample material variability expressed by the coefficient of variation (ratio between standard deviation and average value), which is less than 20 p.c.

Because of the fact that physical properties of wood are strongly dependent on the orientation of the reference coordinates, it may be of interest to investigate several sets of technical constants deduced from one or another optimization procedure of the non-diagonal terms of the stiffness matrix .

TABLE 4 : C_{12} on poplar wood for several angles

Angle	C_{12} (10^8 N/m2) calculated from			
	quasi-longitudinal waves	quasi-transversal waves	both waves	average
15°	complex	20.80	45.64	33.22
30°	6,53	23.96	37.43	22.64
45°	complex	27.75	35.91	31.83
60°	complex	19.48	30.33	24.91
75°	3.41	16.36	41.05	20.27

C_{12} optim = 20.27 x 10^8 N/m²

TABLE 5 : Optimum off-diagonal terms

	Stiffness matrix 10^8 N/m2			Compliance matrix 10^{-11} m2/N		
	C_{12}	C_{13}	C_{23}	S_{12}	S_{13}	S_{23}
Spruce	34.66	19.73	14.42	-12,59	2.40	-177.26
Poplar	20.27	8.41	6,03	-16,09	-18.69	-136.92
Maple	31.97	21.48	13,75	-12.01	-11.05	-18.57

TABLE 6 : Technical constants

	Young' moduli			Poisson's ratios			Reference
	E_L	E_R	E_T	ν_{RT}	ν_{RL}	ν_{LT}	
Spruce	129,41	6,01	3.45	0,61	0,07	-0.31	this article
Poplar	57.73	6.84	2.47	0.94	0.11	1,07	
Maple	55.21	21,67	13,56	0.40	0,26	0,61	
Poplar	83,24	6.11	2,44	-	-	-	BODIG, 1972
Spruce	166.00	8,50	6,90	0.43	0,02	0.52	HEARMON, 1948

Table 7 gives results concerning the technical constants calculated for two hypotheses: for the first hypothesis we accept the minimum value for C_{ij} calculated from quasi-longitudinal and quasi-transversal velocities, Eqs (12,13). For the second hypothesis we use the procedure described before for both of them. The choice of one set of technical constants was made by selecting those values corresponding to the highest E_L modulus. The argument in favour of this depends on previous measurements of longitudinal wave propagation in cylindrical bars on the same species, using the same ultrasonic equipment. We found that the E_L on cylindrical bars was considerably higher than the E_L in the present experiments on cubes. The realistic values of Poisson's ratios are also to be considered.

Those considerations obviously do not cover all aspects concerning the calculation of technical constants from ultrasonic measurements, but it seems that for each species a suitable combination of theoretical and experimental data may allow the determination of 9 elastic constants. The corresponding stiffness terms should be used furthermore for velocity or slowness surface calculations as well as for studies on the relation between propagation direction and polarization of waves.

5.2. The eigenvectors

We now proceed to the study of real eigen vectors (p_k) corresponding to the real values of velocities. For this calculation we used Eq. (18) and the polarizations corresponding to the same sign. The off diagonal terms Γ_{ij} were calculated from shear stiffness C_{44}, C_{55}, or C_{66} and optimized C_{ij}, which give us real components of displacement.

The dependence of $\delta(\theta)$, the deviation of QL and QT displacement from wave normal in three wood species is plotted in figure 2. The displacement angle δ is shown to be positive in the same sense as $\theta°$. The results obtained indicate how displacements excited in the interior of wood will be propagated within the medium. It is obviously of considerable interest to note the particular geometrical features of QL and QT curves in various planes. Spruce has a particular behaviour and we did not observe an intersection in the RT plane.

TABLE 7 : Technical constants calculated in two hypothesis
1) minimum value for all C_{ij}; 2) minimum average for C_{ij}

		Units	Acer 1)	Acer 2)	Poplar 1)	Poplar 2)	Spruce 1)	Spruce 2)
Stiffness	C_{12}	10^8 N/m2	4.51	41.95	16.40	20.27	24.73	34.66
	C_{13}	"	7.19	21.48	4.64	8,41	19.73	19.73
	C_{23}	"	12,37	13.75	3,38	6,03	3.75	14.42
Compliances	S_{12}	10^{-11} m2/N	-48.43	-27.36	-14.71	-16.09	- 6,15	-12,59
	S_{13}	"	- 4.05	- 9.22	- 4,17	-18,69	-10,38	+ 2,40
	S_{23}	"	-21.81	-16,29	-48.55	-136,92	- 5.07	-177,26
Youngs'moduli	E_L	10^8 N/m2	86,97	38.15	66,58	57,73	132.46	129,41
	E_R	"	27.98	14.98	11,23	6,84	20,51	6,01
	E_T	"	15.58	14.24	4.05	2,47	9,90	3.45
Poisson's ratios	ν_{LR}	-	*	1.04	0.16	0,12	0.13	*
	ν_{RL}	-	*	0,41	0.98	1,04	0.81	*
	ν_{LT}	-	0,06	0,13	0,02	0,05	0,10	-0,01
	ν_{TL}	-	0.35	0,31	0,28	1,07	1,37	-0,31
	ν_{RT}	-	0,34	0,23	0,19	0.33	0.05	0.61
	ν_{TR}	-	0.61	0,24	0,55	0,94	0,10	1.06

*been very erratic, the value was rejected because $1 - \nu_{ij}.\nu_{ji} > 0$

TABLE 8 : Wood anisotropy and ultrasonic velocity

	Velocities ratios		Anisotropy V_{QT}/V_T in					
	$V_{LL}:V_{RR}:V_{TT}$	$V_{LR}:V_{LT}:V_{RT}$	LR		RT		LT	
			0°	90°	0°	90°	0°	90°
Spruce	1 :2.5:3.6	1 ; 1 ;0,65	1,00	0.66	1.52	1.53	0.99	0.66
Maple	1 :1,6:2.1	1 ;1.2:2.5	1,22	2,47	0,57	0.49	0.82	2.01
Poplar	1 ;2.3:2.4	1 ;1.2:2.4	1,23	2,39	0,42	0,51	0,81	1,95

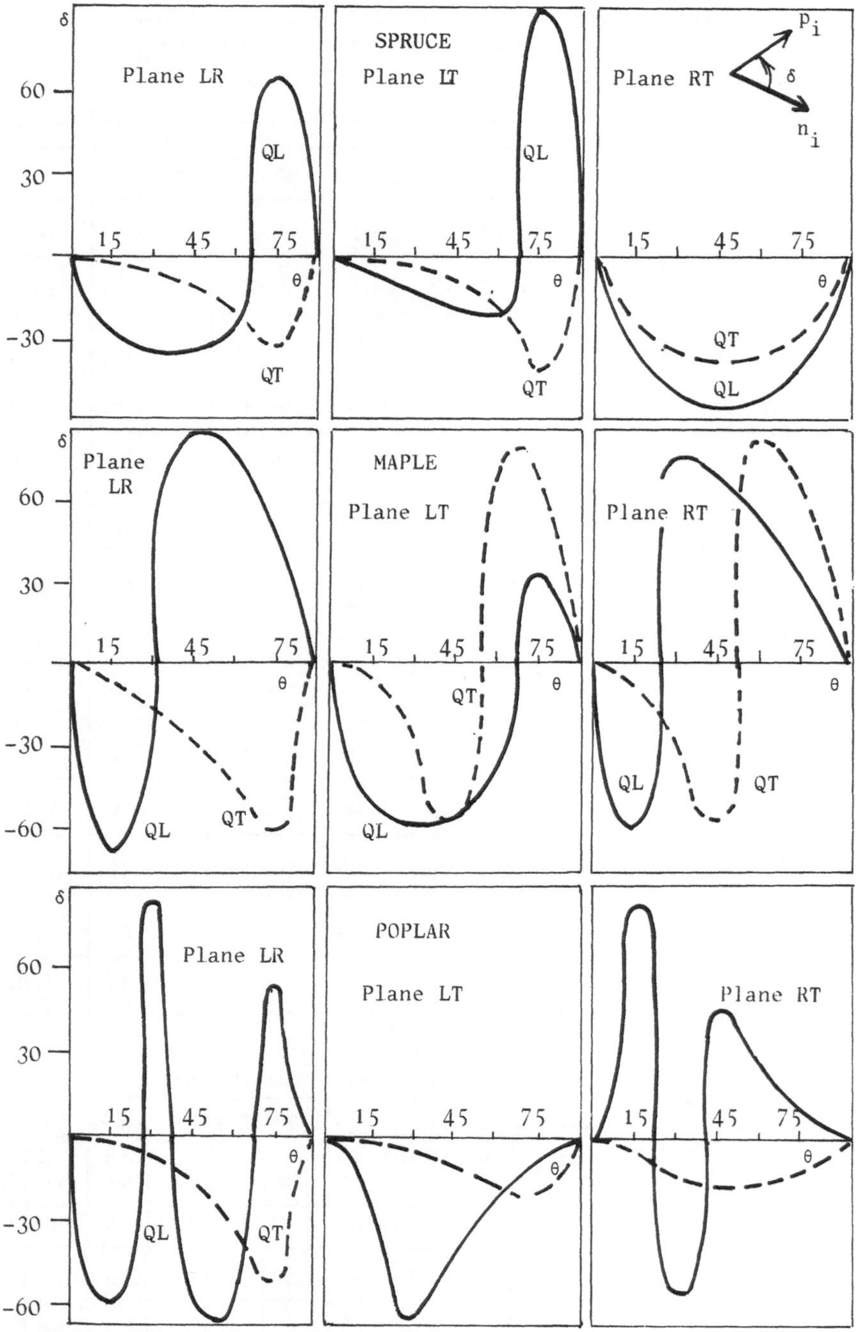

FIGURE 2: Dependence of $\delta(\theta)$, the deviation of QL and QT displacement from wave normal

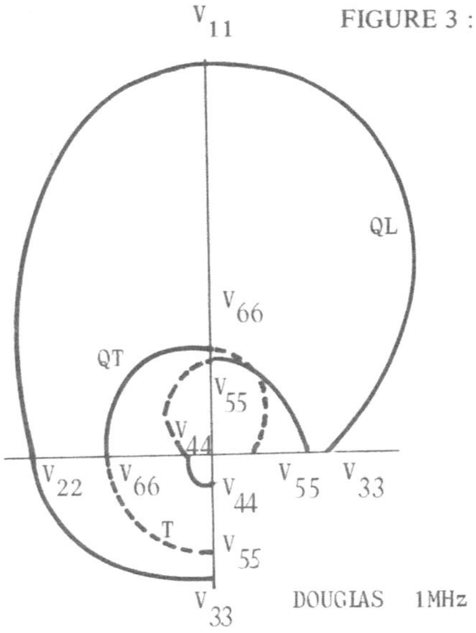

FIGURE 3 : Velocity surface on Douglas

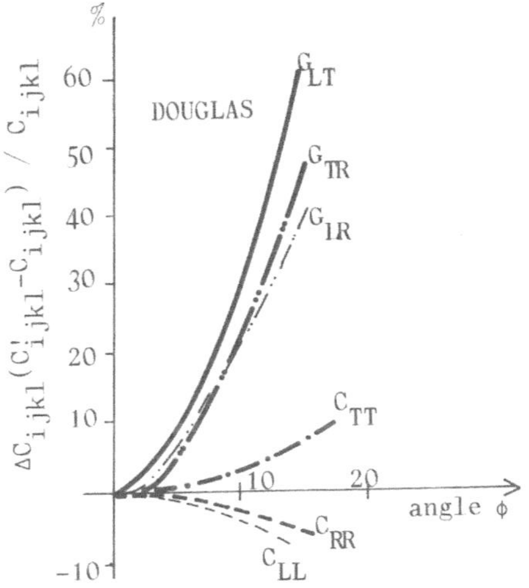

FIGURE 4 : Stiffness and grain angle

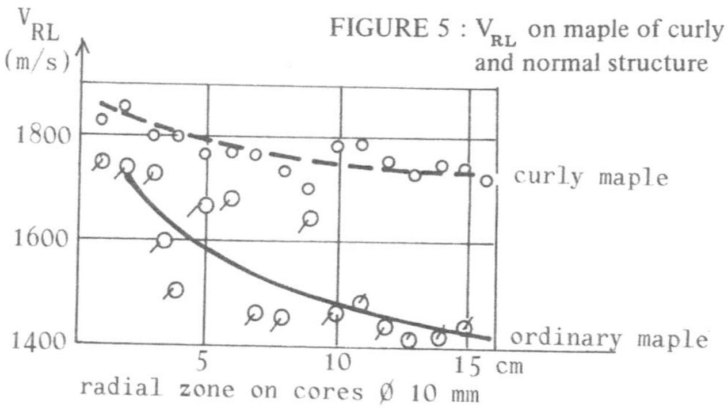

FIGURE 5 : V_{RL} on maple of curly and normal structure

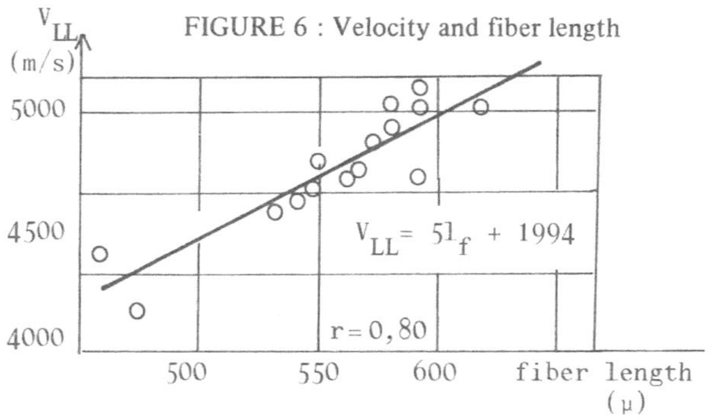

FIGURE 6 : Velocity and fiber length

The existence of similarly shaped dependence $\delta(\theta)$ has been established from ultrasonic measurements for crystals and composite materials. It is appropriate to observe that these results on wood reported here are primarily of mathematical significance in relation to the methodology adopted for the calculation of optimum C_{ij}, since the accuracy with which any propagation direction in wood can be uniquely established is not so high as for crystals.

It is also apparent from the intersections of the inner slowness sections of spruce that the classification of waves in quasi-longitudinal and transversal becomes meaningless and we must be careful in identifying the displacement vectors associated with different regions of the slowness sheets.

In subsequent papers it is hoped to examine the wave surface for various wood species and also to study the reflection - refraction and boundary problems which are highly dependent on the hypothesis assumed and partially verified here, because the particular shape of the slowness surface for spruce does not occur for all woods.

5.3. Wood anisotropy

Figure 3 shows the velocity for infinite plane waves deduced from Eq. (2) for douglas fir.

The anisotropy of wood can be expressed as the ratio between quasi-transversal and pure transversal waves in principal axes (Table 8).

In addition to velocity ratioes, the velocity surface can give a graphical representation of wood anisotropy. For elastically isotropic solids, the theory predicts that the velocity surfaces are concentrical where T and QT coincide and cannot intersect the QL surface. For anisotropic materials these surfaces can intersect. In wood the most common intersection occurs between T and QT surfaces (8) in the LT plane. The angle is about 56° for European softwoods and varies between 47° and 64° for hardwoods.

For species studies in this report the intersection angle between T and QT curves is given in table 9.

5.4. Some practical applications in forestry science

Some practical application of ultrasonic velocity methods used to study wood quality on non-destructive specimens like increment cores bored from living trees, are presented here. It seemed appropriate to show that wood quality can be discriminated by factors depending on its macroscopic structure (grain angle, curly structure, sapwood-heartwood, adult-juvenil wood), its microscopic structure as a tension wood or fiber length and on the other hand by factors related to forest treatment, i.e. pruning. Measurements on standing trees are also reported.

5.4.1. Relationships between wood grain angle and ultrasonic velocity

It is obvious that grain angle is an important factor affecting wood stiffness. In order to calculate the influence of the deviation of grain direction from theoretical symmetry directions, we propose a simulation leading to a calculation of the rotation of the axes. The grain deviation ranges from 0° to 15°.

It is well know that the values of stiffness terms depend on the orthotropic orthogonal references axes. If these references change, the stiffness matrix must be calculated using composite equations for transforming stiffnesses (10). For a rotation in 12 plane from 1 to 2 around the axis 3, through an angle ϕ, the equations relating stiffnesses in old C_{ijkl} and new C'_{ijkl} references axes are the following:

$$C'_{11} = \cos^4\phi C_{11} + \cos^2\phi\sin^2\phi\, C_{12} + \sin^4\phi\, C_{22} + 4\cos^2\phi\sin^2\phi\, C_{66}$$

$$C'_{22} = \sin^4\phi C_{11} + 2\cos^2\phi\sin^2\phi\, C_{12} + \cos^4\phi\, C_{22} + 4\cos^2\phi\sin^2\phi\, C_{66} \quad (21)$$

$$C'_{66} = \cos^2\phi\sin^2\phi C_{11} - 2\cos^2\phi\sin^2\phi C_{12} + \cos^2\phi\sin^2\phi C_{22} + (\cos^2\phi - \sin^2\phi)^2\ C_{66}$$

The expression of C_{16} and C_{26} are small enough to be neglected here.

The difference $\Delta C_{ijkl} = (C'_{ijkl} - C_{ijkl}) / C_{ijkl}$ between stiffness in the old and new reference axes versus the rotation angle of wood specimens is given in figure 4. If $C_{ijkl} > C'_{ijkl}$ then $\Delta C_{ijkl} < 0$. It can be observed that the curve of ΔC_{ijkl} (for C_{LL}, C_{RR}) corresponds inversely with an increase of the rotation angle.

If $C_{ijkl} < C'_{ijkl}$ then $\Delta C_{ijkl} > 0$ and the curve ΔC_{ijkl} for C_{TT}, G_{LR}, G_{LT}, G_{TR} increases, and increase of the rotation angle occurds.

The most important variation is given by the shear moduli G_{LT}. In douglas-fir a deviation of 5° in the rotation angle may induce a 8 p.c. increase in ΔC_{ijkl} and a 15 p.c. deviation in rotation angle may cause a 62 p.c. increase in ΔC_{ijkl}. In conclusion, we can say that the ultrasonic technique demonstrates that there is a close relationship between velocity and the grain angle of a wood specimen. The stiffness terms which are the most affected by the variation of the grain angle are the shear moduli.

5.4.2. Ultrasonic velocities on wood of particular structure

The experimental values (fig. 5) of ultrasonic velocity V_{LR} show some differences

TABLE 9 : Intersection between QL, QT and T in 3 anisotropic planes

Species	Spruce				Maple	Poplar		
Plane	LR	RT		LT	LT	RT	LT	
Surface	QT, T	QL, T	QT, T	QL, T	QT, T	QL, T	QL, T	QT, T
Angle	6°	87°	23 et 78°	84°	51°	85°	89°	55°

N.B. : the angle is measured versus the first axis of the plane

TABLE 10 : Transversal waves velocities on sapwood and heartwood of douglas

		Velocities of transversal waves					
		V_{LR}		V_{LT}		V_{TR}	
		average	c.v.	average	c.v.	average	c.v.
Pruned tree	S	1405	0.11	1342	0.09	519	0.17
	H	1620	0.10	1448	0.10	615	0,18
	%	13.27	-	7.32	-	15.60	-
Control tree	S	1471	0.10	1430	0.10	508	0.13
	H	1623	0.08	1524	0.08	580	0.21
	%	9.36	-	6.17	-	12.41	-

N.B. : S = sapwood H = heartwood c.v. = coefficient of variation

TABLE 11 : Ultrasonic velocity on wood of particular structure

	Structure	Ultrasonic velocities m/s					Observation
		V_{LL}	V_{TT}	V_{LR}	V_{LT}	V_{TR}	
Sitka spruce	adult wood	4928	1677	1605	1450	592	32 trees
	juvenile wood	4894	1758	1535	1448	521	32 trees
Beech	tension wood	4404	1567	1450	1190	838	5 specimens
	opposit wood	3977	1871	1434	1180	964	5 specimens

between curly and normal structure of maple.

In table 10 the transversal velocities measured on sapwood and heartwood of a pruned and a control douglas tree show the most discriminating velocity is V_{TR}.

In table 11 the velocities on adult and juvenile wood of Sitka spruce as well as those of tension and opposite beech wood are analysed. Statistical tests show that significant differences (at 1 p.c.) exist between V_{TT} or V_{TR} on adult and juvenile wood of sitka spruce and between the same velocities of tension and opposite wood of beech (5 p.c.).

5.4.3. Measurements on standing trees

Surface waves were used to measure the ultrasonic velocity on the circumference of trees in longitudinal anisotropic direction of wood. The results were compared with a rapid hardness test using a pilodyn instrument which gives the depth of penetration of a pin in to wood at 6 J. The results obtained were also compared with static E_L measured on standard specimens.

Like the pilodyn, surface wave velocities give a good idea of the variation of wood quality on a stem (11). The correlation coefficient between these two variables significant at 1 p.c. is r = 0.67. Another interesting correlation coefficient, also significant at 1 p.c., was established between surface velocity on the tree and E_L on small clear specimens (r = 0.58).

With these brief considerations, we can state that the ultrasonic velocity method carried out on living trees is a useful tool for wood quality analysis.

CONCLUDING REMARKS

A complete set of elastic constants was obtained for wood using Christoffel's equations for an orthotropic solid. The calculated elastic constants are in good agreement with values obtained from ultrasonic velocity measurements. Using arguments more or less plausible according to mathematical taste, an optimisation procedure of non-diagonal terms of the stiffness matrix has been generalized on 3 wood species. The limits of its validity appear to be quite wide. Mathematically speaking, it appears that a full knowledge of the stiffness matrix would embody the solutions to many problems connected with wave propagation in wood. Some practical applications of ultrasonic velocity methods to forestry science were found. Measurements on standing trees seem to be of major interest for the practitioner.

Wood is not a modern material but makes you pose certain questions. Hence it serves and stimulates the practical and scientific interest of the user and fundamentalist of ultrasonic velocity method.

REFERENCES

1. Auld B.A.: Acoustical fields and waves in solids.Vol. 1, Wiley Interscience Pub., New York, 1973.
2. Dieulesaint E., Royer D : Ondes élastiques dans les solides. Masson & Cie, Paris, 1974.
3. Musgrave M.J.P.:Crystal acoustics, Holden-Day, San Francisco, 1970.
4. Jayne B.A.(ed) : Theory and design of wood and fiber composite materials. Syracuse University Press, 1972.
5. Bucur V.: Ultrasonic velocity, stiffness matrix and elastic constants of wood. Journal of Catgut Acoustical Society (in press). 1986.
6. Hearmon R.F.S.: The elasticity of wood and plywood. London, H.N.S.O., 1948.

7. Hearmon R.F.S. :The assessment of wood properties by vibrations and high frequency acoustic waves. Proc. 2nd Symposium fo the NDT od wood, Washington State University, 1965.

8. Bucur V. : Terms of stiffness matrix, ultrasonic velocity and anisotropy of wood. Communication Ultrasonics International 85, King's College, London, 2-4 July 1985.

9. Bodig J.,Goodman J.R.: Prediction of elastic parameters for wood. Wood Science, 5 (4), 249-264, 1973.

10. Hearmon R.F.S.: An introduction to applied anisotropic elasticity. Oxford, University Press, 1961.

11. Bucur V. : Ultrasonic, hardness and X-ray densitometric analysis on wood. Ultrasonics (in press), 1986.

ACKNOWLEDGEMENTS

This work has been partially supported by the Ministère de la Recherche et de la Technologie (France).

The author wishes to express her gratitude to her colleagues: D. Aubert for assistance in preparing the programs and handling the data, to P. Gelhaye and J.R. Perrin for technical assistance, to G. Jacquemot for typing the manuscript and to E. Drevet and J.A. Evertsen for correcting the English revision. Also we wish to thank the Catgut Acoustical Society, C. Hutchins, for sending us samples from Soundboards of particular interest for wood science.

DISCUSSION

Comment: Chivers
How are the measurements of velocity affected by the moisture content of the wood? Secondly, what precautions are needed in the handling of the specimens in the laboratory?

Reply: Bucur
The ultrasonic measurements are affected by the moisture of the wood: in the present report, all the measurements on cubic specimens were performed at 12% m.c. The sensibility of ultrasonic velocity to the moisture content is in the range of 0% to more than 100%.

Comment: Alippi
The dispersion of points in the velocity curves you showed, could it be due to the obvious interaction between wavelength and average distance between fibers in the wood? That is a case not considered in the lecture by Auld.

Reply: Bucur
In biological materials like wood, the natural variability is the principal factor of the dispersion of the experimental measurements. Expressed in statistical terms (coefficients of variation), the dispersion of the longitudinal wave velocity measurements is 10%, and for shear wave velocity, it could be even 20%.
Certainly, the anatomical variability at macroscopic level (as the width of annual rings) or at microscopic level (as the fiber length, the diameter of cells, the microfiber angle, the presence of reactive wood, etc.) can be considered as sources of the discussed variations. In this sense, some similarities can be found with the layered composites discussed previously by Auld.

Comment: Sliwinski
Have you observed any relation between the direction of cross section of velocity surface characteristics and the anatomy of the tree? How does it depend upon the place where the tree has been cut?

Reply: Bucur
Generally speaking, on the velocity surface, the intersection between QT and T curves is in the LT plane. The angle of the intersection depends upon the optimization procedure accepted for the calculation of the off diagonal terms of Γ_{ij}. For the moment, no anatomical measurements were performed in the view of establishing a possible relationship between this angle and some anatomical characteristics, as the microfiber angle, for example.

Comment: Busse
As the wood is piezoelectric, it should be possible to generate ultrasonic waves by applying

an rf field with no contact; also, it should be possible to detect acoustic waves in a remote way. That technique would eliminate boundary or surface problems and allow at the same time non destructing testing.

Reply: Bucur
The non-contact methods are the techiniques of the future for detection of defects in wood; it is a line of research to be developed, specially for industrial NDT controls on planks.

MEASUREMENT OF THE PERMEABILITY OF A POROUS PLATE BY AN ULTRASONIC TEST

G. BONNET*, D. SCHMITT**, J. EL MAAROUFI*

1. INTRODUCTION

The theory of dynamic poroelasticity for fluid-saturated porous media was first presented by Biot (1). This first model was next confirmed by some recent results in the theory of homogenization (2) (3) (4). In addition, some recent experimental results confirmed some important effects predicted by the theory, particularly the existence of the second (slow) compressional wave (5).

The theory for fluid saturated or nearly saturated porous media seems to be quite well established.

It seems therefore to be quite reasonable to study practical applications of such a poroelastic model. The coupling between the two phases being directly related to the permeability, the purpose of this paper is to study the possibility of the measurement of the permeability by using ultrasonic waves propagating through a porous plate.

In a first step, the equations of the classical poroelasticity are recalled, including partial differential equations and continuity relations.

The equations for the transmission of waves through a porous plate are next written, and these equations are solved numerically in some practical cases of interest.

2. DYNAMIC POROELASTICITY

The motion of a deformable saturated porous medium is usually described by taking as variables the mean displacements of the solid and of the fluid, denoted below u and U, respectively.

2.1 Stress-deformation relations

For small displacements, the deformation tensors of each phase may be written:

$$e_{ij}=\frac{1}{2}(u_{i,j}+u_{j,i}) \qquad \text{(solid)}$$

$$\epsilon_{ij}=\frac{1}{2}(U_{i,j}+U_{j,i}) \qquad \text{(fluid)}$$

The partial stress tensors are related to these deformation tensors by:

$$\sigma_{ij}= 2N\, e_{ij} +\delta_{ij}(Ae+Q\epsilon) \qquad \text{(solid)}$$

$$\Sigma_{ij}=(Qe+R\epsilon)\delta_{ij} =s\delta_{ij} \qquad \text{(fluid)}$$

where $e=e_{ii}$ $\qquad \epsilon=\epsilon_{ii}$

with the elastic coefficients A, Q, R, N used by Biot (1).

2.2 Equations for the displacements

The equations which govern the wave propagation in a saturated porous medium were written by Biot (1) in the form

$$N\, u_{i,kk} + \left| (N+A)\, e + Q\epsilon \right|_{,i} - b(u_i - U_i) = \rho_{11}\ddot{u}_i + \rho_{12}\ddot{U}_i$$
$$(Qe+R\epsilon)_{,i} + b(u_i - U_i) = \rho_{12}\ddot{u}_i + \rho_{22}\ddot{U}_i \qquad (1)$$

where

$$\rho_{11}+\rho_{12} = (1-\phi)\rho_s$$
$$\rho_{12}+\rho_{22} = \phi\rho_f$$

ρ_s and ρ_f are specific mass of the solid and of the fluid,
ϕ is the porosity,
u_i (resp. U_i) are the components of the solid (resp. fluid) displacement (i = 1, 3).

The equations are confirmed by the results obtained by using the theory of homogenization (2) (3) (4), with a slight difference described below.
b and ρ_{ij} must be considered as functions of the frequency, the equations (1) being valuable only for a harmonic motion (or in Fourier domain).

The equations in the time domain are then obtained with the form of a convolution product and not in the form (1).

In the following, we assume that the motion is a sinusoïdal one, and thus the equations (1) are valid.

2.3 Equations for the potentials

To solve these equations, we use the potentials introduced by

$$u = (\text{grad}\,\phi^{(s)} + \text{curl}\,\psi^{(s)})e^{i\omega t}$$
$$U = (\text{grad}\,\phi^{(1)} + \text{curl}\,\psi^{(1)})\, e^{i\omega t} \qquad (2)$$

P waves

The P waves are obtained by applying the divergence operator to (1), thus

$$P\Delta\phi^{(s)} + Q\Delta\phi^{(1)} = -\omega^2(\gamma_{11}\phi^{(s)} + \gamma_{12}\phi^{(1)})$$
$$Q\Delta\phi^{(s)} + R\Delta\phi^{(1)} = -\omega^2(\gamma_{12}\phi^{(s)} + \gamma_{22}\phi^{(1)}) \qquad (3)$$

where

$$\gamma_{ii}=\rho_{ii}-ib(\omega)/\omega \qquad \gamma_{12}=\rho_{12}(\omega)+ib(\omega)/\omega$$

and P = A + 2N.

Δ= Laplace operator

The expression of $\phi^{(1)}$ as a function of $\phi^{(s)}$ is given by solving (3):

$$\phi^{(1)}=\frac{PR-Q^2}{\omega^2(Q\gamma_{22}-R\gamma_{12})}\phi^{(s)}+\frac{R\gamma_{11}-Q\gamma_{12}}{Q\gamma_{22}-R\gamma_{12}}\phi^{(s)} \tag{4}$$

(3) and (4) give:

$$(PR-Q^2)\Delta^2\phi^{(s)}+\omega^2(P\gamma_{22}+R\gamma_{11}-2Q\gamma_{12})\Delta\phi^{(s)}+\omega^4(\gamma_{11}\gamma_{22}-\gamma_{12}^2)\phi^{(s)}=0 \tag{5}$$

The relation (5) may be written as

$$(PR-Q^2)\left(\Delta\phi^{(s)}+\frac{\omega^2}{\alpha_1^2}\phi^{(s)}\right)\left(\Delta\phi^{(s)}+\frac{\omega^2}{\alpha_2^2}\phi^{(s)}\right)=0 \tag{6}$$

where α_1^2, and α_2^2 are the two solutions of

$$\alpha^4(\gamma_{22}\gamma_{11}-\gamma_{12}^2)-\alpha^2(P\gamma_{22}+R\gamma_{11}-2Q\gamma_{12})+PR-Q^2=0 \tag{7}$$

$\phi^{(s)}$ may be therefore written in the form

$$\phi^{(s)}=\phi_1^{(s)}+\phi_2^{(s)} \tag{8}$$

$\phi_1^{(s)}$ and $\phi_2^{(s)}$ are the solutions of the Helmholtz equations which appear in (6).

α_1^2 and α_2^2 are the phase velocities of the two P waves in the Biot Model.

$\phi^{(1)}$ is next obtained by (4), which leads to:

$$\phi^{(1)}=\xi_1\phi_1+\xi_2\phi_2 \tag{9}$$

$$\xi_j=-\frac{1}{\alpha_j^2}\frac{PR-Q^2}{Q\gamma_{22}-R\gamma_{12}}+\frac{R\gamma_{11}-Q\gamma_{12}}{Q\gamma_{22}-R\gamma_{12}} \tag{10}$$

S waves

In the same way, by applying the curl operator to (1), one obtains:

$$N\Delta\psi^{(s)}=-\omega^2(\gamma_{11}\psi^{(s)}+\gamma_{12}\psi^{(1)})$$
$$\psi^{(1)}=-\frac{\gamma_{12}}{\gamma_{22}}\psi^{(s)}=\chi\psi^{(s)} \tag{11}$$

The velocity of the S wave is therefore:

$$\beta = \left| N/(\gamma_{11} - \frac{\gamma_{12}^2}{\gamma_{22}}) \right|^{1/2}$$

u and U are then obtained by:

$$u = \text{grad}\phi_1 + \text{grad}\phi_2 + \text{curl}\ \psi$$

$$U = \xi_1 \text{grad}\phi_1 + \xi_2\ \text{grad}\phi_2 + \chi \text{curl}\ \psi$$

3. CONTINUITY RELATIONS AT AN INTERFACE

The continuity relations may be obtained by using the following results (8).

The balance of mass may be written, at an interface between two porous media with the same pore-filling fluid (13) (8)

$$[\phi(u\text{-}U).N] = [w.N] = 0 \qquad (12)$$

where the brackets denote the jump of the enclosed quantity through the interface, with $w=\phi(u\text{-}U)$. N is a unit normal vector to the interface.

When the fluid is not the same on each side of the interface, the balance of mass is no more valid, but the relation above may be obtained by writing the balance of volume in a channel of the porous medium. The equation (12) is still valid.

The balance of momentum leads to

$$[((\Sigma_{ij}+\sigma_{ij}) + \alpha\delta_{ij})N_j] = 0 \qquad (13)$$

where α contains only terms with the form $\rho\dot{u}_i\dot{u}_j$, where ρ is a specific mass, $\dot{u}_i$ and $\dot{u}_j$ are velocities of the two constituents. It is assumed in the following that such terms may be negligible compared to σ_{ij} and Σ_{ij}.

Therefore: $$[(\Sigma_{ij}+\sigma_{ij})\ N_j] = 0 \qquad (14)$$

Continuity of the skeleton displacement

If the solid skeleton remains continuous without slip, the solid displacement must be continuous, therefore

$$[u_i] = 0 \qquad (15)$$

Continuity of the pressure

Deresiewicz & Skalak (11) introduce a continuity relation with the form:

$$[p+\beta\phi(\dot{u}_i\text{-}\dot{U}_i)N_i)] = 0 \qquad (16)$$

where β is a "resistance coefficient".

The homogenization theory (14) leads to

$$[p]=0 \qquad (17) \qquad \text{and not (16)}$$

Without clear justification of the "resistance" term we assume that the pressure is continuous at the interface.

4. TRANSMISSION THROUGH AN INTERFACE BETWEEN A DRY AND A SATURATED POROUS MEDIUM

Before presenting a computation of the transmission of waves through a porous plate, it is useful to present a computation which is not directly applicable to our purpose, but which gives first indications of a part of the physical phenomena which appear at an interface between an elastic and a porous medium.

Some transmission coefficients for the case of a P wave in a dry porous medium which is incident on a dry medium-saturated medium interface are shown in figure 1 (8).

It appears that the low frequency results are practically identical to the elastic computation, but for higher frequencies, it is no longer true. The lower frequency where the effects of the porous medium are visible is approximately equal to $0.01\, f_c = 0.01\, \dfrac{\mu\phi}{2\pi k \rho_f}$ where f_c is the characteristic frequency defined by Biot, which is function of the dynamic viscosity μ, the permeability k, the specific mass of the fluid ρ_f. It appears that the difference between an elastic and a porous case is due to the following reason: for low frequencies, the P2 wave does not propagate and is a diffusive type, but for higher frequencies, it is no longer true.

It appears therefore that the threshold of $0.01\, f_c$ must be exceeded if an effect due to the wave propagation in a porous medium is in view.

5. TRANSMISSION THROUGH A POROUS PLATE

5.1 Equations for the problem

We consider a fluid plane wave incident on a porous plate (figure 2). The problem in the following is to compute the amplitude of the wave which is transmitted through the plate as a function of the characteristics of the porous plate.

The plane P wave $e^{i(\omega t-kx)}e^{i\upsilon_\alpha z}$ is incident on the porous plate with an angle of incidence θ with $\sin\theta = k\alpha/\omega$.

A reflected P_R wave $A_R e^{i(\omega t-Kx)}e^{i\upsilon_\alpha z}$ and a transmitted P_T wave (coef A_T) are generated at the two faces of the plate.

In the porous plate, the potentials are

$$\phi_i(z)=\phi_i(z)+\phi_i(z)=A_i\, e^{i\upsilon_{\alpha i} z}+B_i e^{-i\upsilon_{\alpha i} z} \qquad (i=1,2)$$

$$\psi(z)=\psi(z)+\psi(z)=C\, e^{i\upsilon_\beta z}+D\, e^{-i\upsilon_\beta z}$$

The wave field in the plate is obtained by superposition of 6 plane waves.

P1 waves : an upward wave $A_1 e^{i(\omega t-Kx)} e^{i\upsilon_{\alpha 1} z}$

a downward wave $B_1 e^{i(\omega t-Kx)} e^{-i\upsilon_{\alpha 1} z}$

P2 waves : an upward wave $A_2 e^{i(\omega t-Kx)} e^{i\upsilon_{\alpha 2} z}$

a downward wave $B_2 e^{i(\omega t-Kx)} e^{-i\upsilon_{\alpha 2} z}$

S waves : an upward wave $C\, e^{i(\omega t-Kx)} e^{i\upsilon_\beta z}$

a downward wave $D\, e^{i(\omega t-Kx)} e^{-i\upsilon_\beta z}$

The expressions for $\upsilon_{\alpha 1}$, $\upsilon_{\alpha 2}$; υ_β are obtained by replacing these expressions in the wave equation.

Continuity equations

The boundary equations may be written:

a) for $z = z_0$

- mass conservation

$$U_z(P_R, z_0) + U_z(P_I, z_0) = u_z + \phi(U_z - u_z)$$

where U_z (resp. u_z) is the vertical component of the fluid (resp. solid) displacement.

- normal traction vector on the face of the plate

$$0 = \sigma_{xx} \qquad \sigma_{zz}(P_R, z_0) + \sigma_{zz}(P_I, z_0) = \sigma_{zz} + s$$

- continuity of pressure

$$-\sigma_{zz}(P_R) - \sigma_{zz}(P_I) = -s/\phi$$

$s = -\phi p$ is the partial stress in the fluid.

b) similar conditions are applied for $z = z_1$

Finally, a linear 8x8 system allows computation of the 8 constants A_T, A_R, A_1, A_2, B_1, B_2, C, D, which may be written

$$M\, N = P$$

where

$$^tN = \{A_R, A_1, A_2, B_1, B_2, C, D, A_T\}$$

$$^{t}P = \left\{ -i\, \nu_{\alpha} e^{i\nu_{\alpha} z_{o}},\ \rho\omega^{2} e^{i\nu_{\alpha} z_{o}}, 0, \rho\omega^{2} e^{i\nu_{\alpha} z_{o}}, 0,0,0,0 \right\}$$

The matrix M, function of elastic coefficients, wave numbers and geometrical parameters,is not given here.

5.2 Solution

The linear system previously obtained has been solved in a practical case where the porous plate is a paper sheet immersed in water. The low thickness of a paper sheet makes it difficult to perform a usual quasi static measurement of the permeability. In addition, an ultrasonic method would be helpful in the control of the process of papermaking.

a) Effect of the frequency

The result of the computation for the transmission coefficient is given in figure 3 as a function of the frequency, for a normal incident wave for two different values of the permeability of the plate. As it may be seen, the low thickness of the paper sheet makes it practically transparent to the low frequency waves. However, for frequencies greater than 1 MHz, it is no more true. In addition, the transmission coefficient is not the same for the two values of the permeability for frequencies higher than 1 MHz.

The threshold previously defined for this case by 0.01 f_c is equal to 50 kHz for the case under consideration. It is surprising to notice that the effect of the porous medium does not appear for frequencies between 50 kHz and 1 MHz.

Such a result may be explained in the following way: on the upper interface, the incident P wave is split into three waves in the plate, but the phase difference between the two faces of the plate is very small, which allows a complete recomposition of these waves and makes the plate practically transparent to the wave, even if a P2 wave appears in the plate.

b) Effect of the angle of incidence

The results of the computation are shown on the figure 4 for an angle of incidence $\simeq 60°$. It appears that the threshold for which the sheet is no more transparent appears for lower frequencies, which is an obvious consequence of a longer path of the wave in the direction of the incident wave. As a consequence, the effect of permeability appears at a lower frequency.

6. CONCLUSIONS

The purpose of the paper was to show in which conditions the permeability of a porous plate could affect the transmission of waves through it. It was shown that the low frequency threshold of such an effect is higher than the one which would be expected from a simple transmission through one interface. It is a consequence of the recomposition of the waves on the output side of the plate.

A further numerical study is being performed to obtain the best conditions for the practical realisation of the test.

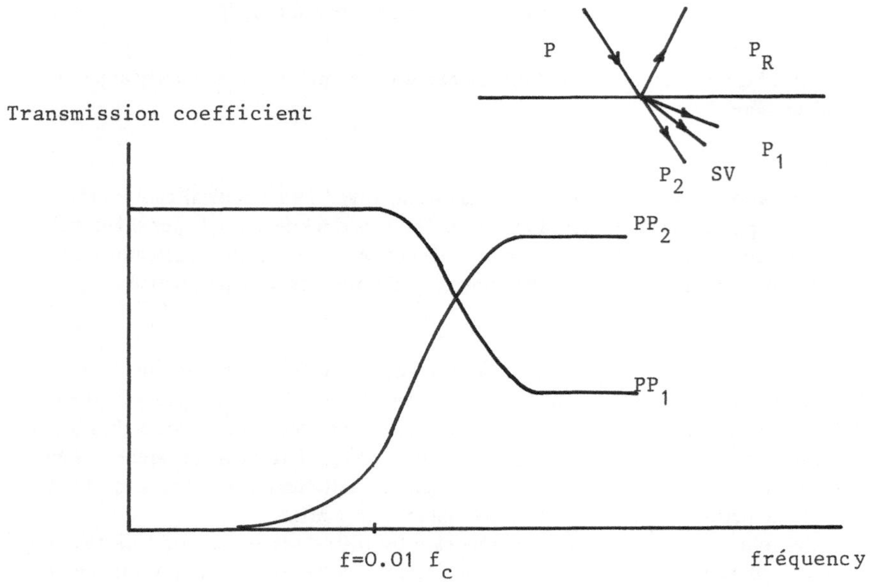

FIGURE 1 Transmission through an interface dry medium/saturated medium (ref 8)

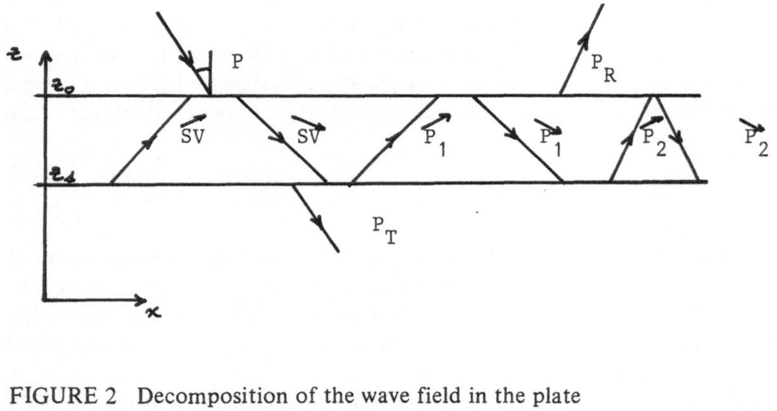

FIGURE 2 Decomposition of the wave field in the plate

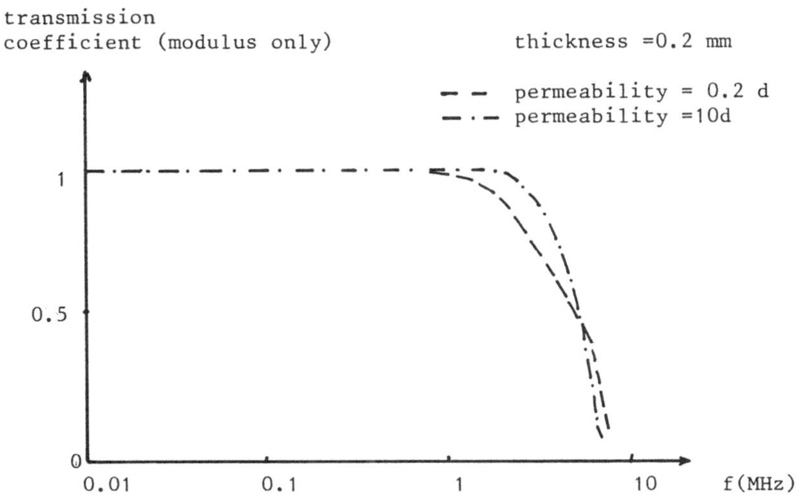

FIGURE 3 Transmission through the plate. Normal incidence

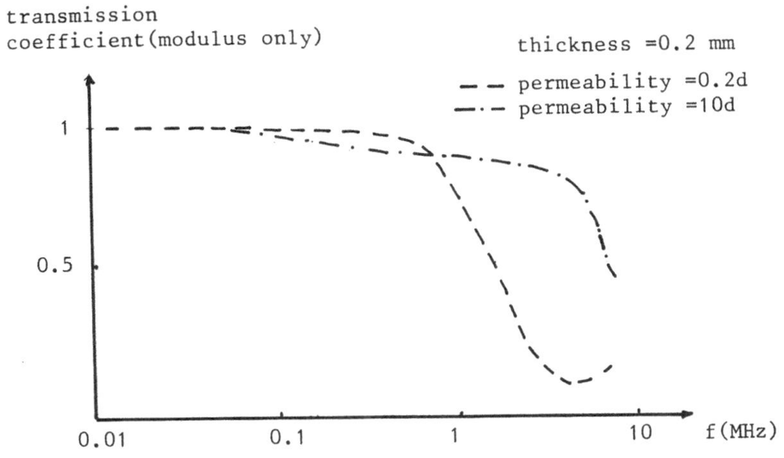

FIGURE 4 Transmission through the plate. Incidence 60°.

REFERENCES

1. Biot M.A., Theory of propagation of elastic waves in a fluid saturated porous solid. J. Acoust. Soc. Am., vol. 28, n. 2, pp. 168-178 (1956).
2. Levy T., Propagation of waves in a fluid saturated porous elastic solid. Int. J. Eng. Sci. Vol. 17, pp. 1005-1014 (1979).
3. Auriault J.L., Dynamic behaviour of a porous medium saturated by a newtonian fluid. Int. J. Eng. Sci., vol. 18, pp. 775-785 (1979).
4. Burridge R., Keller J. B., Poroelasticity equations derived from microstructure. J. Acoust. Soc. Am. vol. 70 (4) pp. 1140-1146.
5. Plona T.J., Observation of a second bulk compressional wave in a porous medium at ultrasonic frequencies. Appl. Phys. Lett., 36 (4) pp. 259-261 (1979).
6. Dutta N.C., Ode H., Seismic reflections from a gas water contact, Geophysics, vol. 48 n. 2, pp. 148-162 (1983).
7. Stoll R.D., Kan T.K., Reflection of acoustic waves at a water sediment interface, J. Acoust. Soc. Am. vol. 70 (1) pp. 149-156 (1981).
8. Bonnet G., Boutin C., Transmission of compressional waves through a water table (subm. to J. Geot. Eng. Div. ASCE).
9. Deresiewicz H., Wolf B., Rice J.T., The effect of boundaries on wave propagation in a liquid filled porous solid. Bull. Seism. Soc. Am. vol. 52, pp. 595-638, (1962), vol. 54, pp. 409-423, pp. 1537-1961 (1964), vol. 55, pp. 919-923 (1965).
10. Bonnet G., Contribution à l'étude des milieux poreux en régime dynamique. Application à la reconnaissance par pompage harmonique et à la reconnaissance sismique. Thèse D.ès.Sc. Montpellier (France) (1985).
11. Deresiewicz H., Skalak F., On uniqueness in dynamic poroelasticity. Bull. Seism. Soc. Am., vol. 53 n. 4, pp. 783-789 (1963).
12. Coussy O., Bourbie T., Propagation des ondes acoustiques dans les milieux saturés. Revue de l'I.F.P. vol. 39 n. 1 (1984).
13. Truesdell C., Toupin R., Classical theory of fields. Handbuch der Physik, vol. III/1, Springer Verlag ed. Berlin (1960).
14. Auriault J.L., Pers. Comm. (1985).

LASER-GENERATED ULTRASONIC WAVES FOR THE INVESTIGATION OF POROUS SOLIDS

D.A. HUTCHINS*, R.P. YOUNG† AND J. UNGAR*

1.0 INTRODUCTION

The use of a pulsed laser to generate well characterised acoustic sources has certain advantages over traditional methods in that it is reproducible and results in a wide bandwidth. Further, it is not necessary to couple the source transducer to the sample by immersion techniques or under pressure, as is frequently the case when the attenuation properties of porous solids are studied. An investigation into the propagation of ultrasonic transients within porous solids using the pulsed laser technique is described. The mechanisms leading to ultrasonic generation by irradiation of solids with pulsed lasers is reviewed in Section 1.1, followed by a brief outline of elastic wave propagation through porous media. A description of the apparatus used is given in section 2 and the results obtained by laser irradiation are compared to those obtained using PZT transducers in section 3.

1.1 Wave propagation in porous media

Propagation of acoustic waves in porous media has been the subject of several recent studies in which the theories of Biot [1,2] have been applied extensively in investigations of solids with fluid-filled pores. These theories are, however, applicable to a wide range of media [3].

Biot's theory describes a class of porous solid in which both the solid phase and any fluid held within its pores are continuously connected. These structures may be represented schematically, as in Fig. 1(a) [4]. Many porous solids, however, do not have interconnecting pores and have negligible permeability, as is the case with porous metals, and they are represented in Fig. 1(b). In this situation, modified theories are required.

It is assumed in all the theories described here that both pore and grain sizes are much smaller than the wavelength of interest. These theories are used to predict the velocity of longitudinal and shear waves

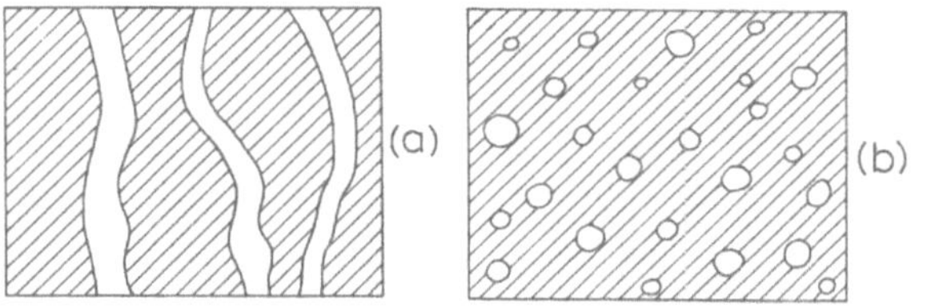

Fig. 1 Schematic diagram of porous solids, (a) pores connected. (b) pores not connected.

in the composite material, with the solid containing air in its pores. If the pores contain fluid, however, interconnection leads to the propagation of a third wave mode of a compressional nature, whose maximum velocity is that of the bulk fluid. Plona [5] was the first to observe such waves at ultrasonic frequencies in fluid saturated porous media. Other authors [4] have shown that as porosity reduces, the observed longitudinal and shear wave velocities approach those of the bulk solid, as might be expected; the slow compressional wave velocity, however, approaches zero. These results appear to be consistent with predictions derived from Biot's theory [7].

Biot's approach allows the porosity ϕ of a material to be evaluated using measurement of ultrasonic wave velocities. In brief, these velocities may be represented in a simplified form as:

$$v(\text{shear}) = \left[\frac{N}{(1-\phi)\rho_s + (1-(1/\alpha))\phi\rho_s}\right]^{\frac{1}{2}} \qquad (1a)$$

$$v(\text{long}) = \left[\frac{K_b + (4/3)N}{(1+\phi)\rho_s + (1-(1/\alpha))\phi\rho_s}\right]^{\frac{1}{2}} \qquad (1b)$$

$$v(\text{slow}) = \frac{v(\text{fluid})}{\sqrt{\alpha}} \qquad (1c)$$

The various quantities referred to in the equations above are derived from properties of both the fluid, and the solid matricies (or frames). ϕ is the porosity, as judged by the fraction of the volume not occupied by solid, and ρ_s and ρ_f are the density of the bulk solid and pore fluid. κ_b and N

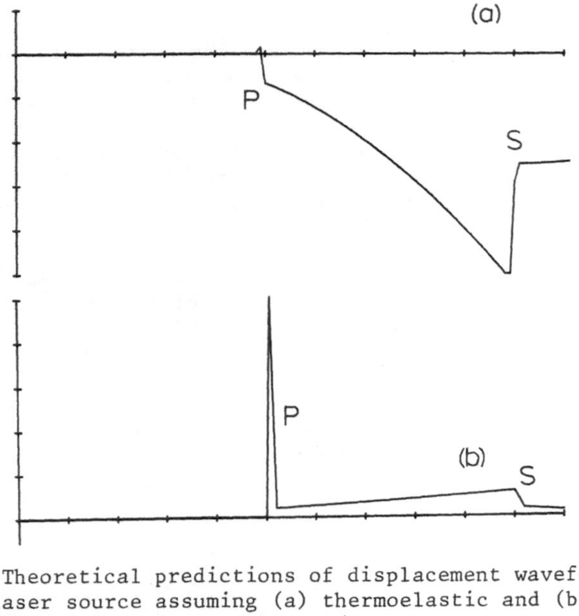

Fig. 2 Theoretical predictions of displacement waveforms from a pulsed laser source assuming (a) thermoelastic and (b) evaporation mechanisms. Poisson's ratio $\nu = 1/3$.

are properties of the skeletal frame (i.e. the solid frame with no fluid), and are the bulk and shear moduli respectively. The quantity α is related to the geometry of the solid particles in the porous solid, and is one of the most critical quantities to evaluate. It is described in greater detail elsewhere [6-8].

The expressions in (1) only hold under conditions such that the fluid has negligible viscosity or the frequency is high enough, and where both κ_b and $N >> \kappa_f$, where κ_f is the bulk modulus of the fluid, Under such limits, attenuation (which is frequency dependent) has little affect on the velocities, but the theory can also be used to predict attenuation due to viscous effects in the fluid.

In many solid materials, the pores contain gases such as air. In such cases, the velocities can still be estimated, although the slow wave is likely to be negligible. Equations (1) can still be used, however, to estimate material parameters by an experimental measurement of the shear and longitudinal velocities. For instance, the porosity ϕ may be estimated if the other parameters are known.

As stated above, a fundamental precondition of the use of the Biot theory is that the pores are interconnected. If this is not so, as shown in Fig. 1(b), then other approaches must be used, such as those invoking the Self-Consistent Theory (SCT) [9,10]. This has been used recently to explain scattering in glass beads [11] and other such materials [12]. Under this scheme, it is possible to have a vacuum in the pores, as well as a wide range of fluids. In effect, the skeletal frame of the porous material is represented as a two-phase composite, where one phase has the shear and bulk moduli of the solid material (N and κ_b as above), with the other phase assumed to be a vacuum. The Self Consistent Theory estimates the elastic moduli of the "effective medium" of a composite material, under the assumption that the wavelengths considered are much larger than the inclusions of the various components. This type of theory is used in this paper to estimate longitudinal and shear velocities, for a comparison to experimental data obtained from the laser generation technique.

1.2 Ultrasonic generation by pulsed lasers.

Pulsed lasers have been used in many experiments for the generation of ultrasonic transients in solids, and several reviews (e.g. [13,14])have been published which describe the generation mechanisms. The laser technique has several advantages for the investigation of solids, in that it leads to reproducible, wide bandwidth sources whose characteristics are known [15]. Two types of acoustic source formed by laser irradiation have been used in

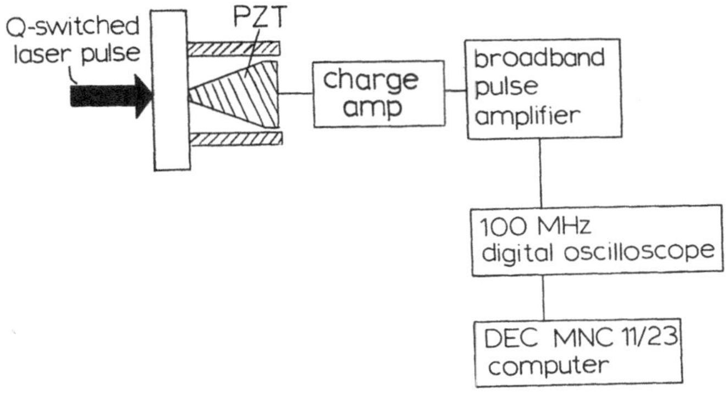

Fig. 3 Schematic diagram of apparatus.

Sample	Average sphere size (μm)	% wt of resin content
AB	229	2.3
AD	148	3.27
AG	75	3.63

Table 1: Properties of porous solids in the form of resin-coated silica spheres. The bulk silica had P and S wave velocities of 5,600 ms^{-1} and 3,300 ms^{-1} with a density of 1.5 gm/cc.

this investigation. The first involves thermoelastic expansion, which leads to a source characterized by horizontal, dipolar stresses, whose time dependence is close to a step function (in practice, it is the integral of the laser pulse shape). A second source is that created by the evaporation of a coating, previously applied to the surface. Momentum transfer from the departing material causes the formation of forces normal to the surface, whose time dependence is pulse-like. This latter mechanism usually requires a higher optical power density, conveniently obtained by focusing the laser pulse.

Wave propagation theory may be used to predict the displacement waveforms expected from such sources on-epicentre. This is demonstrated in Fig. 2 for (a) a thermoelastic and (b) an evaporation source. Note that in the former, the longitudinal (P) signal is smaller than the shear (S), with both being step-like as expected; in the evaporation source, however, the longitudinal signal is step-like, with the shear arrival being of reduced amplitude.

It is possible to use pulsed laser generation in conjunction with interferometric detection [16], leading to an all-optical measurement system. Due to the limited sensitivity of interferometers, however, such an approach has not been used in the experiments to be detailed below, as the ultrasonic attenuation was too high. Hence, wide bandwidth PZT detectors were used in this study as will now be described.

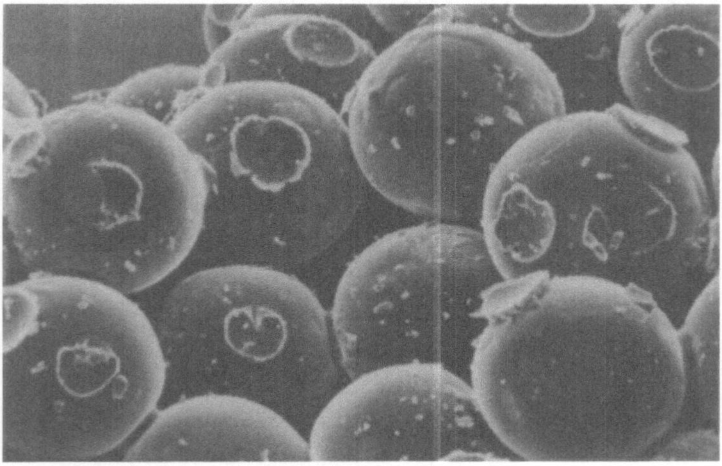

Fig. 4 Scanning electron micrograph, showing structure of typical porous solid, fabricated from resin-coated porous spheres. This sample was not formed under pressure.

2. APPARATUS AND EXPERIMENT

A schematic diagram of the apparatus is presented in Fig. 3. A frequency doubled, Q-switched ruby laser was used to provide 30ns pulses at a wavelength of 347nm in the UV. Typical pulse energies were 150 mJ. The optical wavelength quoted was chosen to enhance absorption at the surface of our samples, which were in the form of consolidated silica spheres. The ultrasonic waves generated by a single laser pulse propagated through the sample, and were detected at the far surface using a thick conical PZT detector. This device was designed to have a wide bandwidth, by reducing resonance effects, and was held against the sample within a shielded, spring loaded mount.Detected waveforms were amplified using a wide bandwidth charge amplifier and subsequent voltage amplifiers, and digitized using a transient waveform recorder. Permanent storage and replotting was then achieved by transfer to a DEC11/23 computer.

One aim of the present research was to compare results from the apparatus above to those obtained with a more conventional approach, using specially-designed broadband PZT transducers as both source and receiver [17]. These were constructed so as to be incorporated conveniently within a compression testing machine, allowing estimates of velocity and attenuation in porous samples to be obtained under various states of deformation.

In this paper, the results obtained using both of the above techniques will be compared for porous samples in the form of resin-coated silica spheres, the coating being polymerized under pressure. Three such samples were investigated, with different sphere sizes, whose physical properties are outlined in Table 1. A scanning electron micrograph of a typical sample is presented in Fig. 4, where the polymer coating, of average thickness 10 μm, is clearly visible. It should be noted that these samples are a much more realistic model material for rock types such as sandstone than other samples used by other workers (e.g. [5]), where sintered specimens were used.

In both techniques, attenuation in each sample was esimated by a comparison of the received power spectra with that following propagation

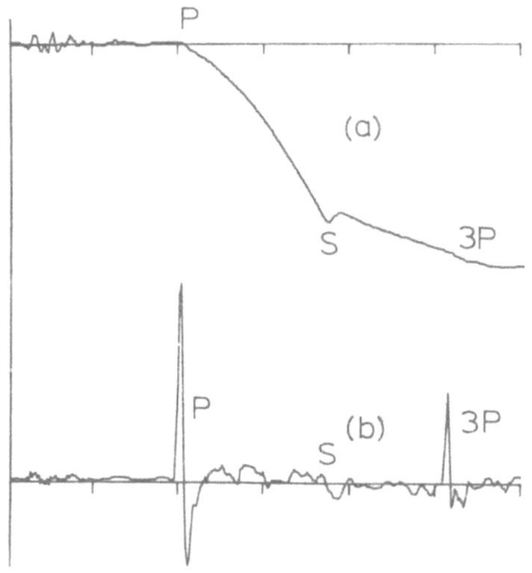

Fig. 5 Experimental displacement waveforms in aluminum, following generation by pulsed laser irradiation of a 10mm thick sample. (a) thermoelastic generation, (b) evaporation of a silicone grease coating.

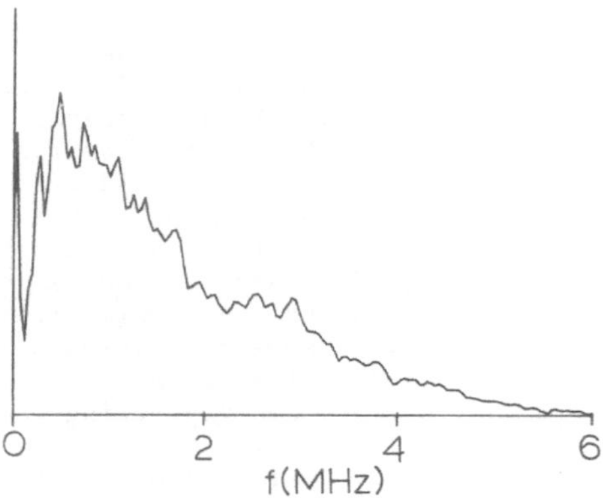

Fig. 6 Amplitude spectrum of waveform shown in Fig. 5 (b).

through aluminum. Velocities of longitudinal and shear waves were measured directly from received waveforms. In all cases, the thickness of the samples was 10mm.

3. RESULTS AND DISCUSSION

3.1 Waveforms in Aluminum

Displacement waveforms generated within aluminum samples may be compared directly to those predicted by wave propagation theory, which requires a knowledge only of longitudinal and shear velocities. Experimental waveforms following generation by laser irradiation in a 10mm thick aluminum sample are presented in Fig. 5 for (a) thermoelastic generation and (b) evaporation of silicon grease coating. Comparison to Fig. 2, in which such waveforms were predicted theoretically, leads to some interesting conclusions. Consider first the thermoelastic waveform, Fig. 5(a), where it will be seen that the expected step-like arrivals were not reproduced. This was the result of using an unfocused laser beam, of ~6mm x 4mm cross-section, the extended source area causing a range of time delays at the receiver. A smaller area could not be used, because of the high attenuation of porous solids which necessitated a high incident energy. A focused beam, of 1mm diameter, was used for the evaporation source however, and as is evident from Fig. 5(b) the longitudinal signal was pulse-like with a fast rise and fall. This waveform gave good agreement with theory, confirming the stated source mechanism.

A spectrum of Fig. 5(b) is presented in Fig. 6. Note the wide bandwidth, of use in studies which required the waveform in aluminum to be used as a reference for comparitive attenuation studies.

To illustrate the similarities between the laser techniques and that using PZT for generation, a waveform obtained in aluminum for the PZT system is shown in Fig. 7(a), and its power spectrum in Fig. 7(b). Note that in this case also a wide bandwidth is possible, but the need for a couplant to porous samples under pressure would lead to a reduction in reproducibility, and problems arise when a measurement is required with a minimal applied load.

3.2 Waveforms in porous solids

The displacement waveforms obtained using the laser techniques in one of the porous samples (AG) fabricated from resin-coated silica spheres is shown in Fig. 8, following generation by (a) thermoelastic mechanisms and (b) evaporation of a silicon grease coating. Comparison to theoretical waveforms (Fig. 2) and those in aluminum (Fig. 4) indicates that the bandwidth is much reduced due to attenuation. Note in particular that the prominent longitudinal (P) transient is reduced in amplitude and frequency content. Waveforms generated by evaporation in samples AB and AD are presented in Fig. 9, and show similar features to that obtained in sample AG.

It was observed that the longitudinal and shear velocities differed little in each of the porous samples examined, representative velocities being 2,500 ms^{-1} for longitudinal and 1,300 ms^{-1} for shear waves. Actual values for each sample are included in Table 2, which includes data from both the laser technique and the conventional PZT appraoch.

Attenuation in the samples was estimated in the form of power spectra, the attenuation spectrum being presented relative to that in aluminum. Fig. 10 shows results for sample AG. Note the rapid attenuation to a frequency of ~2MHz, whereafter the attenuation seemed to go through a maximum. This behaviour was also noted in the results obtained using the PZT arrangement. A waveform obtained after passage through sample AG, at an applied load 1MPa, is presented in Fig. 11(a) for this case, and the power spectrum derived from this waveform is shown in Fig. 11(b). Comparison to Fig. 7, that for propagation through aluminum, indicates that marked attenuation was again present at high frequencies. The attenuation of the sample is plotted in Fig. 11(c), and comparison to that of Fig. 10, obtained using the laser technique, demonstrates that the spectra show similarities, especially the discontinuity in the spectra at ~2MHz.

The velocities quoted in Table 2 have been compared to the predictions of the Self-Consistent Theory, assuming a composite solid comprised of silica spheres with vacuum in the pores. As was expected from the arguments indicated previously, it was found that the predicted velocities were a function of the porosity ϕ, as shown in Fig. 12. Note that for the velocities measured experimentally, the predicted porosity would correspond to ~44%. This is close to the value that would be expected from close packing of the spheres, and indicates that the model is performing in a

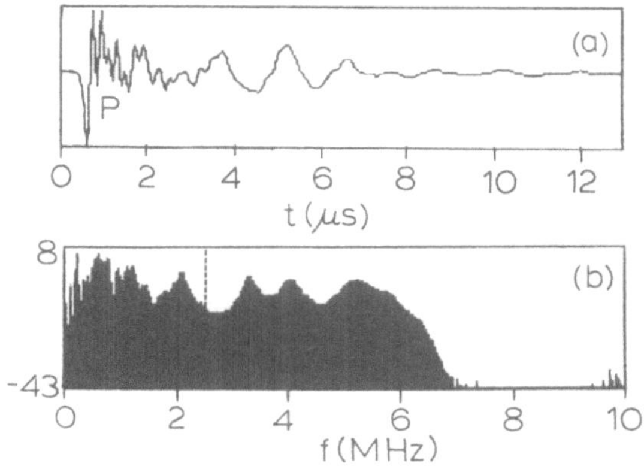

Fig. 7 Results in aluminum, following experiments with a system using PZT transducers for both generation and detection. (a) received waveform, (b) the resulting power spectrum.

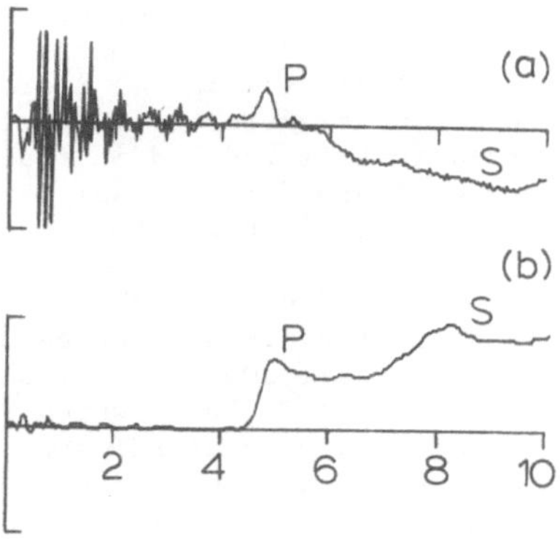

Fig. 8 Laser-generated waveforms in the porous sample AG, following (a) thermoelastic generation and (b) generation by evaporation.

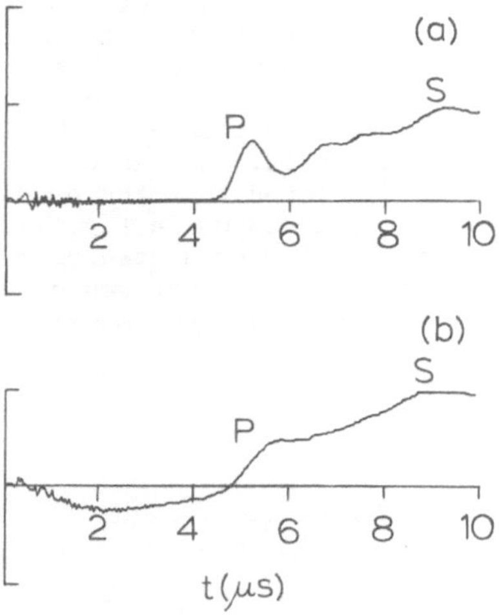

Fig. 9 Waveforms generated by evaporation in (a) sample AD and (b) sample AB.

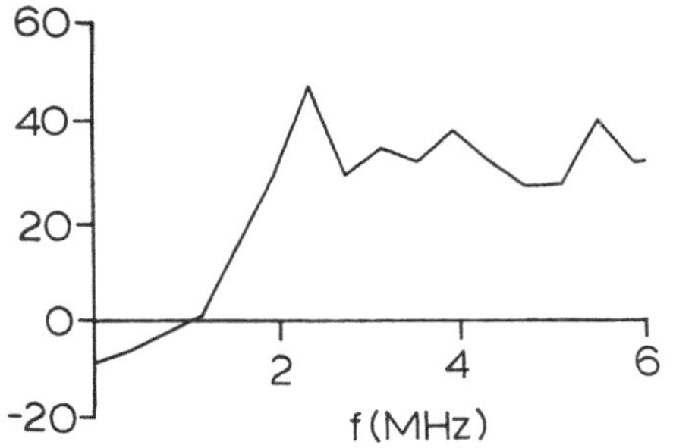

Fig. 10 Attenuation in dB of sample AG, as a function of frequency, using the laser technique.

Sample	Poisson's† Ratio ν	Shear† Modulus (x1000 MN m^{-2})	Bulk† Modulus	C_t/C_ℓ †	C_t/C_ℓ *	C_t * ms^{-1}	C_ℓ * ms^{-1}
AB	0.226	3.54	5.28	0.595	0.48	1,221	2,540
AD	0.239	3.71	5.89	0.58	0.55	1,337	2,445
AG	0.361	3.09	10.06	0.47	0.53	1,316	2,500

Table 2: Measured values, obtained by the laser technique (*) and the conventional PZT method (†).

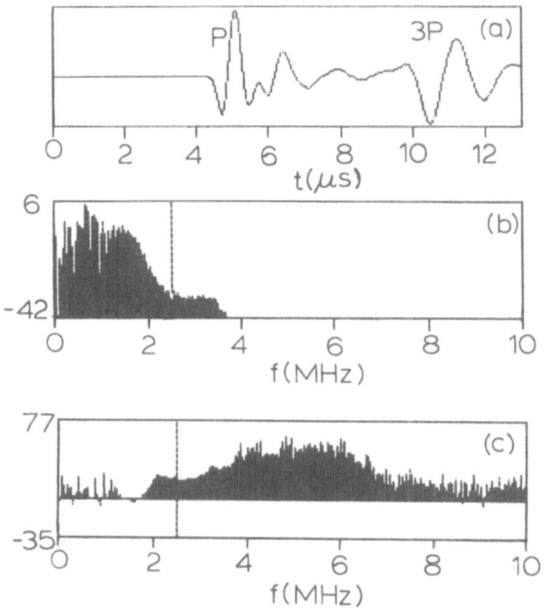

Fig. 11 (a) Waveform following passage through sample AG, using PZT transducers for both generation and dectection. (b) Corresponding power spectrum, vertical scale in dB. (c) Resulting attenuation spectrum of sample.

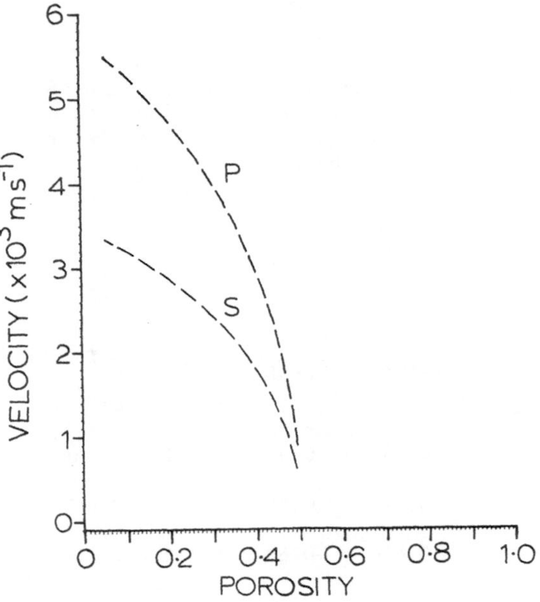

Fig. 12 Theoretical velocities of longitudinal (P) and shear (S) waves in sample AG, using the Self-Consistent Theory approach, as a function of porosity ϕ.

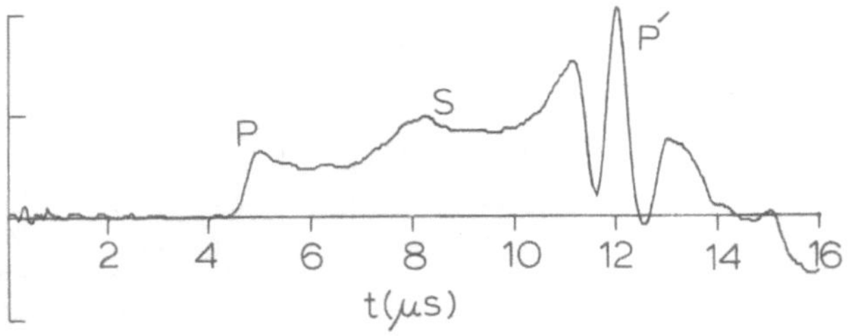

Fig. 13 As Fig. 8(b), but showing presence of P' wave, travelling at ~1/3 the P wave velocity.

satisfactory manner. The limited variation in velocity with sphere size in the AB, AD and AG samples is likely to be due to the fact that the porosities were very similar, as each was formed under pressure to result in a close packing of the silica spheres.

The above discussion has treated the primary longitudinal and shear arrivals only. As was outlined in the Introduction, it is also possible to propagate a slow compressional wave in any fluid contained in the pores, provided they are interconnected (as they are in the present samples). An interesting phenomenon was observed in the silica samples, in that an unexpected arrival was detected on many of the samples investigated. This was observed in sample AG, and an extension of the waveform of Fig. 8 is presented in Fig. 13, where the signal (P') is shown to be oscillatory in nature and of a larger amplitude than the original compression (P) wave. If it is a true wave, and not an artefact of the thick PZT detector, then its velocity would be approximately one third of that of the direct P wave. We do not have an explanation for this signal, and further work is underway with a different detector to investigate whether such a signal is always present in such samples when a laser is used for generation. It is interesting to note, however, that the P' signal is of greater amplitude and at a higher frequency when smaller sphere sizes are used.

3.3 Waveforms in porous rocks

The laser has been used to generate waveforms in a sandstone sample of 23mm thickness, the results for (a) thermoelastic and (b) evaporation mechanisms being presented in Fig. 14. Note the similarities with those obtained in the artificial samples (e.g. Fig. 8) following laser generation. In the case of these real materials, the particles comprising the composite solid are characterized by an extended size distribution; however, it can be seen that the laser technique may potentially be useful for the investigation of such materials in the laboratory. It is interesting to note that the P' wave of Fig. 13 was not observed in sandstone samples, suggesting that this arrival may depend on the presence of a narrow range of particle sizes and geometries.

4. CONCLUSIONS

The pulsed laser generation technique has been used to estimate both the velocity of propagation of longitudinal and shear waves, and the attenuation as a function of frequency, within artificial porous samples. These samples were prepared using known size distributions of silica spheres, and formed under pressure to result in a well-characterized solid. The measured velocities correlated well with the predictions of theory, using the Self Consistent Theory approach.

The results obtained also showed a reasonable correlation with those derived from the specially-designed PZT transducer system, applied to the specimen under load. This supports the conclusion that the laser technique shows promise as a reproducible method for the measurement of material parameters.

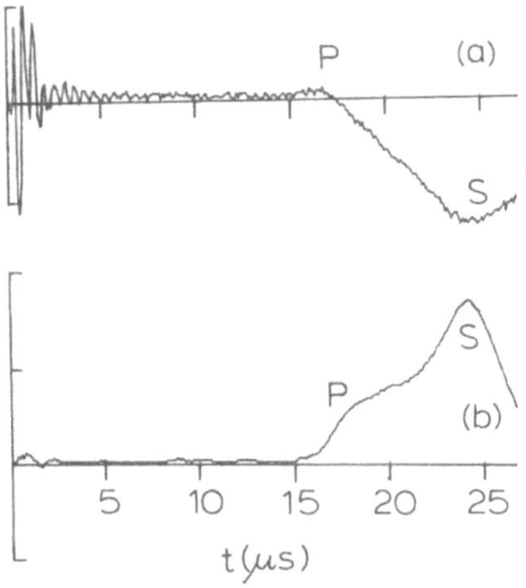

Fig. 14 Laser-generated displacement waveforms in a sandstone sample of 23 mm thickness. (a) thermoelastic waveform, (b) waveform generated by evaporation.

5. ACKNOWLEDGEMENTS

This work was funded by the Province of Ontario, and the National Science and Engineering Research Council of Canada.

6. REFERENCES.

1. M.A. Biot, J. Acoust. Soc. Am. 28, 168 (1956).
2. M.A. Biot, J. Acoust. Soc. Am. 28, 179 (1956).
3. K. Attenborough, Phys. Rep. 82, 179 (1982).
4. T.J. Plona, Proc. 1982 IEEE Ultrasonics Symp., pp.1044-1048.
5. T.J. Plona, Appl. Phys. Lett. 36, 259 (1980).
6. J.A. Berryman, Appl. Phys. Lett. 37, 382 (1980).
7. N.C. Dutta, Appl. Phys. Lett. 37, 898 (1980).
8. D.L. Johnson, Appl. Phys. Lett. 37, 1065 (1980).
9. J.A. Berryman, J. Acoust. Soc. Am. 68, 1809 (1980).
10. J.A. Berryman, J. Acoust. Soc. Am. 68, 1820 (1980).
11. K.W. Winkler and W.F. Murphy, J. Acoust. Soc. Am. 76, 820 (1984).
12. P.R. Ogushivitz, J. Acoust. Soc. Am. 77, 429 (1985).
13. C.B. Scruby, R.J. Dewhurst, D.A. Hutchins and S.B. Palmer, In "Research techniques in Nondestructive Testing" Vol. V(R.S. Sharpe, ed. Academic Press, N.Y., 1981).
14. D.A. Hutchins, to be published in a forthcoming volume of "Physical Acoustics" (Academic Press, N.Y., 1986).
15. D.A. Hutchins, R.J. Dewhurst and S.B. Palmer, J. Appl. Phys.
16. D.A. Hutchins and F. Nadeau, Proc. 1983 IEEE Ultrasonics Symp.
17. R.P. Young and H.I. Alinossawi, to be published in Proc. 1985 Soc. Exploration Geophys. Annual Meeting, Washington, U.S.A.

DISCUSSION

Comment: Cheeke
Are you able to extract any physically interesting information from the wash signal?

Reply: Hutchins
The wash signal arises because the surface has momentum imparted upon it by the longitudinal wave. Hence, it may be possible to derive some elastic information concerning the solid material (such as the Poisson's ratio).

ULTRASONIC DETERMINATION OF TEXTURE AND RESIDUAL STRESS IN POLYCRYSTALLINE METALS

C M SAYERS

1. INTRODUCTION

In a polycrystalline aggregate the elastic constants in the specimen reference frame vary from grain to grain due to the random orientation of the grains. Polycrystalline metals are therefore elastically inhomogeneous, and the elastic constant mismatch at the grain boundaries leads to scattering of the ultrasonic wave. In the long wavelength limit, however, the metal can be modelled as an elastic continuum with elastic constants determined by the elastic constants of the grains and the crystallite orientation distribution function (CODF). This function gives the probability of a crystallite having a given orientation with respect to the specimen frame, and gives a quantitative description of the texture, or crystallographic alignment, of the material. In a strongly textured metal the yield stress varies as a function of direction and this can lead to non-uniform flow in deep drawing for example. As a result there is a need for a non-destructive measurement of texture in process control, and there is considerable interest in the use of ultrasonics for this purpose. In section 2 it is shown how information on the CODF can be obtained from ultrasonic velocity measurements.

Also of interest in process control are internal stresses arising from the manufacturing process which must be added to the external stress when determining the response of the structure to an applied load. In the presence of a stress the ultrasonic velocity in the material depends on the propagation and polarisation directions. Figure 1a shows the fractional change in ultrasonic velocity for the three modes of propagation along the plate normal direction in a mild steel specimen subjected to a compressive stress applied along the rolling direction (1). The fractional change in velocity for the two shear waves when the stress is applied along the transverse direction is shown in figure 1b. It is seen that the change in velocity is different for the shear waves polarised parallel and perpendicular to the stress axis, and this difference may be used in principle to determine the stress present. This is the basis of the shear wave birefringence technique (2) which involves the measurement of the relative velocities of two orthogonally polarised shear waves travelling in the same direction.

In metals with a texture or non-random distribution of crystallite orientations, the ultrasonic velocity depends on the orientation of the propagation and polarisation directions with respect to the principal texture axes, even in the absence of stress. In general, therefore, a measurement of the shear wave birefringence alone is not sufficient for determining the stress, except in samples in which the texture does not vary and in which only inhomogeneous or localised stresses are present. For such samples, a stress-free region can be identified allowing the measurement of the stress-free birefringence and hence the correction for

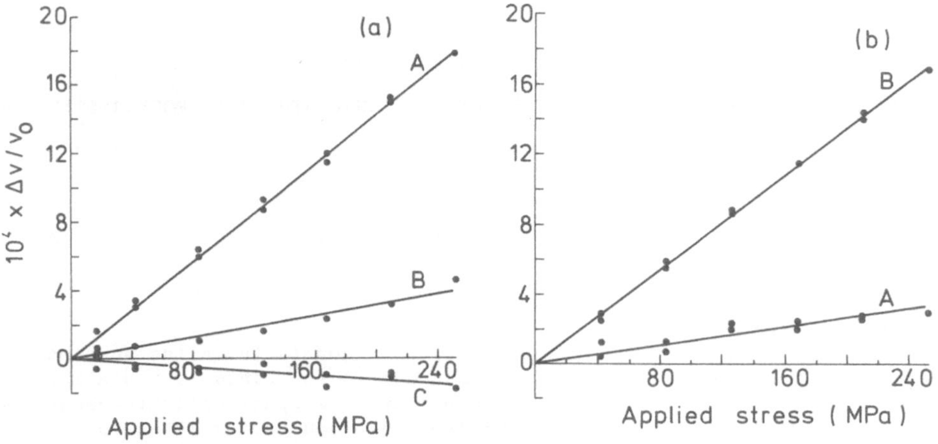

Figure 1. Fractional change in ultrasonic velocity for propagation along the normal direction of a mild steel plate for an applied uniaxial compressive stress acting (a) parallel to rolling direction and (b) parallel to the transverse direction (1). Curves: A, shear wave polarised parallel to the rolling direction; B, shear wave polarised perpendicular to the rolling direction; C, longitudinal wave.

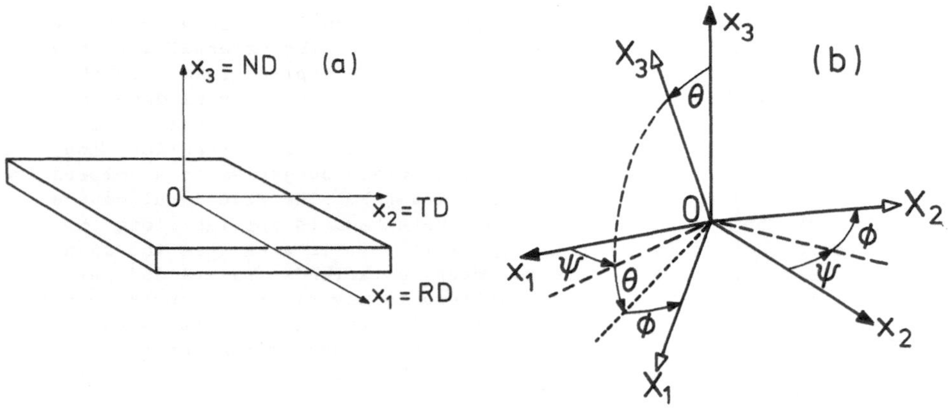

Figure 2(a) Choice of specimen axes for a rolled sheet, with Ox_1 along the rolling direction (RD), Ox_2 along the transverse direction (TD) and Ox_3 along the normal direction (ND). (b) The orientation of the crystallite coordinate system $OX_1X_2X_3$ with respect to the sample coordinate system $Ox_1x_2x_3$ specified by the Euler angles ψ, θ and ϕ (5).

the effect of texture to be made in the region of interest. For samples in which the texture varies over a length scale similar to that of the variation in stress, or for samples in which neither the texture nor the stress vary over the accessible area of measurement, shear wave birefringence does not give a unique determination of the stress. In section 3 methods for the separate measurement of both stress and texture are discussed. These rely on the theory of ultrasonic propagation in textured polycrystalline solids presented in section 2.

2. ULTRASONIC PROPAGATION IN POLYCRYSTALLINE AGGREGATES IN THE ABSENCE OF STRESS

Let $Ox_1x_2x_3$ be an orthogonal set of reference axes fixed in the sample. For a rolled plate these axes could be chosen as the rolling, transverse and normal directions as shown in figure 2a. It is assumed that the sample has orthorhombic symmetry, ie it is assumed to possess three orthogonal mirror planes given by the planes x_1x_2, x_2x_3 and x_3x_1 in figure 2a. This is the symmetry of rolled plate. It will be assumed, in addition, that the sample is an aggregate of crystallites of cubic crystallographic symmetry, as is the case for steel and aluminium. The crystallographic alignment, or texture, of the plate is most conveniently described by the crystallite orientation distribution function (CODF). This method of analysis has been developed extensively by Roe (3-5) and by Bunge (6,7). In this paper the notation of Roe is used.

Let $OX_1X_2X_3$ be an orthogonal set of axes for a crystallite given by the (100), (010) and (001) crystallographic directions. The orientation of a given crystallite with respect to the sample axes $OX_1X_2X_3$ can be specified by the three Euler angles ψ, θ and ϕ shown in figure 2b. The CODF is denoted by $W(\xi,\psi,\phi)$ with $\xi = \cos\theta$. $W(\xi,\psi,\phi)\, d\xi d\psi d\phi$ gives the fraction of crystallites with orientations between ξ and $\xi + d\xi$, ψ and $\psi + d\psi$, and $\phi + d\phi$. It is convenient to expand $W(\xi,\psi,\phi)$ as

$$W(\xi,\psi,\phi) = \sum_{l=0}^{\infty} \sum_{m=-1}^{1} \sum_{n=-1}^{1} W_{lmn}\, Z_{lmn}(\xi) \exp(-im\psi) \exp(-in\phi) \tag{1}$$

where the Z_{lmn} are the generalised Legendre functions defined by Roe (5). For an orthorhombic aggregate of crystallites of cubic crystallographic symmetry, the elastic constant tensor depends only on the coefficients W_{400}, W_{420} and W_{440} in equation (1). Explicit expressions for the elastic constants in terms of these coefficients have been given elsewhere (8-10).

Using these expressions for the elastic constants, the ultrasonic wave velocities in aggregates with small anisotropy can be calculated using Hamilton's principle (11). For orthorhombic aggregates of cubic crystallites, the change $\delta v_R(\theta)$ in the Rayleigh wave velocity for propagation on the x_1x_2 surface from its isotropic value v_R^o is given by

$$\delta v_R(\theta) = (R_1 + R_2 \cos 2\theta + R_4 \cos 4\theta)/ 2 v_R^o \tag{2}$$

in the absence of stress, where θ is the angle between the propagation and rolling directions (11). The amplitudes R_1, R_2 and R_4 are functions of Poisson's ratio ν and are proportional to the coefficients W_{400}, W_{420} and W_{440} respectively. These coefficients may therefore be obtained from a fit of the measured angular variation of the Rayleigh wave velocity to equation (2) and used to plot ultrasonic pole figures as described in reference (9).

The texture coefficients W_{400}, W_{420} and W_{440} may also be determined from bulk wave velocity measurements. For propagation in the plane of the plate

$$\delta v_{SH}(\theta) \equiv v_{SH}(\theta) - v_s^o = \frac{2\sqrt{2}\pi^2 c}{35\rho\, v_s^o} (W_{400} - \sqrt{70}\, W_{440} \cos 4\theta) \tag{3}$$

$$\delta v_{SV}(\theta) \equiv v_{SV}(\theta) - v_s^o = \frac{-8\sqrt{2}\pi^2 c}{35\rho\, v_s^o} (W_{400} - \sqrt{\tfrac{5}{2}}\, W_{420} \cos 2\theta) \tag{4}$$

$$\delta v_1(\theta) \equiv v_1(\theta) - v_1^o = \frac{6\sqrt{2}\pi^2 c}{35\rho\, v_1^o} (W_{400} - \frac{2\sqrt{10}}{3} W_{420} \cos 2\theta + \frac{\sqrt{70}}{3} W_{440} \cos 4\theta) \tag{5}$$

where $c = c_{11} - c_{12} - 2c_{44}$, the c_{ij} being the single crystal elastic constants. $v_{SH}(\theta)$, $v_{SV}(\theta)$ and $v_1(\theta)$ are the velocities of the shear horizontal, shear vertical and longitudinal waves propagating in the plane of the plate at an angle θ to the rolling direction. v_s^o and v_1^o are the isotropic shear and longitudinal velocities.

For propagation in the through thickness direction Ox_3

$$\delta v_{33} \equiv v_{33} - v_1^o = \frac{16\sqrt{2}\pi^2 c}{35\rho v_1^o} W_{400} \tag{6}$$

$$\delta v_{31} \equiv v_{31} - v_s^o = \frac{-8\sqrt{2}\pi^2 c}{35\rho\, v_s^o} (W_{400} - \sqrt{\tfrac{5}{2}}\, W_{420}) \tag{7}$$

$$\delta v_{32} \equiv v_{32} - v_s^o = \frac{-8\sqrt{2}\pi^2 c}{35\rho\, v_s^o} (W_{400} + \sqrt{\tfrac{5}{2}}\, W_{420}) \tag{8}$$

where v_{ij} is the velocity of the wave propagating in the direction Ox_i with polarisation in the Ox_j direction.

3. MEASUREMENT OF STRESS IN THE PRESENCE OF TEXTURE

3.1 The velocity combinations approach

Let λ and μ be the second order and l,m, and n be the third order elastic constants of the aggregate in the absence of texture. The shear wave birefringence $\Delta v_s/v_s^o$ is given in the presence of texture and stress by

$$\frac{\Delta v_s}{v_s^o} \equiv \frac{v_{31} - v_{32}}{v_s^o} = \left.\frac{(v_{31} - v_{32})}{v_s^o}\right|_{\text{texture}} + \left.\frac{(v_{31} - v_{32})}{v_s^o}\right|_{\text{stress}} \tag{9}$$

where v_s^o is the shear wave velocity in the absence of texture and stress and v_{ij} is the velocity of a wave propagating in the direction x_i and polarised in the direction x_j. From equations (7) and (8)

$$\left.\frac{v_{31} - v_{32}}{v_s^o}\right|_{\text{texture}} = \frac{16\pi^2\, c\, W_{420}}{7\sqrt{5}\,\mu} \tag{10}$$

the stress dependent term being given (2) by

$$\left.\frac{v_{31} - v_{32}}{v_s^o}\right|_{\text{stress}} = \left(1 + \frac{n}{4\mu}\right) (\varepsilon_{11} - \varepsilon_{22}) \tag{11}$$

Here ε_{ij} are components of the infinitesimal strain tensor, $(\varepsilon_{11} - \varepsilon_{22})$ being the difference in principal strains in the plane of the plate.

It is therefore not possible to obtain the difference in the principal strains $(\varepsilon_{11} - \varepsilon_{22})$ in the plane of the plate without a separate determination of the texture parameter W_{420}. The aim of the velocity combination approach is to use the additional information provided by the compressional phase delay to separate the effects of texture and stress in the birefringence measurement (12).

If the sample thickness can be measured, the three velocities v_{31}, v_{32} and v_{33} can be calculated from the measured times of flight. To first order in the texture and stress

$$\rho_o \Sigma v^2 = \rho_o (v_{31}^2 + v_{32}^2 + v_{33}^2) = (\lambda + 4\mu) + P (T_1 + T_2) / 3K_o \tag{12}$$

Here $K_o = \lambda + 2\mu/3$ is the bulk modulus, ρ_o is the density of the material in the unstressed state, and

$$P = (21 + 2m + 2\mu - 5\lambda) - (\lambda n + 8\lambda m + 8\lambda^2 + 2\mu n)/ 4\mu \tag{13}$$

It is assumed that the principal stress directions coincide with the principal texture axes $Ox_1x_2x_3$, and that the stress component in the x_3 direction is zero, the components along Ox_1 and Ox_2 being denoted by T_1 and T_2. Equation (12) is independent of the texture or crystallographic alignment and may therefore be used for a texture independent determination of the sum of the principal stresses $(T_1 + T_2)$. Using the elastic constants of mild steel measured by Sayers and Allen (1), equation (13) gives $P = -55.38 \times 10^{10}$ pa. For an applied stress $(T_1 + T_2) = 100$ MPa equation (12) gives a change in $\rho_o \Sigma v^2$ of 0.0246%.

Although equation (12) is independent of texture, it is often difficult to obtain an accurate measurement of the path length in practice. A second texture independent combination of the ultrasonic velocities would allow the thickness to be eliminated, but depends on some knowledge of the texture. For example if $W_{400} = 0$, $\rho_o v_{33}^2$ and $\rho_o (v_{31}^2 + v_{32}^2)$ are independent of texture and give the sum of principal stresses $(T_1 + T_2)$. For fibre symmetry about Ox_1, $W_{420} = -\sqrt{10}\, W_{400}/3$, independently of the magnitude of the texture. $\rho_o (v_{31}^2 + 4 v_{32}^2)$ is therefore independent of texture to first order. If $W_{420} = A\, W_{400}$ so that the texture coefficients W_{420} and W_{400} vary in a proportionate manner then $v_{33}^2 / \Sigma v^2$ will vary linearly with $(v_{31}^2 - v_{32}^2)/\Sigma v^2$:

$$\frac{v_{33}^2}{\Sigma v^2} = \frac{(K + 4\mu/3)}{(K + 10\mu/3)} + \frac{1}{A}\sqrt{\frac{2}{5}}\, \frac{(v_{31}^2 - v_{32}^2)}{\Sigma v^2} \tag{14}$$

in the absence of stress (12). Deviations from this line can then be used to evaluate the level of stress present. Figure 3a shows a plot of equation (14) for the case of steel with fibre texture, the propagation direction being perpendicular to the fibre axis Ox_1. The effect of an applied stress of 300 MPa is also shown for a point with texture induced

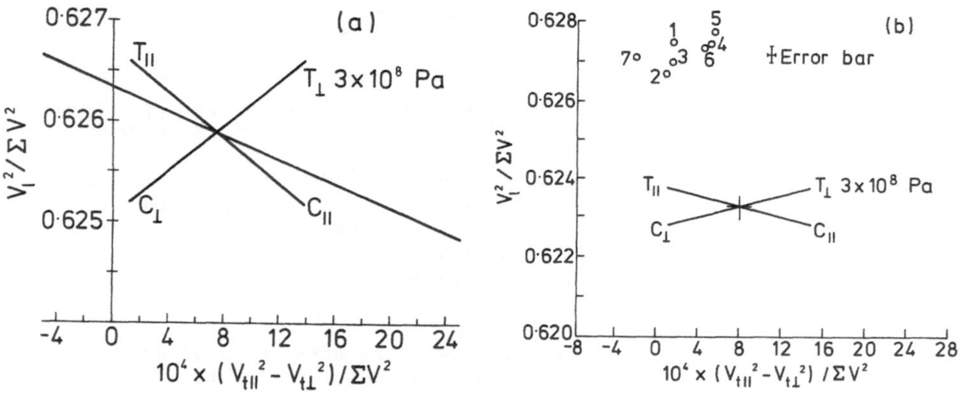

Figure 3(a) Velocity combinations plot for steel with fibre texture, the propagation direction being perpendicular to the fibre direction Ox_1. The effect of an applied stress of 300 MPa is shown for a point with a stress free birefringence of 0.2% T - tensile stress; C - compressive stress; // - parallel to rolling direction; ⊥ - perpendicular to the rolling direction. (b) Velocity combinations plot for the seven measurement positions on the compact tension specimen shown in figure 4a (12). The effect of an applied stress of 300 MPa on a point + is indicated.

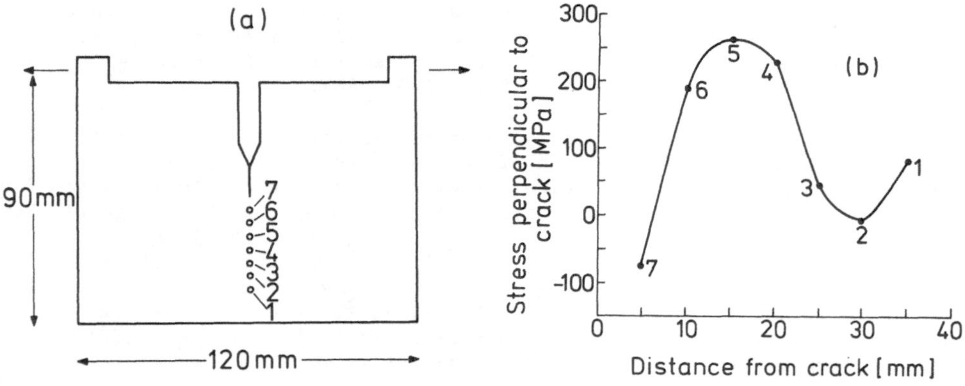

Figure 4(a) Schematic diagram of the compact tension specimen of 22 NiMoCr 3,7 steel (12,13) with the seven measurement points indicated. (b) Plot of residual stress as a function of distance from the crack tip for the compact tension specimen deduced from the velocity combinations plot (12).

birefringence of 0.2%. Measurements of the three times of flight on a large number of steel samples (12) showed the ratio W_{420}/W_{400} to vary from sample to sample. This method is therefore most suitable for determining the variation of stress across a sample or between specimens made by the same process for which the assumption $W_{420} = A\ W_{400}$ is valid. Figure 3b shows a velocity combinations plot for a scan near the crack in a 19mm thick compact tension specimen of 22 NiMoCr 3.7 steel with small anisotropy (12). Also shown in the figure are the displacements associated with different types of stress of 3×10^8 Pa in magnitude, calculated from the measured elastic constants λ, μ, l, m, and n (1). This provides a calibration for the determination of the stress present. Figure 4a shows a schematic diagram of the sample. Taking account of the measurement error shown, the distribution of data points suggests that the main component of stress lies in a direction perpendicular to the line of the crack. This information is unobtainable with the birefringence measurement alone, since the birefringence can only give the difference in principal stresses. The velocity combinations plot can be considered to provide a two-dimensional map of the stress and texture states in the sample, whereas the birefringence is a projection of this map onto one dimension with a resultant loss of information. The calculated stress distribution is shown in figure 4b and is in reasonable agreement with theory (13).

3.2 The SH wave method

In the absence of stress it is seen from equation (3) that an ultrasonic SH wave propagating in the plane of the plate has the same velocity for propagation parallel($\theta = 0°$) and perpendicular ($\theta = 90°$) to the rolling direction. This corresponds to an interchange of the propagation and polarisation directions of the wave. For an orthorhombic material in the presence of stress with components σ_{ij}.

$$\rho\ (v_{ij}^2 - v_{ji}^2) = \sigma_{ii} - \sigma_{jj} \qquad (15)$$

Here ρ is the density and v_{ij} is the phase velocity of a shear wave propagating in the direction x_i and polarised in the direction x_j, x_i and x_j being the principal axes of the elastic constant tensor (14-16). It is assumed that the principal stresses are parallel to the principal texture axes. This relationship is independent of the magnitude of the texture of the material. Its use for separating texture and stress was suggested by MacDonald (14) following earlier work of Biot (15) and Thurston (16).

A disadvantage of the method is that propagation of waves along the stress axis is necessary. However Thompson et al (17-19) have overcome this problem for thin plates by using electromagnetic acoustic transducers (EMATs) to excite the fundamental horizontally polarised shear (SH) wave of the plate travelling in the plane of the plate. The velocity of this mode is frequency independent, and is identical to the SH wave velocity in an unbounded medium with the same texture and stress distribution. Measurements on 6061-T6 aluminium, 304 stainless steel and commercially pure copper plate gave essentially equal velocities for propagation along the rolling and transverse directions as predicted by theory. The stress predicted using equation (15) was found to be in good agreement with the uniaxial tensile stress applied. Measurements on a rolled titanium sample did not show the expected behaviour, and this may be due to a failure of the assumption of orthorhombic symmetry for this sample.

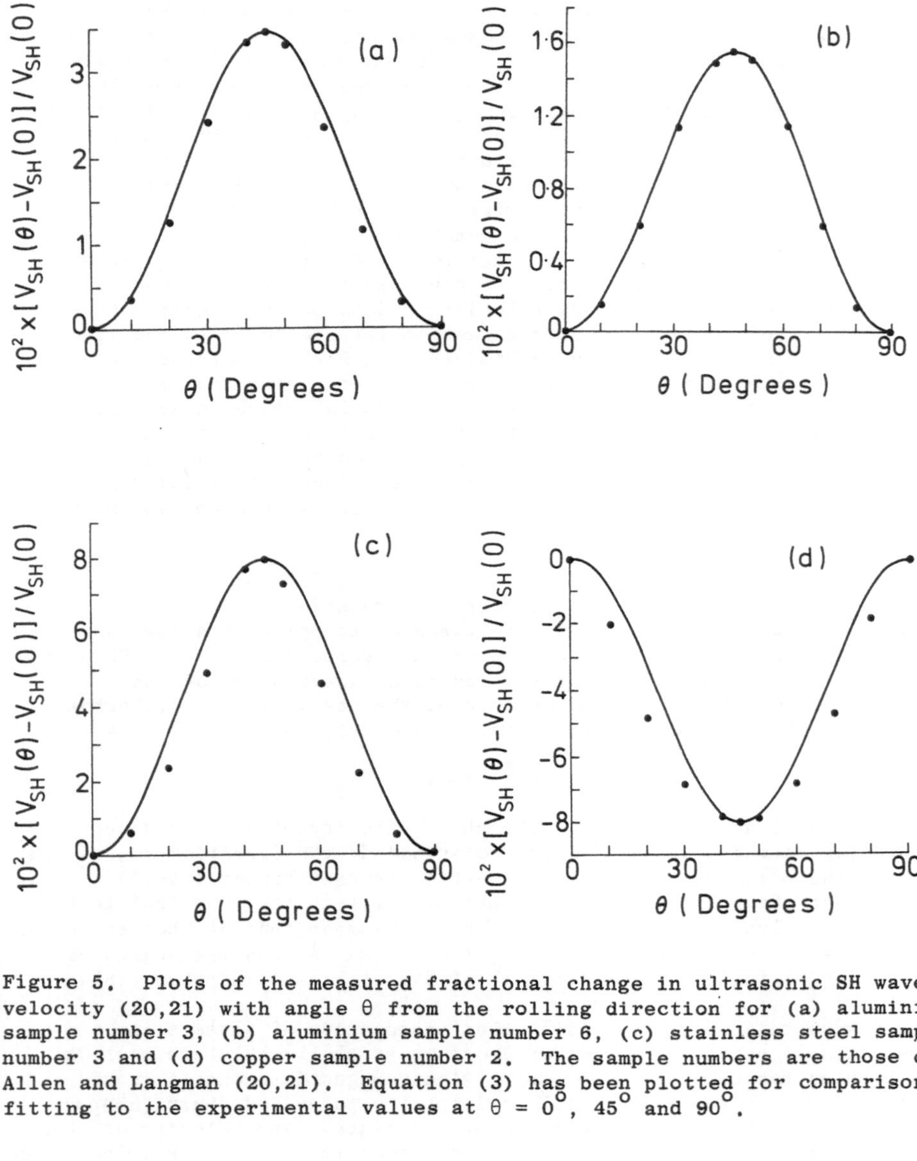

Figure 5. Plots of the measured fractional change in ultrasonic SH wave velocity (20,21) with angle θ from the rolling direction for (a) aluminium sample number 3, (b) aluminium sample number 6, (c) stainless steel sample number 3 and (d) copper sample number 2. The sample numbers are those of Allen and Langman (20,21). Equation (3) has been plotted for comparison by fitting to the experimental values at $\theta = 0^\circ$, 45° and 90°.

Allen and Langman (20,21) generated the zero-order SH plate wave using a conventional piezoelectric shear wave transducer aligned to produce horizontally polarised waves. Two EMATs mounted in a rigid base were used as receivers. The three transducers lie along the same axis and the phase delay associated with the propagation of the elastic wave between the two EMATs arising from a stimulus of the piezoelectric transducer is measured. Their results are in good agreement with theory at zero stress as may be seen in table 1 which summarises the results obtained on all the samples measured. The angular dependence of the SH wave velocity measured on four of the plates is shown in figure 5. Deviations from the theoretical curves shown in figure 5 occur because in anisotropic media the phase and group velocities are not parallel, this phenomenon being referred to as beam skewing. The effect of this on the angular variation of the SH wave velocity has been discussed elsewhere (22).

3.3 Rayleigh wave method

Let v_{Ri} denote the velocity of the Rayleigh wave propagating on the x_1 x_2 surface in the direction x_i. In the presence of texture and stress

$$\frac{\Delta v_R}{v_R^o} \equiv \frac{v_{R1} - v_{R2}}{v_R^o} = \left.\frac{(v_{R1} - v_{R2})}{v_R^o}\right|_{\text{texture}} + \left.\frac{(v_{R1} - v_{R2})}{v_R^o}\right|_{\text{stress}} \qquad (16)$$

where, from equation (2)

$$\left.\frac{v_{R1} - v_{R2}}{v_R^o}\right|_{\text{texture}} = \frac{16\pi^2 c\, W_{420}\, \beta}{7\sqrt{5}\, \mu} \qquad (17)$$

β is a function of Poisson's ratio and is given in reference (11).

The stress dependent is given by

$$\left.\frac{v_{R1} - v_{R2}}{v_R^o}\right|_{\text{stress}} = \frac{\alpha_1}{2\alpha_o} (\varepsilon_{11} - \varepsilon_{22}) \qquad (18)$$

α_o and α_1 being functions of λ, μ and the third order elastic constants l, m and n (23, 24).

It is seen from equations (10), (11), (17) and (18) that equations (9) and (16) depend only on two unknowns W_{420} and $(\varepsilon_{11} - \varepsilon_{22})$. The texture parameter W_{420} may therefore be eliminated to give the difference in principal strains $(\varepsilon_{11} - \varepsilon_{22})$:

$$\varepsilon_{11} - \varepsilon_{22} = \frac{\Delta v_R/v_R^o - \beta\, \Delta v_s/v_s^o}{\alpha_1/2\alpha_o - (1+n/4\mu)\beta} \qquad (19)$$

Equation (19) depends, for its validity, on the accuracy of equation (2). This has been checked by determining the parameters W_{420} and W_{440} which determine the angular dependence of the Rayleigh wave velocity using bulk waves (25) and comparing the predicted variation with that measured. For a stress free plate W_{440} can be determined from the angular dependence of the SH wave velocity (equation (3)), whilst the measured shear wave birefringence gives W_{420} (equation (10)). Two samples were used. The first was a 99.5% pure aluminium plate (BS 1470/1050A) and the second an aluminium 4.5% Mg alloy plate (BS 1470/5083/0). Both plates were 50mm

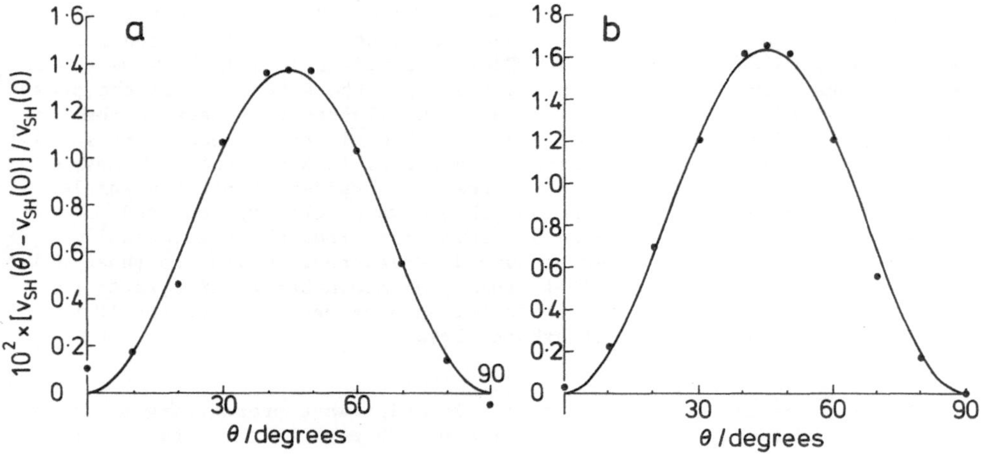

Figure 6. Variation of $(v_{SH}(\theta) - v_{SH}(0))/v_{SH}(0)$ with angle θ from the rolling direction for (a) a 99.5% pure aluminium plate and (b) an aluminium 4.5% magnesium alloy (25).

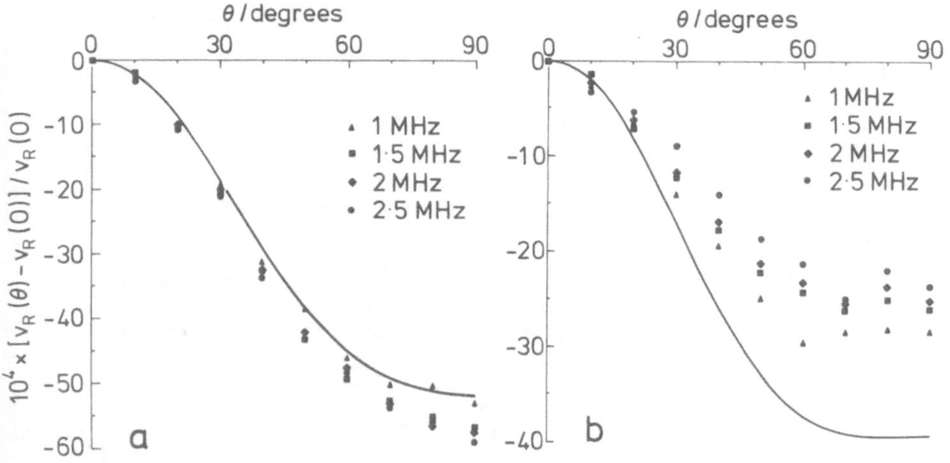

Figure 7. Variation of $(v_R(\theta) - v_R(0))/v_R(0)$ with angle θ from the rolling direction for (a) a 99.5% pure aluminium plate and (b) an aluminium 4.5% magnesium alloy at several different frequencies (25). The curve shows the prediction of equation (2).

thick. Figure 6 shows a fit of equation (3) to the data for the aluminium and aluminium alloy plates giving $W_{440} = -0.00256$ for the pure aluminium plate and $W_{440} = -0.00306$ for the alloy. The measured shear wave birefringence for the two plates were $\Delta v_s/v_s^o = 92.62 \times 10^{-4}$ and $\Delta v_s/v_s^o = 70.32 \times 10^{-4}$ respectively giving $W_{420} = -0.002267$ for the pure aluminium plate and $W_{420} = -0.001721$ for the alloy. Figure 7 shows the prediction of equation (2) using these values compared with the measured Rayleigh wave velocity made at several different frequencies (25). The main angular dependent term is seen to arise from the term R_2 involving W_{420}, which was obtained from the shear wave birefringence. Since this represents a through thickness average of the texture of the plate, the low frequency results should be in better agreement with the prediction of equation (2), and this is verified by the results shown in figure 7. Assuming the plates to contain zero stress, the results show that both plates have a surface texture different to the bulk average. However, in contrast to the pure aluminium plate which contains a texture gradient over a thin layer only, the aluminium alloy plate has a texture which varies over a longer length scale.

The theoretical model presented in section 2 was also checked by comparing the values of W_{420} and W_{440} obtained from ultrasonic bulk wave velocity measurements with values obtained by a neutron diffraction measurement of the texture in the sample. The values obtained from the (200) and (111) pole figures are in reasonable agreement with each other and with the values determined using ultrasonics as seen in table 2. This confirms that the main effect of the texture on the ultrasonic velocities is due to the crystallographic alignment of the grains and not due to other factors such as any grain shape anisotropy, etc.

4. Conclusion

In this paper the theoretical basis for treating the propagation of ultrasound in textured polycrystalline metals has been outlined. It has been demonstrated how quantitative information concerning the crystallite orientation distribution function can be obtained from the angular variation of ultrasonic velocities. Three methods for the texture independent measurement of stress in anisotropic polycrystalline metals have been described. Of these the Rayleigh wave method is capable of determining the variation of stress with depth by using the frequency dependence of the Rayleigh wave velocity. However, in the presence of a depth dependent texture, equation (19) is only approximately texture independent. As a result the combination of the Rayleigh wave velocity with that of either a skimming SV or longitudinal wave would be more satisfactory (equations (4) and (5)). The velocities of these waves need to be measured at a frequency at which the waves average over the depth in the same way as the Rayleigh wave. This should allow the depth dependence of both texture and stress to be obtained from the frequency dependence of the Rayleigh wave velocity.

REFERENCES

1. Sayers, C M and Allen, D R, J Phys D 17 (1984) 1399.
2. Hsu, N N, Exp Mech 14 (1974) 169.
3. Roe R J and Krigbaum, W R, J Chem Phys 40 (1964) 2608.
4. Roe R J, J Appl Phys 36 (1965) 2024.

5. Roe R J, J Appl Phys 37 (1966) 2069.
6. Bunge H J Z Metallk 56 (1965) 872.
7. Bunge H J Krist Tech 3 (1968) 431.
8. Morris P R, J Appl Phys 40 (1969) 447.
9. Sayers C M, J Phys D 15 (1982) 2157.
10. Allen A J, Hutchings M T, Sayers C M, Allen D R and Smith R L, J Appl Phys 54 (1983) 555.
11. Sayers C M, Proc Roy Soc A400 (1985) 175.
12. Allen D R and Sayers C M Ultrasonics 22 (1984) 179.
13. Schneider E and Goebbels K Periodic Inspection of Pressurised Components. Conference at Inst Mech Engs London (1982).
14. MacDonald D E, IEEE Trans on Sonics and Ultrasonics SU-28 (1981) 75.
15. Biot M A, J Appl Phys 11 (1940) 522.
16. Thurston R N, J Acoust Soc Am 37 (1965) 348.
17. Thompson R B, Smith J F and Lee S S in 'Review of Progress in Quantitative NDE 2' ed. D O Thompson and D E Chimenti, Plenum Press, New York (1983).
18. Thompson R B, Lee S S and Smith J F in 'Review of Progress in Quantitative NDE 3' ed. D O Thompson and D E Chimenti, Plenum Press, New York (1984).
19. Lee S S, Smith J F and Thompson R B in 'Review of Progress in Quantitative NDE 4' ed. D O Thompson and D E Chimenti, Plenum Press, New York (1985).
20. Allen D R and Langman R (1985) AERE-R11573.
21. Langman R and Allen D R (1985) AERE-R11598.
22. Allen D R, Langman R and Sayers C M Ultrasonics 23 (1985) 215.
23. Hayes M and Rivlin R S, Arch Rat Mech Anal 8 (1961) 358.
24. Hirao M, Fukuoka H and Hori K, J Appl Mech 48 (1981) 119.
25. Sayers C M, Allen D R, Haines G E and Proudfoot G G, Phil Trans Roy Soc (1986).

TABLE 1 Comparisons between ultrasonic phase delays for various samples of thin sheet measured by Allen and Langman (13,14). t_θ is the phase delay for propagation of SH waves at an angle θ° to the rolling direction, and t_{av} is the average of t_0 and t_{90}.

Material	Sample number	Thickness/mm	$\frac{t_0 - t_{90}}{t_{av}} \times 10^4$	$\frac{t_{45} - t_{av}}{t_{av}} \times 10^4$
Aluminium	1	1.60	1.8	374
	2	1.20	-1.3	-219
	3	1.00	-2.7	-335
	4	1.00	0.9	-231
	5	0.70	0.0	122
	6	0.50	0.4	-154
Stainless Steel	1	1.19	-4.9	-331
	2	0.91	-1.3	-537
	3	0.56	0.0	-733
Copper	1	1.17	-1.3	595
	2	0.98	-4.4	870
	3	0.60	-1.4	1684
Brass	1	1.00	0.6	-369

TABLE 2 Comparison of values of W_{400}, W_{420} and W_{440} determined by ultrasonics with those determined by neutron diffraction.

99.5% pure Al	W_{400}	W_{420}	W_{440}
Ultrasonics	-	-0.002267	-0.002560
(200)	0.005671	-0.0025925	-0.0027206
(111)	0.0060636	-0.0024745	-0.0026664
Al 4.5% Mg	W_{400}	W_{420}	W_{440}
Ultrasonics	-	-0.001721	-0.003060
(200)	0.0073897	-0.0015506	-0.002420
(111)	0.0072983	-0.0016496	-0.0024196

HARMONIC WAVES IN THREE-DIRECTIONAL CARBON-CARBON COMPOSITES

S. BASTE and A. GERARD

Summary

The propagation of harmonic elastic waves through three-directional composites with a periodic structure is analysed. Methods using the Floquet or Bloch theory common in the study of the quantum mechanics of crystal lattices are applied. Mixed variational principle in the form of integrals over a single cell of composite are developed, and applied for a 3D carbon-carbon composite.

1. INTRODUCTION

An important motivation for a detailed study of the propagation of harmonic waves in composites is that effective elastic constants can conveniently be measured by ultrasonics testing techniques /1/. These procedures require a good understanding of the dynamic behavior of composites. In the 3D carbon-carbon case, the wavelength is of the same order of magnitude as the characteristic length in the texture of this material, and so, the dispersion (i.e. the dependance of the phase velocity and the group velocity on the wavelength) is emphasised.Our wave propagation experiments on 3D-CC have also revealed higher modes of propagation, sometimes calledoptical modes, in addition to the three lowest modes (which are usually called the acoustical modes).

A composite consists in a collection of identical unit cells which repeat themselves in all directions, and hence form a periodic structure. The analysis of the propagation of harmonic waves through composite material is mathematically closely analogous to theory of propagation solution of the Schrodinger equation in periodic lattices /2/ ; it is natural to examine the composite problem by a variational approach used in solid-state physics. Since the elasticity tensor and the mass-density in elastic composites admit large discontinuities, the usual Rayleigh quotient /3/ or the stress Rayleigh quotient /4/ are not quite effective, whereas a mixedvariational method (i. e. a modified Reissner's variational principle)in which the displacement and the stress are given independent variations have been proposed by S. NEMAT-NASSER /5/. This author demonstratedthe accuracy and effectiveness of this method for layered and one-directional fiber-reinforced composites /6-7/.

In this paper, the problem of harmonic waves in a three-directional elastic composite will be examined from this standpoint, and explicit numerical results for a three-directional carbon-carbon composite in which the fibers lie in the three directions.

2. THEORETICAL BACKGROUND

2.1. Variational statement

Using a rectangular cartesian coordinates system we denote the displacement, the strain and the stress components by u_j, ε_{jk} and σ_{jk}, respectively, and assume that the body forces f_j measured by unit mass are given throughout the volume V occupied by an elastic body with mass-density ρ and elasticity coefficients $C_{jklm}(x)$, where x is the position vector of a point in V. The mass-density $\rho(x)$ and the elasticity tensor $C_{jklm}(x)$ are continuous and continuously differentiable functions of x in the subregions occupied by the matrix and by the fibers, but, in general, admit discontinuities at the boundaries between these two constituents.

The field equations are :

$$\sigma_{jk,k} + \rho f_j = 0 \quad ; \quad \sigma_{jk} = C_{jklm} \varepsilon_{lm} \; ; \; \varepsilon_{jk} = 1/2 \; (u_{j,k} + u_{k,j}), \tag{1}$$

with boundary conditions on S :

$$\sigma_{jk} n_k = T_j \quad , \quad u_j = U_j \, , \tag{2}$$

where n_j is the exterior unit normal on the regular surface S which bounds the body, T_j is the prescribed traction components and U_j the prescribed displacement components.

The Reissner's functional, in which the displacements, and the stress can be given arbitrary variation, is expressed in the following way :

$$J_r(u_j, \sigma_{jk}) = \int_V \{ 1/2 \, D_{jklm} \sigma_{jk} \sigma_{lm} + \rho f_j u_j - 1/2 \, \sigma_{jk} (u_{j,k} + u_{k,j}) \} \, dV$$

$$\int_{Sf} T_j u_j \, dS + \int_{Su} T_j (U_j - u_j) \, dS \quad , \tag{3}$$

where D_{jklm} is the elastic compliance matrix, and Sf and Su are two distinct parts of S.

2.2. Application to composites

A composite consists of hexahedron unit cells defined by means of three base vectors 1^β, $\beta = 1,2,3$. Because of the periodic structure :

$$\rho(x + 1^\beta) = \rho(x) \quad , \; C_{jklm} (x + 1^\beta) = C_{jklm} (x) \; . \tag{4}$$

For harmonic waves of pulsation ω and wave vector k_j, the field equations have periodic coefficients, and therefore, according to the Floquet theory, they admit solutions of the form :

$$u_j (x+1^\beta) = u_j (x) \exp (i \, k_l \, 1^\beta_l) \; , \; \sigma_{jk} (x+1^\beta) = \sigma_{jk} (x) \exp (i \, k_l \, 1^\beta_l) \, . \tag{5}$$

In particular, for x on the surface S, (5) defines the quasi-periodicity conditions for the displacement or the stress :

$$u_j (x+1^\beta) = u_j (x) \exp (i \, k_l \, 1^\beta_l) \; , \; T_j (x+1^\beta) = - T_j (x) \exp (i \, k_l \, 1^\beta_l), \tag{6}$$

where x on S. To arrive at a variational statement in which the displacement and the stress are varied independently, we identify the body forces by $1/2 \; \omega^2 u_j$, and from (3), obtain :

$$L = \int_V 1/2 \; \{ D_{jklm} \; \sigma_{jk} \sigma^*_{lm} + \rho \, \omega^2 u_j u^*_j - \sigma_{jk} (u^*_{jk} + u^*_{k,j}) + c.c.\} \; dV$$

$$+ \int_S \{ T_j \; (x+l^\beta) \; (u_j^* \; (x+l^\beta) - u_j^* \; (x) \exp \; (-i \; k_k \; l_k^\beta) \;) + c.c.\} \, dS \quad , \tag{7}$$

where the superscript star denotes the complex conjugate and the term c.c. stands for the complex conjugate of the quantities which precede it.

2.3. Approximation by exponential functions

For approximation, consider the test functions :

$$u_j = \sum_{\alpha,\beta,\tau=0}^{\pm N} U^{\alpha\beta\tau} \; f^{\alpha\beta\tau} \quad , \quad \sigma_{jk} = \sum_{\alpha,\beta,\tau=0}^{\pm N} S_{jk}^{\alpha\beta\tau} \; f \tag{8}$$

and the coordinate functions :

$$f^{\alpha\beta\tau}(k,x) = \exp \; \{ \; i(\; (K_1 + 2\pi\alpha)\frac{x_1}{a_1} + (K_2 + 2\pi\beta)\frac{x_2}{a_2} + (K_3 + 2\pi\tau)\frac{x_3}{a_3}) \; \}. \tag{9}$$

where a_j is the cell dimension, and $K_j = k_j . a_j$ is the dimensionless wave vector (no sum) and $U_j^{\alpha\beta\tau}$ and $S_j^{\alpha\beta\tau}$ are the Fourier coefficients. Note that these test functions are continuous and hence no jump conditions is necessary. Moreover, since the quasi-periodicity conditions (6) are also satisfied, the last term of (7) drops out, and we obtain the functional :

$$L = \int_V \{ \; D_{jklm} \, \sigma_{jk} \; \sigma_{lm}^* + \rho \, \omega^2 \; u_j \, u_j^* - \sigma_{jk} \, u_{j,k}^* - \sigma_{jk}^* \, u_{j,k} \} \, dV \; . \tag{10}$$

2.4. Matrix organization

We introduce the notation :

$$U = \{U_1, \; U_2, \; U_3\}^T \quad , \; S = \{S_{11}, \; S_{22}, \; S_{33}, \; S_{23}, \; S_{13}, \; S_{12}\}^T \quad ,$$

where

$$|U_j| = \{U_j^{-N \; -N \; -N}, \; U_j^{-N+1 \; -N \; -N}, \; \ldots \; \ldots, \; U_j^{N \; N \; N}\} \quad ,$$

$$|S_{jk}| = \{S_{jk}^{-N \; -N \; -N}, \; S_{jk}^{-N+1 \; -N \; -N}, \; \ldots \; \ldots, \; S_{jk}^{N \; N \; N}\} \; , \tag{11}$$

and write (10) as :

$$L = \begin{vmatrix} U^* \\ S^* \end{vmatrix} \begin{vmatrix} \Omega & H \\ H^* & \Phi \end{vmatrix} \begin{vmatrix} U \\ S \end{vmatrix} \; . \tag{12}$$

The stationary condition imposed on (12), leads to :

$$\Omega \, U + H \, S = 0 \quad , \quad H^* \, U + \Phi \, S = 0 \quad , \tag{13}$$

which yield :

$$S = - \; \Phi^{-1} \, H^* \, U \quad , \quad (\Omega - H \; \Phi^{-1} \, H^*) \, U = 0 \tag{14}$$

The second equation provides the eigenfrequencies and the corresponding eigenvectors, and the first equation gives the Fourier coefficients for stress fields. The characteristic equation is then :

$$\det (\Omega - H \quad \Phi^{-1} H^{*}) = 0 \quad . \tag{15}$$

For a given value of the wave vector K_j, the roots of this equation give the corresponding frequencies.

2.5. Ω,Φ and H matrices for a 3D composites

Reorganizing the eigenvalues system (14) in a dimensionless form, and calculating the Ω, Φ and H matrices for the 3D-carbon-carbon cell (see fig. 1), we find :

$$\{ \frac{\nu^2}{d} \quad \Omega - H \, \Phi^{-1} H^{*} \} U = 0 \quad . \tag{16}$$

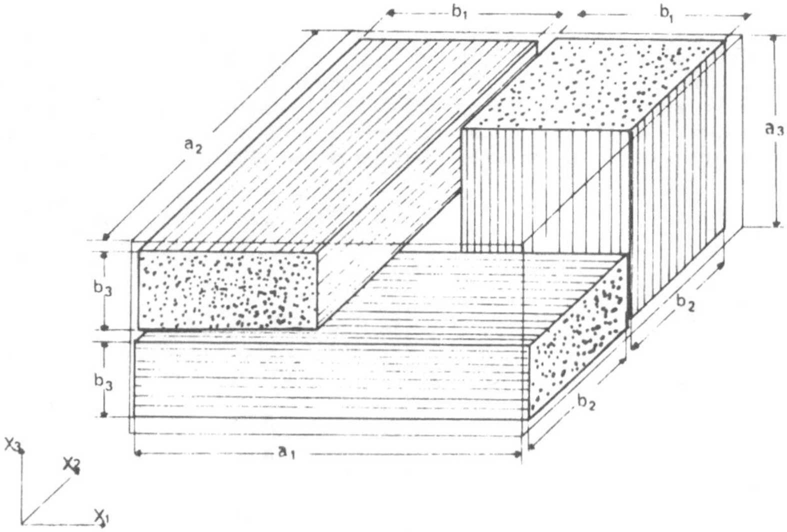

FIGURE 1. The 3D-CC cell

where matrices H, Ω and Φ are defined as follows :

$$H = \begin{vmatrix} H_1 & 0 & 0 & 0 & H_3 & H_2 \\ 0 & H_2 & 0 & H_3 & 0 & H_1 \\ 0 & 0 & H_3 & H_2 & H_1 & 0 \end{vmatrix} , \tag{17}$$

where H_1, H_2 and H_3 are $(2N+1)^3 \times (2N+1)^3$ matrices defined in the following manner :

- for $\alpha=\delta$ $\beta=\mu$ and $\tau=\gamma$, and with $K_j = k_j a_j$ (no sum)

$H_1 (I_1,J_1) = -i(K_1+2\pi\alpha)$, $H_2(I_1,J_1) = -i(K_2+2\pi\beta) n_0$, $H_3(I_1,J_1) = -i(K_3+2\pi\tau)m_0$

- for $\alpha \neq \delta$, $\beta \neq \mu$ or $\tau \neq \gamma$

$$H_1(I_1,J_1) = H_2(I_1,J_1) = H_3(I_1,J_1) = 0 \tag{18}$$

where

$$I_1 = (\alpha+1+N)+(\beta+N)\ (2N+1) + (\tau+N)\ (2N+1)^2,$$

$$J_1 = (\delta+1+N)+(\mu+N)\ (2N+1) + (\gamma+N)\ (2N+1)^2.$$

$\alpha,\beta,\tau = 0, \pm 1, \pm 2, \ldots\ldots, \pm N \qquad \delta,\mu,\gamma = 0, \pm 1, \pm 2, \ldots\ldots, \pm N$

$$\Phi = \begin{vmatrix} \Delta_{11} & \Delta_{12} & \Delta_{13} & 0 & 0 & 0 \\ \Delta_{12} & \Delta_{22} & \Delta_{23} & 0 & 0 & 0 \\ \Delta_{13} & \Delta_{23} & \Delta_{33} & 0 & 0 & 0 \\ 0 & 0 & 0 & \Delta_{44} & 0 & 0 \\ 0 & 0 & 0 & 0 & \Delta_{55} & 0 \\ 0 & 0 & 0 & 0 & 0 & \Delta_{66} \end{vmatrix} \qquad \Omega = \begin{vmatrix} \underline{\Omega} & 0 & 0 \\ 0 & \underline{\Omega} & 0 \\ 0 & 0 & \underline{\Omega} \end{vmatrix} \tag{19}$$

where $\underline{\Omega}$ and Δ_{jk} are $(2N+1^3) \times (2N+1^3)$ matrices defined in the following manner :

- for $\alpha \neq \delta$, $\beta \neq \mu$ and $\tau \neq \gamma$:

$$\Delta_{jk}\quad (I_1,J_1) = 0$$

- for $\alpha = \delta$, $\beta \neq \mu$ and $\tau \neq \gamma$:

$$= \Gamma_{jk}^{(1)} \exp \left| -i\pi/2\ \{(\beta-\mu) + (\tau-\gamma)\} \right| \frac{\sin\pi(\beta-\mu)m_2}{\pi(\beta-\mu)} \quad \frac{\sin\pi(\tau-\gamma)l_2}{\pi(\tau-\gamma)}$$

- for $\alpha \neq \delta$, $\beta = \mu$ and $\tau \neq \gamma$:

$$= \Gamma_{jk}^{(2)} \exp \left| i\pi/2\ \{(\tau-\gamma) - (\alpha-\beta)\} \right| \frac{\sin\pi(\alpha-\delta)n_2}{\pi(\alpha-\delta)} \quad \frac{\sin\pi(\tau-\gamma)l_2}{\pi(\tau-\gamma)}$$

- for $\alpha \neq \delta$, $\beta \neq \mu$ and $\tau = \gamma$: (20)

$$= \Gamma_{jk}^{(3)} \exp \left| i\pi/2\ \{(\alpha-\delta) + (\beta-\mu)\} \right| \frac{\sin\pi(\alpha-\delta)n_2}{\pi(\alpha-\delta)} \quad \frac{\sin\pi(\beta-\mu)m_2}{\pi(\beta-\mu)}$$

- for $\alpha = \delta$, $\beta = \mu$ and $\tau \neq \gamma$:

$$= \left| \Gamma_{jk}^{(1)} n_2 \exp \left| i\pi/2\ (\tau-\gamma) \right| + \Gamma_{jk}^{(2)}\ m_2 \exp \left| -i\pi/2\ (\tau-\gamma) \right| \right| \frac{\sin\pi(\tau-\gamma)l_2}{\pi(\tau-\gamma)}$$

- for $\alpha = \delta$, $\beta \neq \mu$ and $\tau = \gamma$:

$$= \left| \Gamma_{jk}^{(1)} n_2 \exp \left| i\pi/2\ (\tau-\gamma) \right| + \Gamma_{jk}^{(3)} l_2 \exp \left| -i\pi/2\ (\beta-\mu) \right| \right| \frac{\sin\pi(\beta-\mu)m_2}{\pi(\beta-\mu)}$$

- for $\alpha \neq \delta$, $\beta=\mu$ and $\tau=\gamma$:

$$= \left| \Gamma_{jk}^{(2)} m_2 \exp \left| i\pi/2 \ (\alpha-\delta) \right. + \Gamma_{jk}^{(3)} l_2 \exp \left| -i\pi/2 \ (\alpha-\delta) \right| \right| \frac{\sin\pi(\alpha-\delta) n_2}{\pi(\alpha-\delta)} ,$$

- for $\alpha=\delta$, $\beta=\mu$ and $\tau=\gamma$:

$$= \frac{D_{jk}^{(1)}}{\overline{D}_{11}} + \Gamma_{jk}^{(1)} m_2 l_2 + \Gamma_{jk}^{(2)} n_2 l_2 + \Gamma_{jk}^{(3)} n_2 m_2 .$$

Ω is obtained if one substitutes $\theta - 1/(\overline{n}_1 + \overline{n}_2\theta)$ for $\Gamma_{jk}^{(1)}$, $\Gamma_{jk}^{(2)}$ or $\Gamma_{jk}^{(3)}$ and $\overline{1}/(\overline{n}_1 + \overline{n}_2\theta)$ for $D_{jk}^{(1)}/\overline{D}_{11}$ in the expression for $\Delta_{jk}(I_1,J_1)$.

in the above expressions the following notations is used :

$$\nu^2 \ \frac{\omega^2 a_1^2 \overline{\rho}}{\overline{C}_{11}} , \quad \overline{\rho} = \rho^{(1)} \overline{n}_1 + \rho^{(2)} \overline{n}_2 , \quad \overline{C}_{11} = C_{11}^{f_1} \overline{n}_2 + C_{11}^{(1)} \overline{n}_1 ,$$

$$\overline{D}_{11} = D_{11}^{f_1} \overline{n}_2 + D_{11}^{(1)} \overline{n}_1 , \quad \overline{n}_1 = 1 - \overline{n}_2 , \quad \overline{n}_2 = \frac{b_1 b_2 b_3}{a_1 a_2 a_3} , \quad \theta = \frac{\rho^{(2)}}{\rho^{(1)}}$$

$$n_2 = \frac{b_1}{a_1} , \quad m_2 = \frac{b_2}{a_2} , \quad l_2 = \frac{b_3}{a_3} , \quad n_0 = \frac{a_1}{a_2} , \quad m_0 = \frac{a_1}{a_3} , \qquad (21)$$

$$d = 1/(\overline{C}_{11} \overline{D}_{11}) , \quad \Gamma_{jk}^{(1)} = \frac{D_{jk}^{f_1} - D_{jk}^{(1)}}{\overline{D}_{11}} , \quad \Gamma_{jk}^{(2)} = \frac{D_{jk}^{f_2} - D_{jk}^{(1)}}{\overline{D}_{11}} , \quad \Gamma_{jk}^{(3)} = \frac{D_{jk}^{f_3} - D_{jk}^{(1)}}{\overline{D}_{11}}$$

where ν is the dimensionless frequency, a_i the cell dimension, b_i the fiber dimension, $\rho^{(1)}$ the matrix mass-density, $D_{jk}^{(1)}$ the matrix compliance, $\rho^{(2)}$ the fiber mass-density and D_{jk}^{fi} the compliance of the fiber parallel to the i-axis.

3. NUMERICAL RESULTS

3.1. Test-functions truncations

We have to solve a general eigenvalue problem for $3(2N+1)^3 \times 3(2N+1)^3$ complex matrices. For N=1, our crudest approximation (i.e. 27 plane waves), we meet an eigenvalue problem for 81 x 81 matrices, and for N=2 (i.e. 125 plane waves) the order of the matrices is 375.

Table 1 shows the computation time (CPU time) needed for the resolution of the system (16) for N=1 and N=2 depending on the computer used.

Computations are made with the aid of subroutines DCMINV from the IMSL library and subroutines LZHES and LZIT writing by L. KAUFMAN /8/.

Since no exact solution exist for this problem, it is not possible to easily assess the acuracy of the results.

Mini 6 s.p. N=1	IBM 3080 s.p. N=1	IBM 3080 s.p. N=2	IBM 3080 d.p. N=1	CRAY 1 d.p. N=1	VAX 780 d.p. N=1	VAX 780 d.p. N=2
14 mn	13 s	24mn 45s	18 s	5.38s.	20mn	27h

TABLE 1. CPU time for N=1 or N=2

Tables 2 and 3 show the first three eigenfrequencies, as functions of the wave number K_1. This lowest three branches are the acoustical branches which correspond to the shear and longitudinal waves. For K=0, the corresponding eigenfrequencies will be equal to zero. Note that Ω et $H\ \Phi^{-1}\ H^*$ are hermitian matrices, and their eigenvalues will have real values.

	MINI 6 simple précision N = 1	IBM 3080 simple précision N = 1	IBM 3080 simple précision N = 2	IBM 3080 double précision N = 1	CRAY 1 double précision N = 1
$K_1 = 0.0$	0.017 - 0.04 i 0.017 + 0.04 i 0.0003 + 0.04 i	0.022 + 0.02 i 0.022 - 0.02 i 0.023 - $0.8\ 10^{-4}$ i	0.04 + 0.16 i 0.04 - 0.32 i 0.35 - $0.3\ 10^{-2}$ i	$0.5\ 10^{-10}$ - $0.7\ 10^{-6}$ i $0.3\ 10^{-9}$ + $1\ 10^{-7}$ i $0.1\ 10^{-5}$ + $7\ 10^{-12}$ i	
$K_1 = 0.628$	0.549 + $0.8\ 10^{-3}$ i 0.597 - $0.1\ 10^{-3}$ i 1.029 - $0.2\ 10^{-3}$ i	0.555 - $0.8\ 10^{-5}$ i 0.599 - $0.4\ 10^{-5}$ i 1.025 - $0.4\ 10^{-4}$ i	0.361 - 0.03 i 0.804 - 0.006 i 1.100 0.004 i	0.564 - $7\ 10^{-15}$ i 0.593 + $7\ 10^{-15}$ i 1.025 + $1\ 10^{-16}$ i	0.564 - $9\ 10^{-13}$ i 0.593 + $9\ 10^{-13}$ i 1.025 $7\ 10^{-14}$ i
$K_1 = 1.256$	1.159 - $0.9\ 10^{-2}$ i 1.160 + $0.9\ 10^{-2}$ i 2.033 + $0.9\ 10^{-2}$ i	1.119 - $0.2\ 10^{-3}$ i 1.195 + $0.2\ 10^{-3}$ i 2.032 - $0.2\ 10^{-4}$ i	1.118 - 0.02 i 1.270 + 0.03 i 2.159 0.01 i	1.128 - $4\ 10^{-12}$ i 1.187 + $3\ 10^{-12}$ i 2.032 + $3\ 10^{-13}$ i	

TABLE 2. Eigenfrequencies of harmonic waves in a 3D composites, $D_{11}^{(1)} = 0.235$.

In figure 2, the eigenfrequencies are plotted for $K_2 = K_3 = 0$, as functions of the wave number K_1 for N=1 and N=2 (only the first ten modes are listed).

As is seen from this figure, our crudest approximation which still exhibits dispersion and which corresponds to N=1, (i.e. 27 planes waves), gives for the first eighth, nine modes results which are quite accurate, in comparison with the second order of approximation (N=2). Note that the mixed variational method are neither lower nor upper bounds for the exact eigenfrequencies.

	IBM 3080 double précision N = 1	VAX 11/780 double précision N = 1	VAX 11/780 double precision N = 2
$K_1 = 0.0$	$0.1 + 2\ 10^{-7}\ i$ $0.3\ 10^{-6} - 3\ 10^{-7}\ i$ $0.3\ 10^{-6} + 2\ 10^{-7}\ i$	$0.8\ 10^{-24} + 6\ 10^{-8}\ i$ $0.2\ 10^{-7} + 1\ 10^{-9}\ i$ $0.1\ 10^{-6} + 1\ 10^{-7}\ i$	$0.6\ 10^{7} + 8\ 10^{7}\ i$ $0.9\ 10^{6} + 2\ 10^{6}\ i$ $0.1\ 10^{-5} + 1\ 10^{6}\ i$
$K_1 = 0.628$	$0.594 + 2\ 10^{-14}\ i$ $0.615\ \ 3\ 10^{-14}\ i$ $1.135 + 9\ 10^{-15}\ i$	$0.594 - 2\ 10^{-15}\ i$ $0.615 + 3\ 10^{-15}\ i$ $1.135\ \ 1\ 10^{-15}\ i$	$0.598 + 3\ 10^{-12}\ i$ $0.619\ \ 4\ 10^{12}\ i$ $1.229 + 5\ 10^{13}\ i$
$K_1 = 1.256$	$1.188 + 0\ i$ $1.231 + 0\ i$ $2.234 + 0\ i$	$1.188 + 0\ i$ $1.231 + 0\ i$ $2.234 + 0\ i$	$1.195 + 4\ 10^{-15}\ i$ $1.240 + 3\ 10^{15}\ i$ $2.394 - 1\ 10^{-16}\ i$

TABLE 3. Eigenfrequencies of harmonic waves in a 3 D composites, $D_{11}^{(1)} = 0.133$.

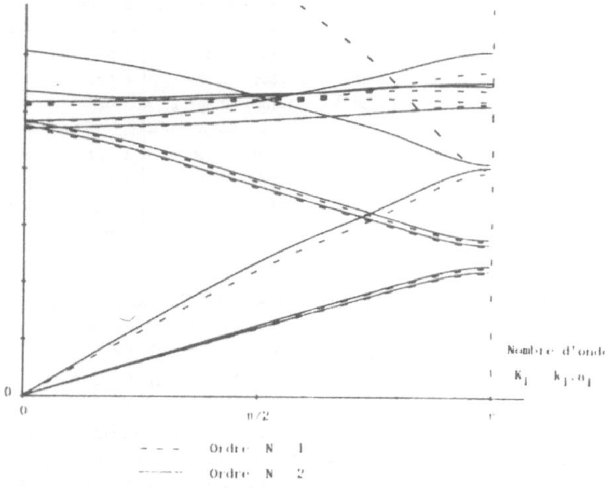

FIGURE 2. Dispersion curves for harmonic waves in a 3 D composite, $(D_{11}^{(1)} = 0.235)$, waves propagating parallel to the 1-axis.

3.2. Dispersion curves

Since the eigenfrequencies are periodic functions of the wave vector, and since the first Brillouin'S zone (i.e. $K_1 \in |0,\pi|$) contains one period of all frequencies, one needs only to estimate the frequencies as function of the wave vector in this Brillouin's zone. All informations for harmonic waves propagating through the composite can be stated in terms of the eigenfrequencies in this zone.

Figure 3 shows the dispersion curve for the waves propagating parallel to the 1-axis. The two lowest curves are for the shear waves, the third one the longitudinal wave. The other curves are the first optical branches. Figure 4 shows the dispersion curves which relate the dimensionless phase velocity to the dimensionless wave number K_1.
In figure 5, the phase velocity for each of the first ten modes are plotted for $K_2 = K_3 = 0$, as functions of the frequency.

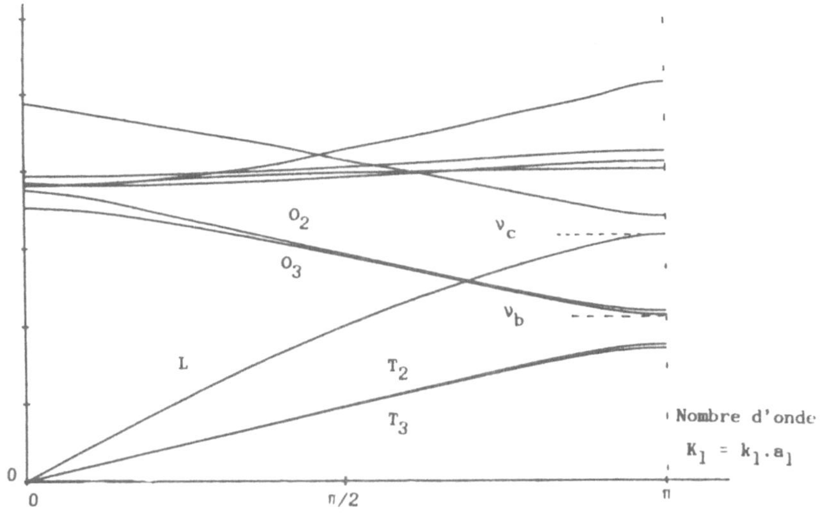

FIGURE 3. Dimensionless frequencies as function of the wave number K_1, ($D_{11}^{(1)} = 0.133$), propagation parallel to the 1-axis, N=1.

This dispersion curves clearly indicate the dispersive character and the band structure of the response associated with the periodicity of the composite. The 3D carbon-carbon acts as waveguide, and the existence of optical modes at low frequencies explain the complexity of the experimental response. The presence, below the cut-off frequency ν_c, of two optical modes with bass cut-off frequency ν_b, and whose spectrum intersects the longitudinal branch, justifies the experimental existence of slow guided mode below and over the cut-off frequency ν_c. The eight branch corresponds to the experimental fast guided mode, sometimes called "precursor", often met in the study of waves propagation through composite materials. Special attention should be paid to the fact that a 3D composite has a periodic structure in the three directions, and hence there are no total stop-bands, i.e. frequency ranges in which propagating harmonic waves are not possible. It has a pulse experimental response with a multi-modal imbricate character.

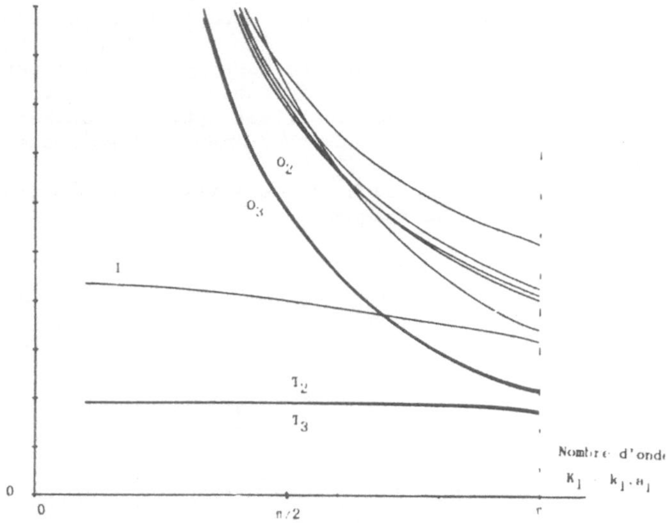

FIGURE 4. Dimensionless phase-velocity vs dimensionless wave number K_1, $D_{11}^{(1)} = 0.133$), propagation parallel to the 1-axis, N=1.

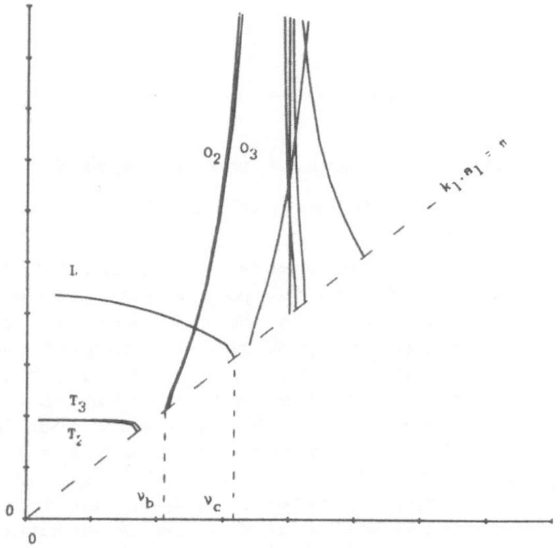

FIGURE 5. Dimensionless phase velocity vs. dimensionless frequency, ($D_{11}^{(1)} = 0.133$), propagation parallel to the 1-axis, N=1.

4. CONCLUSION

This approximate method gives, with acceptable computational effort, a satisfying model of static, and especially dynamic behavior of three-directional composites, in harmony with our experimental data.

ACKNOWLEDGEMENT

This work was supported by DRET/SNIAS n° 83/069 Contract.

REFERENCES

1. Roux J., Hosten B., Castagnède B. and Deschamps M.: Caractérisation mécanique des solides par spectro-interférométrie ultrasonore.:Revue Phys. Appli., 20,1985, 351-358.
2. Brillouin J., Parodi M.: Propagation des ondes dans des milieux périodiques: Masson et Cie, Dunod, 1956.
3. Kohn W., Krumhawsl J.A. et Lee E.H.: Variational methods for dispersions relations and elastic properties of composites materials: J. Appl. Mech., vol. 39, June 1972, 327-336.
4. Lee E.H.: A survey of variational methods for elastic wave propagation analysis in composites with periodic structures: Dynamics of Composite Materials, E.H. Lee, A.S.M.E., N.Y., 1972.
5. Nemat-Nasser S.: General variational methods for waves in elastic composites: J. Elasticity, vol. 2, n°2, June 1972, 73-90.
6. Nemat-Nasser S. et Minagawa S.: Harmonic waves in layered composites : Comparison among several schemes: J. Appl. Mech., vol. 42, Trans.A.S.M.E. vol. 97, Series E, 1975, 699-704.
7. Nemat-Nasser S. et Yamada M.: Harmonic waves in fiber-reinforced orthotropic elastic composites. J. Appl. Mech., vol. 48, A.S.M.E., 1981, 967-971.
8. Kaufman L.C.: The LZ algorithm to solve the generalized eigenvalue problem for complex matrix (F2): A.C.M. Trans. Math. Software, vol. 1, n°3, sept. 1975, 271-281.

DISCUSSION

Comment: Zarembowitch
I presume that it wouln't be very pleasant to introduce attenuation in your calculations. However, is attenuation (particularly for shear waves) high enough at 400 kHz?

Reply: Baste
It is not difficult to introduce attenuation in our calculations: it only results in writing a wave functional with complex terms of propagation. The true difficulty stands in the measurement of the viscoelastic constant of the fiber.

Comment: Sayers
How do you figure out the number of modes? Is there any general experimental method?

Reply: Baste
In the dispersion curves, only the first ten modes could be plotted; the $n = 1$ order gives the correct representation only for the first eight or nine modes.
An experimental method for distinguishing between all these quite imbricate modes, and for measuring them, is in progress.

ULTRASONIC QUANTUM OSCILLATIONS IN SEMIMETALLIC $Bi_{1-x}Sb_x$ ALLOYS*

M.Cankurtaran, H. Çelik and T. Alper

ABSTRACT

Ultrasonic quantum oscillations have been studied in Bi and semimetallic $Bi_{1-x}Sb_x$ alloys ($x < 0.04$). Experiments were carried out in the temperature range 1.2-4.2K using longitudinal ultrasonic waves of frequencies up to 170 MHz in the magnetic fields between 0.05-2.3T. With increasing Sb concentration the oscillation amplitude decreases, the period of the oscillations increases and the peaks become broader. In these alloys the electron Fermi surface becomes smaller isotropically, the electron effective masses and the energy gap between the conduction band and valence band decrease with increasing Sb concentration. The carrier lifetime decreases slightly with increasing Sb concentration which may imply that the relaxation mechanisms in the alloys are due mainly to the phase smearing caused by dislocations.

1. INTRODUCTION

The semimetallic character of bismuth is determined by the overlap of the valence band maximum at the T point with the conduction band minima at the L points of the reduced Brillouin zone. When Bi is alloyed with antimony, this overlap reduces with increasing Sb concentration. At an Sb concentration of $x \simeq 0.07$ the overlap is removed and the $Bi_{1-x}Sb_x$ alloy becomes semiconducting. As the Sb concentration is further increased to about $x \simeq 0.22$ the conduction band minimum at the L point overlaps an additional valence band extremum and the alloy system changes from the semiconducting state to a semimetallic one. The energy gap at the L point between conduction and valence bands (E_G) decreases with increasing Sb concentration. After the inversion of these two bands at $x \simeq 0.04$, this energy gap again increases.

Investigations of electronic properties are usually carried out at low temperatures using the de Haas-van Alphen (DHVA) effect, galvanomagnetic effects, the Shubnikov-de Haas (SDH) effect, cyclotron resonance, magneto-optical measurements and the technique of magneto-plasma waves. The ultrasonic quantum oscillations (UQO) were investigated first in very dilute $Bi_{1-x}Sb_x$ alloys in the composition range $0 < x < 0.001$ by Fukami et al(1). Cankurtaran et al(2) have studied extensively the electronic properties of semimetallic $Bi_{1-x}Sb_x$ alloys up to 4 % Sb by the UQO technique. Dominec et al(3) also reported UQO in the semimetallic $Bi_{1-x}Sb_x$. In this work much emphasis has been put on the effects of alloying on UQO.

The Fermi surface of the semimetallic $Bi_{1-x}Sb_x$ alloys ($0 < x < 0.07$) is similar to that of Bi. The electron Fermi surface consists of three ellipsoids (denoted by a, b and c) at the L points in the Brillouin zone. A principal axis of each of these ellipsoids coincides with the binary axes of the crystal; the other two axes are tilted from the trigonal plane by an angle of about 6°. Thewhole surface consists of a single ellipsoid of revolution about the trigonal (z) axis centred at the T points of the reduced Brillouin zone.

* This work was supported by the Scientific and Technical Research Council of Turkey (TBAG-622).

The conditions $w_c\tau>1$ and $\hbar w_c>kT$, which are necessary for observation of the quantum oscillatory effects introduced by a magnetic field, can be satisfied quite easily in semimetals, due to their small effective masses. Here τ is the mean lifetime of carriers, T is the absolute temperature of the system and w_c is the cyclotron frequency. When ultrasonic waves are propagated through metals at low temperatures the coefficient of ultrasonic wave attenuation shows oscillations with magnetic field known as ultrasonic quantum oscillations. The effect is periodic in reciprocal magnetic field (1/H). The period of the oscillations for an ellipsoidal Fermi surface is given by

$$\Delta(1/H)=\frac{2\pi e}{\hbar}\frac{1}{S(k_{HO})}=\frac{e\hbar}{m_c^*}\left(E_F+\frac{E_F^2}{E_G}-\frac{\hbar^2k_{HO}^2}{2m_H}\right)^{-1} \qquad (1)$$

where $S(k_{HO})$ is the cross-sectional area of the Fermi surface perpendicular to the magnetic field at $k_H=k_{HO}$, m_c^* is the cyclotron mass, E_F is the electron Fermi energy, E_G is the direct energy gap at the L points of reduced Brillouin zone. Since $k_{HO}<<k_F$, $S(k_{HO})$ may be considered to be the extremal cross-sectional area. Therefore the period measured experimentally provides information about the cross sections of the Fermi surface.

In this work we have studied UQO in the DHVA and intermediate regions. An expression for the ultrasonic attenuation coefficient $\alpha(H,T)$ in the DHVA region is given by Matsumoto and Mase[4] as

$$\alpha(T,H)=\alpha(0)\,\frac{2}{\pi}\,q\ell\Big[1+\left(\frac{\hbar w_c}{2E_F}\right)^{1/2}\sum_{r=1}^{\infty}\frac{(-1)^r}{r^{1/2}}\;\frac{2\pi^2 rkT/\hbar w_c}{\sinh(2\pi^2 rkT/\hbar w_c)}\;\cos\left(\frac{2\pi rE_F}{\hbar w_c}-\frac{\pi}{4}\right)$$

$$.\exp\left(-\frac{2\pi^2 rkT_D}{\hbar w_c}\right)\Big]. \qquad (2)$$

Here $\alpha(0)$ is the zero-field attenuation coefficient and $T_D=\hbar/\pi k\tau_D$ is the Dingle temperature; τ_D is the mean lifetime of the electrons which contribute to the attenuation of sound waves. T_D and m_c^* can be obtained from the variation of oscillation amplitude with temperature and magnetic field.

2. EXPERIMENTAL PROCEDURE

The single-crystal Bi was grown by the Bridgman method using 5N purity Bi purchased from Johnson Matthey Chemicals Limited, but the $Bi_{1-x}Sb_x$ alloy single crystals were prepared by the zone-levelling technique using 60 times zone-refined Bi and Sb of 5N purity purchased from the same supplier. A tubular resistance heater furnace with water-cooled ends was used in both purifying and alloying processes. This furnace provided a zone length of 10-15 mm and a temperature gradient of about 20 Kcm^{-1} at the solid-liquid interface. Using the crystal-growth and zone-melting system of Material Research Company (model Z83) a scanning speed of 2 mmh^{-1} was achieved. For the alloys with compositions up to about 3% Sb, this molten-zone speed may be adequate to exclude the cellular substructure which results from constitutional supercooling.

The single crystallinity was investigated by chemical etching and the Laue x-ray technique. The samples were cut from the single-crystal ingots by a Servomet spark cutter and spark planed. The orientations of the single crystals were determined by the Laue x-ray technique and by studying the pattern of etch pits and the slip lines on the cleavage plane (trigonal plane), which guided the identification of the bisectrix (y) and binary (x) axes. The error in the orientation of crystals is less than 1°.

The composition of $Bi_{1-x}Sb_x$ alloys was determined by density measurements and neutron activation analysis. The concentration gradient was measured to be about 3 parts in 1000 per cm.

X-cut quartz transducers with fundamental frequency of 8 and 10 MHz and nominal diameter of 6 and 10 mm were used in the experiments to propagate and detect the longitudinal ultrasonic waves of frequencies 10-170 MHz. Nonaq stopcock grease gave satisfactory transducer bonding. The ultrasonic attenuation coefficient was measured with the pulse-echo method using Matec Company's ultrasonic comparator (model 9000). Changes in the ultrasonic attenuation were detected by Matec's ultrasonic attenuation recorder (model 2470) where the heights of two selected echoes are compared in the differential amplifier. Since the output of this amplifier is proportional to the ultrasonic attenuation, automatic recording of the attenuation coefficient is possible. This system facilitates detecting and recording of as low as 0.02 dB changes in the attenuation coefficient. Static magnetic fields up to 2.3T were obtained by an electromagnet (Varian model 3800). The experiments were performed in the temperature range 1.2-4.2K. Temperatures lower than 4.2K were achieved by pumping the liquid helium. The pumping speed was controlled with a cartesian monostat and the presure over the liquid helium was monitored with an electronic manometer (CGS Scientific Company, type 1018).

For data acquisition an IBM-PC with a data acquisition add on card (Scientifi Solution, Inc., formerly Tecmar Inc., -Lab Master Board) was used. Signals were conditioned prior to A/D converter. The magnet signal was used as trigger to start and terminate the sampling. Within the magnet sweep time, usually 5 minutes, about 2000 data points with equal intervals were collected and saved for further processing. The processed data were plotted on an analog X-Y recorder using Lab Master's two D/A channels. 12 bit resolution of both A/D and D/A converters gave satisfactory accuracy. During the measurements the raw data were also recorded on a X-Y recorder.

The data were collected with four experimental configurations: $\vec{q}//z$ axis $\vec{H}//yz$ plane (case I), $\vec{q}//z$ axis $\vec{H}//xz$ plane (case II), $\vec{q}//z$ axis $\vec{H}//xy$ plane (case III) and $\vec{q}//y$ axis $\vec{H}//yz$ plane (case IV), where $\vec{q}$ is the longitudinal sound wave vector and x,y,z are the crystallographic axes. In the first two cases the magnetic field direction (θ) was measured from z axis but in the last two cases from y axis.

3. RESULTS AND DISCUSSION

The effects of the Sb concentration on the measured attenuation are illustrated in figure 1. The three oscillation curves in this figure were taken under the same experimental conditions. The scale of the ordinate is the same for the three curves and so the differences in the oscillation periods and amplitudes are clear. We select in particular the attenuation data taken at $\theta \simeq 25^{o}$, because in this orientation the three electron pockets become equivalent and so the attenuation curve against 1/H consists of a single period. The oscillation period and the amplitudes depend strongly on the Sb concentration; the period increases while the amplitudes decrease with increasing Sb concentration.

The effect of temperature on UQO is shown in figure 2. Although the peak positions and the period remain unchanged, the amplitude increases and the details become more pronounced with decreasing temperature. The temperature does not have substantial effect on the line widths. This may imply that the collision broadening of the electronic energy levels is much higher than the thermal broadening of the Fermi level.

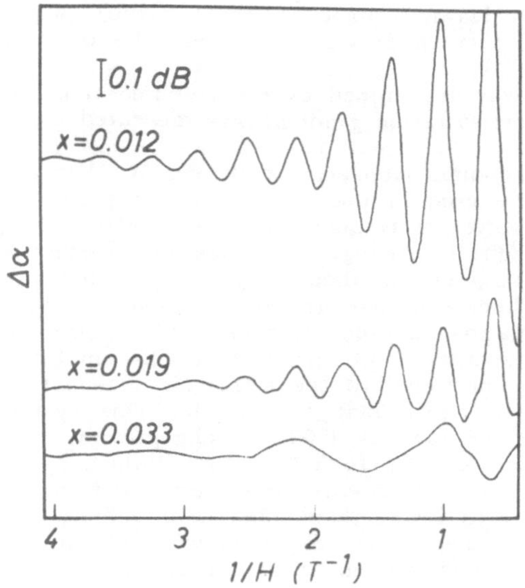

FIGURE 1

Concentration dependence of the oscillations in the attenuation versus 1/H in Case I, θ=25°, f=50 MHz and T=4.1K[2].

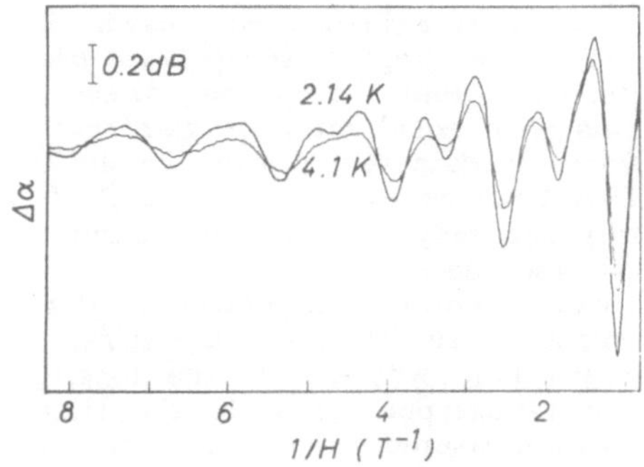

FIGURE 2

The effect of temperature on the attenuation in Case I, x=0.019, θ=100°, f=50 MHz. The ordinate scales are in arbitrary but common units.

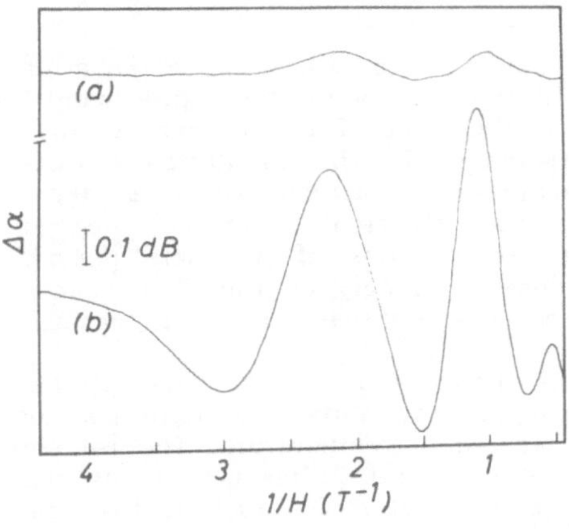

FIGURE 3

The effect of annealing on the attenuation in Case I, x=0.025, θ=90°, f=24 MHz, T=1.34K:
(a) unannealed,
(b) annealed.

Annealing causes substantial increase of the oscillation amplitude (figure 3). This may imply that the crystal defects play an important role in the scattering mechanism of electrons.

The period of the oscillations has been obtained from the slope of the peak number (n) against the reciprocal field (1/H) at which the n th peak occurs. If the electrons in only one carrier pocket participate in the attenuation, the plot of n against 1/H gives a straight line, which is the case, particularly in low magnetic fields. For the cases where more than one pocket contributes to the sound attenuation the periods of the oscillations have been found through Fourier analysis. For Fourier analysis the data havebeen sampled according to the Nyquist sampling criteria. Within the studied magnetic field range there are only several peaks; therefore in most of the cases 128 or 256 data points are adequate, which was achieved by the box-car averaging technique. A typical analysis has been exhibited in figure 4. Here contributions from more than one pocket can be seen (figure 4a). The Fourier spectrum of this data clearly reveals contributions from two pockets (figure 4b).

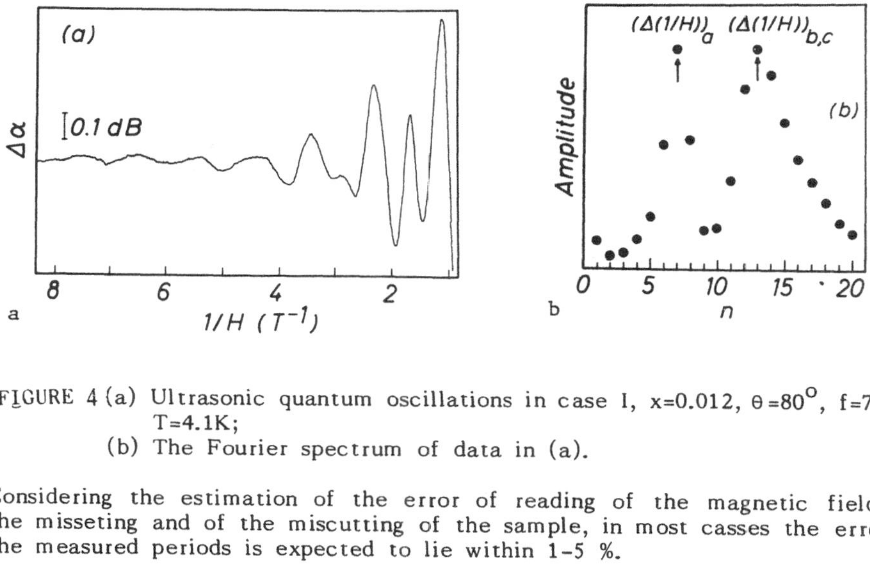

FIGURE 4 (a) Ultrasonic quantum oscillations in case I, x=0.012, $\theta=80^{o}$, f=70 MHz, T=4.1K;
(b) The Fourier spectrum of data in (a).

Considering the estimation of the error of reading of the magnetic field, of the misseting and of the miscutting of the sample, in most casses the error in the measured periods is expected to lie within 1-5 %.

3.1. The electron Fermi surface of semimetallic $Bi_{1-x}Sb_x$ alloys

The angular dependence of the periods for pocket a of the alloys and pure Bi in the yz plane is shown in figure 5. The period increases with increasing Sb concentration for all values of θ, but the shape of the curves of $\Delta(1/H)$ against θ is not dependent on concentration. This implies that the pockets shrink with increasing Sb concentration but keep their shape and orientation.

The effective masses of the principal ellipsoid (pocket a) in the crystal axis system and the values of $\Gamma(E)=E_F(1+E_F/E_G)$ have been obtained from the angular dependence of oscillation period (table 1). The carrier concentration (N) and the tilt angle (φ) of electron ellipsoids(2) are also included in this table. The decrease of N with increasing x is due to the reduction of the overlap of the bands at L and T points. The variations in the tilt angle are within the experimental error of crystal orientation and setting: the tilt angle in semimetallic $Bi_{1-x}Sb_x$ alloys is close to that in pure Bi.

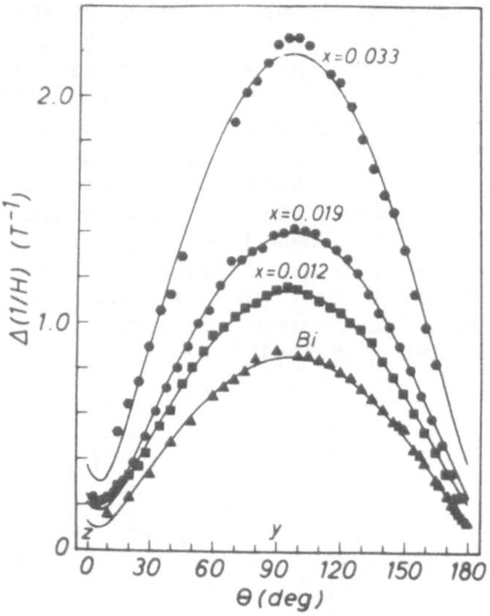

FIGURE 5

The concentration and angular dependence of the oscillation period for a-electrons in the yz plane (Case I)(2).

From table 1 it may be concluded that in semimetallic $Bi_{1-x}Sb_x$ alloys, the electron cyclotron mass decreases with increasing Sb concentration. The ratio of the band edge cyclotron masses of alloys to that of Bi of a-electrons is plotted in figure 6 as a function of Sb concentration. The variation is approximately linear. The extrapolated straight line passing through the experimental points crosses the x-axis at $x \approx 0.04$. In this critical concentration L-bands cross each other ($E_G \approx 0$).

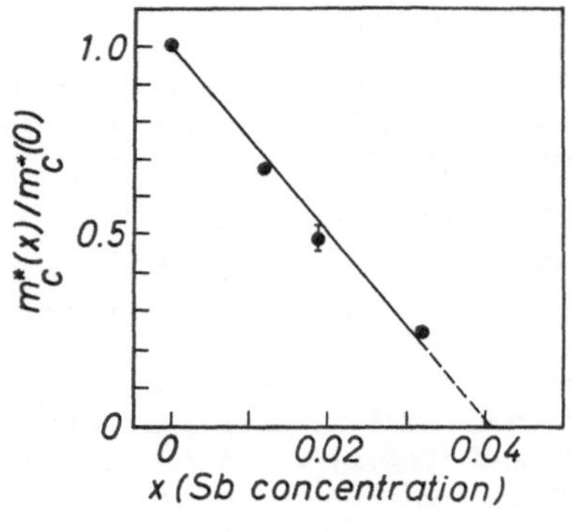

FIGURE 6

The concentration dependence of the electron cyclotron mass at the band edge normalised to that of Bi.

Table 1. Effective masses (in units of m_o) of the principal ellipsoid in the crystal axis system, $\Gamma(E)$, N, and E_F for $Bi_{1-x}Sb_x$ alloys.

x	m_1	m_2	m_3	m_4	$\Gamma(E)$ (meV)	N ($10^{23}m^{-3}$)	φ (deg)	E_F (meV)
0	0.001020	0.224	0.00526	-0.0226	81.5	2.52	-6.1	27.6
0.012	0.000436	0.127	0.00460	-0.0135	87.1	1.47	-6.2	24.6
0.019	0.000434	0.097	0.00318	-0.0121	98.2	1.11	-7.2	22.4
0.033	0.000204	0.056	0.00186	-0.0066	114.8	0.59	-6.9	16.1

It is known that in semimetallic $Bi_{1-x}Sb_x$ alloys ($x<0.07$) the electron Fermi energy decreases with increasing Sb concentration[5,6]. However, as seen from table 1, the value of $\Gamma(E)$ increases with increasing x. This is further evidence that E_G decreases with increasing Sb concentration. Thus the decrease of cyclotron masses in the alloys is correlated with the decrease of E_G.

Using the measured periods in equation (1) and considering the fact that the semimetal-semiconductor transition takes place at $x \simeq 0.07$, the variation of S(x)/S(o) with x was estimated to be

$$S(x)/S(o) \simeq 1-28x-1.9x^2 \tag{3}$$

as seen in figure 7.

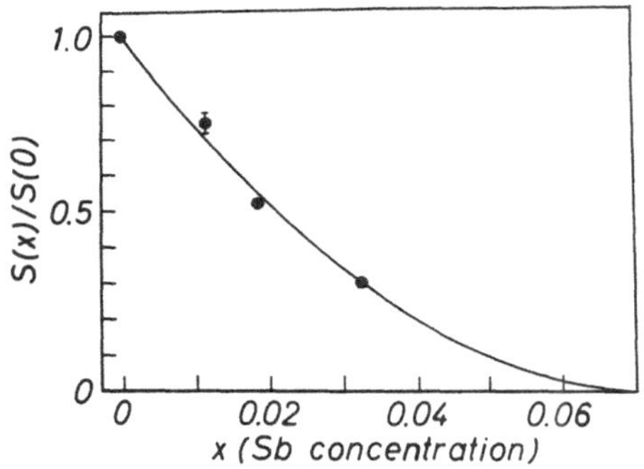

FIGURE 7.
The concentration dependence of the ratio S(x)/S(o).

The electron Fermi energy was determined from the charge neutrality condition using the electron effective masses given in table 1[6]. In the calculations it is assumed that E_G and the overlap energy (E_{ov}) decrease linearly with x, but the hole effective masses do not change with increased values of Sb concentration[5]. For pure Bi we have used E_G=15.3 meV and E_{ov}=38.5 meV. The results are included in table 1; E_F decreases with increasing Sb concentration.

3.2. The variation of the oscillation amplitude with temperature and magnetic field

The variation of the oscillation amplitude with temperature and magnetic field may be estimated from equation (2). However, it is necessary to take into account the effects of non-parabolicity and contributions from various carrier pockets. But equation (2) may be used in the cases when only one pocket participates in the attenuation and only the temperature and magnetic field dependences of the oscillation amplitude are of interest. In low magnetic fields and if only the first harmonic has been considered, equation (2) may be simplified to

$$\ln(\alpha(T)_{amp}/T) = \text{constant} - (2\pi^2 k/\hbar w_c)T \qquad (4)$$

Here it is assumed that T_D and $\hbar w_c$ are independent of temperature. The higher harmonics alter the breadth of the oscillations rather than their amplitude and peak positions. From the slope of $\ln[\alpha(T)_{amp}/T]$ versus T plot, $w_c(=eH/m_c^* c)$, hence the cyclotron mass can be determined. Figure 8 shows examples for the temperature dependence of the oscillation amplitudes. It is clear that, to take only the first harmonic into account may be adequate to describe the temperature dependence of DHVA type oscillations. The effective masses obtained from the temperature dependence of the oscillation amplitudes are consistent with those from period measurements(2). The difference between them may be attributed to neglected effects of the variation of E_F with magnetic field, the errors in estimating both E_F and E_G, and possible contributions from other carrier pockets. However, in the recent work of Cankurtaran et al (6) it was shown that the variation of E_F with magnetic field has negligible effect on the period of light electrons.

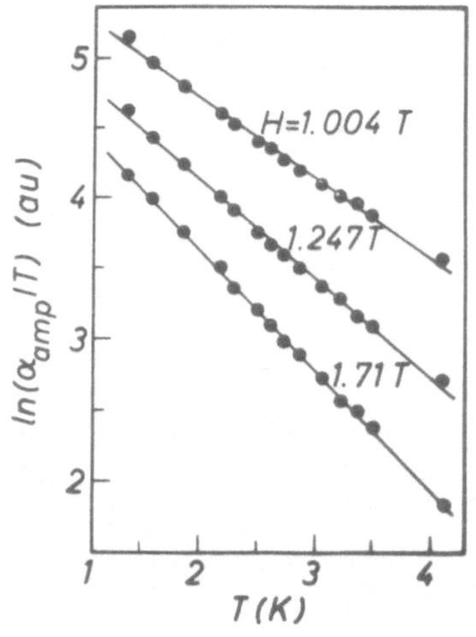

FIGURE 8.

The temperature dependence of the oscillation amplitudes.
x=0.012, Case I, $\theta=7^o$, f=50 MHz, a-electrons.

In the DHVA region it is possible to obtain the mean lifetime of carriers (τ_D) from the variation of oscillation amplitude with magnetic field. At low magnetic fields and at a constant temperature, equation (2) may be approximated by

$$\ln(\alpha(H)_{amp}\sqrt{H})=\text{constant}-[2\pi^2 k m_F^*(T+T_D)/e\hbar]/H \qquad (5)$$

where H is the magnetic field at which a particular peak occurs and m_F^* is the cyclotron mass at the Fermi level. As an example we show $\ln(\alpha_{amp}\sqrt{H})$ against 1/H for Bi and two of the alloys in figure 9. It should be noted that there is a small deviation from linearity which is possibly caused by the effects of neglected higher harmonics and the variation of the relaxation mechanism of carriers with magnetic field(7,8). T_D and τ_D have been obtained (table 2) by using the effective masses from period analysis and the slope of straight lines such as in figure 9. τ_D decreases slightly with increasing Sb concentration. From galvanomagnetic measurements in weak fields, Brandt et al(9) and Red'ko et al(10) deduced that τ_D is of the order of 5×10^{-12} s for semimetallic $Bi_{1-x}Sb_x$ alloys. Similar values of τ_D were reported for pure Bi and Sb(4,11).

The relaxation times for electrons in pure Bi and semimetallic $Bi_{1-x}Sb_x$ alloys up to 2 % obtained from the present magnetoacoustic measurements (table 2) are close to each other. It seems that Sb atoms regarded as impurities do not affect the mean lifetime of carriers substantially.

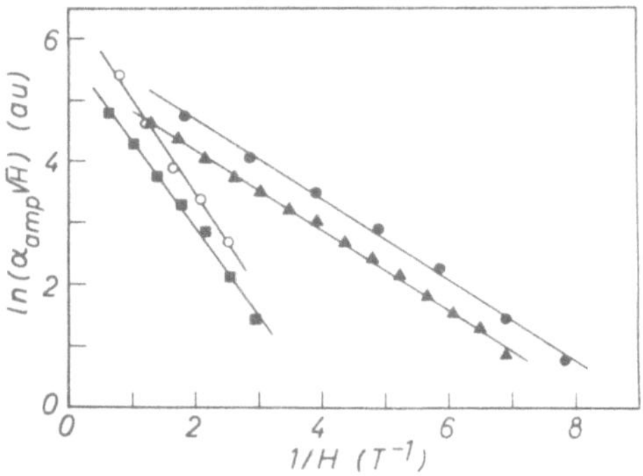

FIGURE 9. The magnetic field dependence of the oscillation amplitudes(2).

▲ , pure Bi, Case I, θ=25°, T=1.75K;
○ , x=0.012, Case I, θ=165°, T=4.1K;
● , x=0.012, Case II, θ=85°, T=4.1K;
■ , x=0.019, Case I, θ=25°, T=2.28K.

The full lines are the best fitsof equation(5).

Table 2. Dingle temperature and mean lifetimes for electrons in semimetallic $Bi_{1-x}Sb_x$ alloys(2).

x	T_D(K)	$\tau_D(10^{-12}s)$
0	0.95	2.56
0.012	1.44	1.68
0.019	1.93	1.26

The diffusion coefficient of Sb in solid Bi is ignorably small. Therefore, annealing has negligible effect on diffusion of Sb atoms in the alloy. However as seen from figure 3 the oscillation amplitudes increase substantially with annealing. In theoretical and experimental studies on dilute alloys (x<0.001) it is proposed that the amplitude of the oscillations is not much sensitive to Sb concentration and by taking into account the phase smearing of the energy levels, the Dingle temperature was found to be 1.7K(12,13). Although we have observed rather different line shape compared to that of very dilute alloys(12), we have predicted similar values for the Dingle temperature for as much as 20 times more concentrated alloys. Thus, the relaxation mechanisms are due mainly to the phase smearing caused by dislocations.

Although Bi and $Bi_{1-x}Sb_x$ alloys have similar band structure and the electron lifetimes are in the same order of magnitude, attenuation curves differ substantially at $\theta \simeq 90^o$, where the sound waves propagate nearly perpendicular to the magnetic field (figures 10,11). To explain the existance of the oscillations at $\theta \simeq 90^o$ in alloys, Fukami et al(1) proposed that Sb atoms distribute inhomogeneously which causes dispersion of sound waves. Therefore even at $\theta \simeq 90^o$ the component of the propagation vector along the magnetic field is not zero.

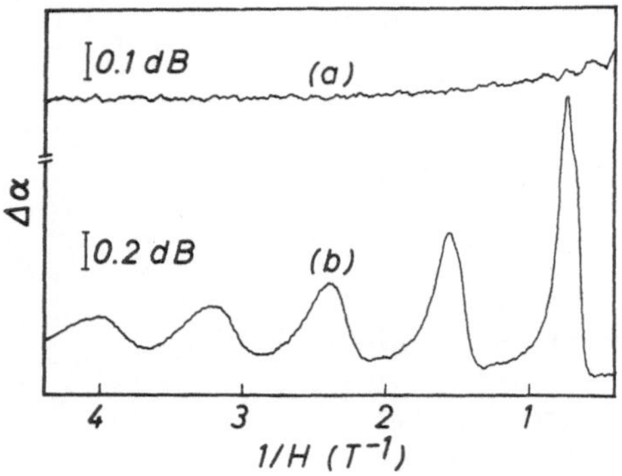

FIGURE 10.

Ultrasonic quantum oscillations in pure Bi in Case IV, f=10 MHz, T=1.38K
(a) $\vec{H} \perp \vec{q}$,
(b) $\vec{H} // \vec{q}$.

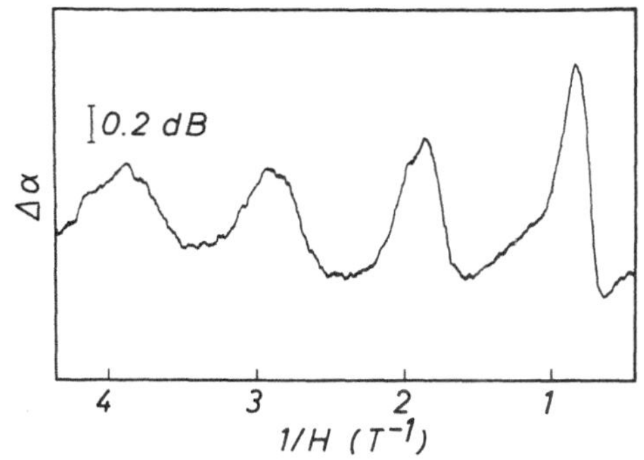

Figure 11.

Ultrasonic quantum oscillations in $Bi_{0.988}Sb_{0.012}$ in Case III, f=170 MHz, T=4.1K, $\vec{H} \perp \vec{q}$.

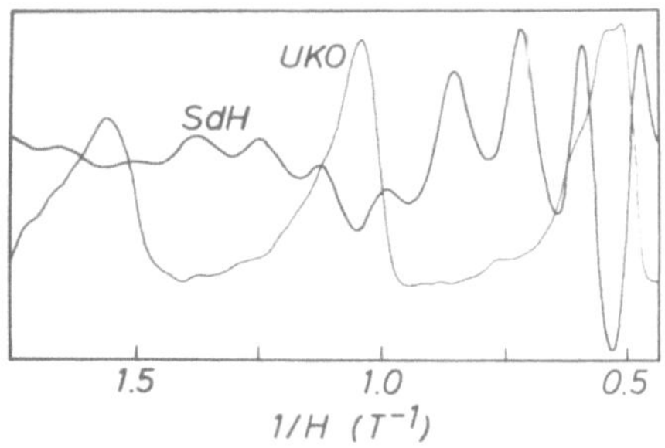

Figure 12.

Comparison of UQO and SDH oscillations in Bi(14).

The results of two different techniques (UQO and SDH effect) are shown in figure 12(14), where ultrasonic oscillations belong to the a-electrons and SDH oscillations belong mainly to the holes. In the investigation of the Fermi surface and the magnetic energy levels of carriers, UQO provide a complementary technique to the others.

REFERENCES

1. Fukami T, Akinaga M, Yamaguchi T and Mase S: J.Phys. Soc. Japan, **47** 435,1979.
2. Cankurtaran M, Çelik H and Alper T: J.Phys F: Metal Phys. **15** 391, 1985.
3. Dominec J, Misek K, Kraak W and Herrmann R: Phys. Status Solidi b **123** 635, 1984.
4. Matsumoto Y and Mase S: J.Phys. Soc. Japan **38** 1328, 1975.
5. Brandt N B and Chudinov S M: Sov. Phys. -JETP **32** 815, 1971.
6. **Cankurtaran M, Çelik H and Alper T: (to be published in J.Phys. F: Metal Phys.) 1985.**
7. Shoenberg D and Vanderkooy J: J. Low Temp. Phys. **2** 483, 1970.
8. Wilde J de and Groot D G de: J. Phys F: Metal Phys. **8** 1131, 1978.
9. Brandt N B, Dittmann Kh and Ponomarev Ya G: Sov. Phys.-Solid State **13** 2408 1972.
10. Red'ko N A, Pol'shin V I and Ivanov G A: Sov. Phys. Solid State **26** 5 1984.
11. Phillips R A and Gold A V: Phys. Rev. **178** 932, 1969.
12. Mase S, Akinaga M and Matsumoto Y: J. Phys. Soc. Japan **50** 3321 1981.
13. Mase S, Fukami T, Akinaga M, Matsumoto Y: J. Phys. Soc. Japan **50** 3329 1981.
14. Çelik H, Alper T, Sümer K, Cankurtaran M: Doğa Bilim Dergisi A1, **9** (inpress), 1985.

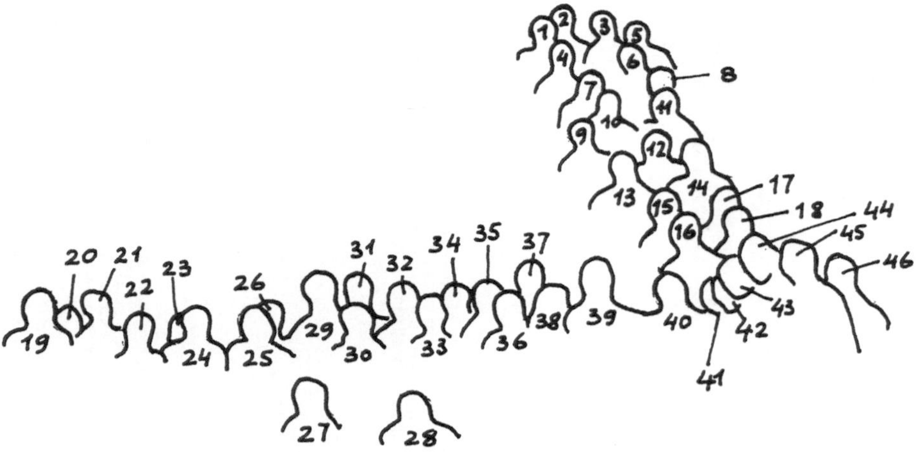

Attendants in the Erice S.Francesco's courtyard

1. A.Zarembowitch
2. G.Bonnet
3. R.Zilber
4. R.C.Chivers
5. E.J.Schmidt
6. B.Terliuc
7. A.Kulik
8. J.F. de Belleval
9. G.Busse
10. G.Gagliardi
11. M.Mezzana
12. A.Sliwinski
13. G.Gremaud
14. J.D.N.Cheeke
15. W.G.Mayer
16. A.Alippi
17. V.Bucur
18. C.M.Sayers
19. C.Sinclair
20. B.Zeqiri
21. G.Carlotti
22. C.Picornell
23. K.Mampaert
24. G.Socino
25. A.Markiewicz
26. D.A.Hutchins
27. D.Rypien
28. L.Palmieri
29. M.Musulluoglu
30. M.A.Neto Coelho
31. J.D.Skinner
32. B.A.Auld
33. L. Covi
34. S.Baste
35. L.Adler
36. J.A.Gallego Juarez
37. J.W.Wolf
38. Mrs W.Madigosky
39. W.Madigosky
40. E.Soczkiewicz
41. H.T.Tran
42. G.R.Laguna
43. A.Marini
44. T.Alper
45. H.Karagülle
46. F.Cordero

NATO Advanced Study Institute
on
ULTRASONIC METHODS IN EVALUATION OF INHOMOGENEOUS MATERIALS
Erice, 15 - 25 October, 1985

LIST OF PARTICIPANTS

Prof. L. Adler
Ohio State University
190 West 19th Av.
Columbus, Ohio 43219, USA

Prof. A. Alippi
Università di Roma "La Sapienza"
Dipartimento di Energetica
Via A. Scarpa, 14
00185 Roma, ITALY

Dr. T. Alper
Hacettepe University
Faculty of Engineering
Dept. Physics Engineering
Beytepe Ankara, TURKEY

Prof. B.A. Auld
Edward Ginzton Laboratory
Stanford University
Stanford, CA 94305, USA

Dr. S. Baste
Université de Bordeaux I
Laboratoire de Mecanique Physique
351, Cours de la Liberation
33405 Talence Cedex, FRANCE

Dr. G. Bonnet
Institut de Mecanique de Grenoble
Domaine Universitaire
B.P. n. 68
38402 Saint-Martin d'Heres Cedex, FRANCE

Dr. V. Bucur
Centre National de Recherches Forestières, Nancy-Amance
Champenoux 54280 Seichampes, FRANCE

Dr. G. Busse
University of Stuttgart
IKP, Postfach 80 11 40
D-7000 Stuttgart 80, FRG

Mr. G. Carlotti
Università di Perugia
Dipartimento di Fisica
Via Elce di Sotto
06100 Perugia, ITALY

Prof. J.D.N. Cheeke
Université de Sherbrooke
Dept. de Physique
Sherbrooke, Quebec J1K 2R1, CANADA

Dr. R.C. Chivers
Dept. Theoretical & Applied Mathematics
Thurston Hall
Cornell University
Ithaca, NY 14853, USA

Dr. F. Cordero
Istituto di Acustica "O.M. Corbino" CNR
Via Cassia 1216
00189 Roma, ITALY

Dr. F. Craciun
Institute of Physics & Technology of Materials
P.O. Box MG-07
Bucharest-Magurele 76900, RUMANIA

Dr. J.F. de Belleval
Université de Technologie de Compiegne
Div. Acoustique et Vibration Ind.
B.P. 233
60206 Compiegne Cedex, FRANCE

Dr. G. Gagliardi
Aeronautica Militare Italiana
Direzione Laboratori
Via Tuscolana, 473
Roma, ITALY

Dr. J.J. Gagnepain
CNRS - LPMO
32, Av. de l'Observatoire
25000 Besançon, FRANCE

Dr. J.A. Gallego Juarez
Instituto de Acustica
Centro de Fisica Aplicada
Serrano 144
28006 Madrid, SPAIN

Dr. G. Gremaud
EPFL - Institut de Genie Atomique
PH - Ecublens
1015 Lausanne, SWITZERLAND

Dr. D.A. Hutchins
Physics Department
Stirling Hall, Queen's University
Kingston, Ontario K7L 3N6, CANADA

Dr. H. Karagülle
Dokuz Eylul Universitesi
Makina Muhendisligi Bolumu
Bornova - IZMIR, TURKEY

Dr. A. Kulik
EPFL - Insitut de Genie Atomique
PH - Ecublens
1015 Lausanne, SWITZERLAND

Mr. G.R. Laguna
Stanford University
Ginzton Laboratories
Stanford, CA 94305, USA

Dr. W. Madigosky
Naval Surface Weapons Center
10901 New Hampshire Av.
Silver Spring, MD 20903, USA

Mrs. K. Mampaert
K.U. Leuven - Campus Kortrijk
Universitaire Campus
8500 Kortrijk, BELGIUM

Mr. A. Marini
Università di Perugia
Dipartimento di Fisica
Via Elce di Sotto
06100 Perugia, ITALY

Dr. A. Markiewicz
Uniwersytet Gdanski
ul. Wita Stwosza 57
Gdansk, POLAND

Prof. W.G. Mayer
Physics Department
Georgetown University
Washington, DC 20057, USA

Mr. M. Mezzana
Aeronautica Militare Italiana
Direzione Laboratori
Via Tuscolana, 473
Roma, ITALY

Mr. M. Musulluoglu
Gazi University
Makina Muh. Bolumu
Maltepe Ankara, TURKEY

M.A. Neto Coelho
Dept. de Engenharia Quimica
Faculdade de Engenharia
Rua dos Bragas
Porto, PORTUGAL

Prof. L. Palmieri
Università di Perugia
Dipartimento di Fisica
Via Elce di Sotto
06100 Perugia, ITALY

Mrs. C. Picornell
Universitat Illes Balears
Fac. Ciènces
Ctra Valldemossa 7,5
Palma de Mallorca, SPAIN

Mr. D.V. Rypien
Ohio State University
Dept. of Welding Eng.
190 19th Ave.
Columbus, Ohio 43210, USA

Dr. C.M. Sayers
AERE Harwell
OX11 ORA, UK

Mr. E.J. Schmidt
Stanford University
Dept. of Geophysics
Stanford, CA 94305, USA

Dr. R. Schmitt
Fraunhofer Institut
IfzP - Abt. Medizintechnik
Universität Gebaude 37
D-6600 Saarbrücken 11, FRG

Dr. J.D. Skinner
General Electric Company
Hirst Research Center
East Lane, Wembley
Middlesex HA97 PP, UK

Prof. A. Sliwinski
Inst. Fyzyki Dosw.
Uniwersytet Gdanski
ul. Wita Stwosza 57
80-952 Gdansk, POLAND

Prof. G. Socino
Dipt. Fisica, Univ. di Perugia
Via Elce di Sotto
06100 Perugia, ITALY

Dr. E. Soczkiewicz
Institute of Physics
Silesian Technical University
ul. Krzywoustego 2
44-100 Gliwice, POLAND

Mr. A.O. Stone
Chelsea College
Polton Place, Folham
London, UK.

Dr. B. Terliuc
Ben-Gurion University
Dept. of Physics
P.O. Box 653
84105 Beer Sheva, ISRAEL

Dr. B.R. Tittmann
Rockwell International
Science Center
1049 Caminos dos Rios
P.O. Box 1085
Thousand Oaks, CA 91360, USA

Mr. H.T. Tran
Università di Perugia
Dipartimento di Fisica
Via Elce di Sotto
06100 Perugia, ITALY

Mr. J.W. Wolf
Georgetown University
Physics Department
Washington, DC 20057, USA

Prof. A. Zarembowitch
Université Pierre et Marie Curie
Dep. de Recherches Physiques
Lab. Dynamique du Reseau
et Ultrasons
Tour 22 - 4 Place Jussieu
75230 Paris Cedex 05, FRANCE

Dr. B. Zeqiri
National Physical Laboratory
Teddington, Middlesex TW1 OLW, UK

Dr. R. Zilber
Israel Atomic Energy Commission
Soreq Nuclear Research Center
N.D.T. Department
Yavne 70600, ISRAEL

INDEX